Coccolithophores are one of the primary algal groups in the oceans. They are the focus of recent research in many disciplines because of their importance in paleoenvironmental reconstruction and stratigraphy. The book starts with a history of coccolithophore studies, followed by chapters discussing their biology, and the composition, function and classification of their skeletal elements. At the heart of the book are taxonomic and atlas chapters with 140 scanning electron micrographs of coccolithophore species. Through a series of contributions from key workers in the field, the reader can follow the path of the organisms from the ocean surface, through the water column to the ocean floor and the addition to the sedimentary rock record. The book concludes with a chapter on geochemical tracers, and the implication of these studies for stratigraphy and paleoenvironmental change.

Coccolithophores

Coccolithophores

Edited by

AMOS WINTER AND WILLIAM G. SIESSER

CAMBRIDGE
UNIVERSITY PRESS

CAMBRIDGE UNIVERSITY PRESS
Cambridge, New York, Melbourne, Madrid, Cape Town, Singapore, São Paulo

Cambridge University Press
The Edinburgh Building, Cambridge CB2 2RU, UK

Published in the United States of America by Cambridge University Press, New York

www.cambridge.org
Information on this title: www.cambridge.org/9780521380508

First published 1994
This digitally printed first paperback version 2006

A catalogue record for this publication is available from the British Library

Library of Congress Cataloguing in Publication data

Coccolithophores / Amos Winter and William G. Siesser, editors.
    p.   cm.
ISBN 0-521-38050-2
1. Coccolithophores. I. Winter, Amos. II. Siesser, William G.
QK569.C63C63   1994
589.4'87–dc20   93-1777 CIP

ISBN-13 978-0-521-38050-8 hardback
ISBN-10 0-521-38050-2 hardback

ISBN-13 978-0-521-03169-1 paperback
ISBN-10 0-521-03169-9 paperback

Contents

Contributors

Larry E. Brand
Rosensteil School of Marine and Atmospheric Sciences, University of Miami,
Miami, FL 33149, USA.

Annelies Kleijne
Duinoord 11, 2224 CA Katwijk, The Netherlands

Ric W. Jordan
Institute of Oceanographical Sciences, Deacon Laboratory, Brook Road, Wormley,
Surrey GU8 5UB, U.K.

Richard N. Pienaar
Department of Botany, University of the Witwatersrand, P.O. Wits 2050,
Republic of South Africa.

Peter H. Roth
Department of Geology, University of Utah, 717 Mineral Sc. Bldg., Salt Lake City,
Utah 84112, USA.

William G. Siesser
Department of Geology, Vanderbilt University, Nashville, Tennessee 37235, USA.

John C. Steinmetz
Marathon Oil, Exploration and Production, P.O.Box 269, Littleton, Colorado 80160-0269, USA.

Amos Winter
Department of Marine Sciences, University of Puerto Rico, P.O.Box 5000,
Mayaguez, Puerto Rico 00681.

Jeremy R. Young
The Natural History Museum, Cromwell Road, London SW7 5BD, UK.

Preface

The idea for this book was conceived several years ago when we realized that there was a need for a multi-disciplinary book on living coccolithophores. In recent years the number of researchers working on coccolithophores has increased greatly. This is in part the result of interest in 'global change' and the important role that coccolithophores probably play in many of the Earth's important biogeochemical cycles. Among many biochemical compounds, coccolithophores produce DMS (a source molecule for cloud nucleation), alkenones (long-chained carbon molecules whose ratios are dependent on ocean temperature) and some unique pigments (used in photosynthetic processes). Coccolithophores, usually at higher latitudes, form communities of extremely high abundances. These are now routinely observed by satellite imagery and often turn the sea surface white. *Emiliania huxleyi* is the main contributing coccolithophore species in 'white water' communities. In some water samples this species reaches over 10 million specimens per liter and is by some accounts the most abundant calcifying species on earth and may also be the number one contributor to the oceanic biomass. Calcite-producing coccolithophores which have fast turnover rates must play an important role in the global cycle of CO_2. There are currently a number of groups examining the effects of coccolithophore blooms on the global biogeochemical cycle including studies of the flux of coccolithophores to the ocean floor using sediment traps. However, it is difficult to estimate changes in global abundances and fluxes of these organisms, especially below the surface.

The large number of coccolithophores present in productive northerly and coastal regions as well as their high diversity in polar areas suggest that their traditional role as indicators of warm, low-productive waters needs to be re-examined. Even though a few species, such as *E. huxleyi*, usually dominate coccolithophore assemblages, other species may be important environmental indicators. Relatively little is known about their life histories, reproductive strategies and ecological preferences. It is hoped that this book will entice researchers to answer some of these questions.

A great number of different (often confusing) taxonomic schemes have been presented for identification of coccolithophores. Often these schemes concerned themselves only with certain classes and not with entire coccolithophore families. We realized that a source book was needed by those involved in identifying coccolithophores. The taxonomic and atlas chapters will hopefully serve as a useful resource for those wishing to identify recent coccolithophores or just admire the shapes and forms of 140 coccolithophore species.

We have tried to arrange the chapters in the book following a logical progression. The book starts with the discovery of coccolithophores in 1836 and offers a chapter on the historical background of their study. This is followed by two chapters discussing aspects of coccolithophore biology (cell ultrastructure and physiology) and four chapters (composition and morphology, function, classification and an atlas) concerning their skeletal elements. After this the reader can follow the path of coccolithophores from the ocean surface (biogeography) through the water column (sedimentation) to the ocean floor (sediments). Methods of preparation and study of coccolithophores are included in the chapter discussing sea-floor sediments. The book concludes with a chapter on a geochemical aspect (stable isotopes) involving coccolithophores and possible implications for stratigraphy and paleoenvironmental reconstruction.

In this book, the living organisms are referred to as 'coccolithophores'. The term 'calcareous nannofossils' is used when specifically referring to extinct forms, which may or may not be biologically related to coccolithophores, but which have long been studied together with coccolithophores and their coccoliths. 'Calcareous nannoplankton' is used here as a collective term for the group when generally referring to both living and fossil forms.

We have purposefully kept the focus of the book on topics covering living coccolithophores and hope that this book will be a useful companion and stimulus for all those interested in these delicate and beautiful organisms. We also hope that this book will help communication between those working on 'modern' topics and those interested in stratigraphy, paleontology, and evolution of 'coccolithophores'.

This book owes much to the untiring efforts, dedication, encouragement and patience of the contributors and the editors of Cambridge University Press.

Amos Winter
William G. Siesser

1 Historical background of coccolithophore studies

WILLIAM G. SIESSER

Christian Gottfried Ehrenberg (1795–1876) has been called the 'Founder of Micropaleontology'. If any one man deserves this title, Ehrenberg must be considered a strong candidate. He was the first to discover and describe fossil silicoflagellates, ebrididans, dinoflagellates, and acritarchs; he did extensive work on diatoms, radiolarians, and foraminifers, and lesser work on sponge spicules, pteropods, ostracodes, phytoliths, and pollen. Ehrenberg's importance to this study is that he was also the first person to discover what would later be called 'coccoliths'.

Ehrenberg made the first recorded observation of coccoliths in 1836, while examining chalk from the island of Rugen in the Baltic Sea. He described the objects as elliptical, flattened discs, having a ring, or a few concentric rings on their surfaces. In contrast to the other microscopic bodies he studied and recognized as being organic, Ehrenberg considered these ovoid 'minerals' to be of inorganic origin. He thought they might be concretions similar to spherulites which form when crystals grow radially around a nucleus. He described, and for the first time illustrated, more of these 'elliptical platelets' in 1840 from a larger collection of chalk samples he had obtained. Ehrenberg spent 14 years preparing his classic work, *Mikrogeologie*, which was published in two volumes in 1854 (see Siesser, 1981, for a review of this monumental book). Volume I contains over 5000 meticulously hand-drawn illustrations, which include drawings of the 'elliptical platelets' he had discovered in the chalk. In *Mikrogeologie*, Ehrenberg called the platelets 'calcareous morpholithe'. His first illustrations of discoasters also appear in *Mikrogeologie* (1854), labelled calcareous 'Crystalldrusen', which he subtitled, 'Scheibensternchen' (and sometimes 'Sternscheibe'). He had originally assigned the 'Scheibensternchen' to the genus *Actiniscus*, but he apparently recognized that *Actiniscus* has a siliceous skeleton (now known to be a siliceous dinoflagellate) and later transferred it to his 'Polygastern' class. He believed that the discoasters, as with the coccoliths, were inorganic in origin.

In 1854, a midshipman in the American Navy, J. M. Brooke, had designed a deep-sea sounding apparatus that could also collect sea floor samples (Thompson, 1874). Brooke's device consisted of a sounding line to which was attached a 65 lb (*c*, 29.5 kg) cannonball with a hole passing though it. An iron rod, hollow at the lower end, passed through and projected below the cannonball. When the ball struck the sea floor, it drove the hollow rod into the sediment with considerable force. A release device then freed the rod, which slipped through the cannonball and was hauled to the surface. The hollow end of the rod was coated with tallow, and, it was hoped, would retain some of the sediment as the rod was pulled to the surface. Samples of sediment could now be routinely retrieved from the deep-sea floor using this device.

In the summer of 1857, HMS *Cyclops* under the command of J. Dayman was engaged in making soundings in the North Atlantic between Ireland and Newfoundland. The purpose of the *Cyclops*' soundings was to ascertain the depth of the sea and the nature of the sea floor along the route proposed for the laying of the first transatlantic telegraph cable. Dayman had modified Brooke's device for use in his sounding operations. He had redesigned the release mechanism, had substituted a lead cylinder for the cannonball (which reduced resistance while the weight was descending) and had constructed an inward-opening valve to prevent the sample washing out (Thompson, 1874). Sea floor samples obtained by Dayman and his crew were sent to the eminent Victorian biologist Thomas H. Huxley for examination; the results of Huxley's investigations appeared as Appendix A of Dayman's report (Huxley, in Dayman, 1858).

In Appendix A, Huxley reported that at depths between 1700 and 2400 fathoms he found 'a multitude of very curious rounded bodies, to all appearances consisting of several concentric layers ... and looking ... some-what like single cells of the plant *Protococcus* ... I will, for convenience sake, simply call them coccoliths.' Huxley attributed an inorganic origin to these 'coccoliths', as had Ehrenberg to the same objects. Huxley apparently did not find coccospheres in the samples he studied.

In 1860, G. C. Wallich was aboard HMS *Bulldog*, which was also engaged in taking soundings for a transatlantic cable. A much improved sounding tool was being used; a device looking more like a modern 'grab sampler' (Thompson, 1874). In the mud samples obtained, Wallich (1861) reported finding not only coccoliths 'in the free state', but also found coccoliths which adhered together to create minute spheres (Fig. 1). Wallich gave these ball-shaped bodies the name 'coccospheres', stating clearly that he found the 'coccoliths' noted by Huxley arranged at regular intervals

Fig. 1. First illustration of a coccosphere. From Wallich (1861).

Fig. 2. Coccospheres and coccoliths from Huxley (1868). Coccospheres of the 'compact' type (Fig. 6) and the 'loose' type (Fig. 7) are shown. His Fig. 1 shows the gelatinous mass, *Bathybius*. Note the coccoliths enclosed in *Bathybius*.

around the surface of each coccosphere. Wallich wrote that the coccoliths were held in position by a delicate gelatinous layer; the fragility of their attachment led to easy detachment and thus was responsible for the vast numbers of free coccoliths in sea floor sediments. Wallich measured the coccospheres in his samples, and found that they ranged from $\frac{1}{1600}$ to $\frac{1}{1250}$ of an inch (c. 16 to 20 µm) in diameter. Wallich (1861) believed that the coccospheres were larvae of foraminifers, probably of Globigerinae. His hypothesis, logical for the time, was based on his observations that: 1) coccospheres were invariably present in greatest numbers wherever

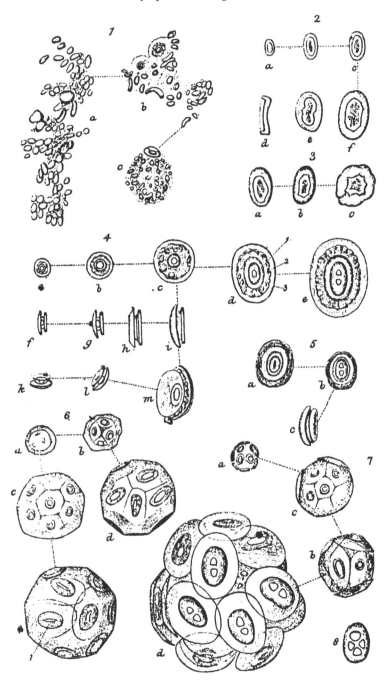

foraminifers were greatest in number, (2) coccoliths were found attached to foraminifer shells, and 3) coccospheres were somewhat similar to the cells of juvenile globigerinids.

At almost the same time, Henry Clifton Sorby (1861) reported finding coccospheres in the English chalk. Sorby stated that he had actually been studying chalk and speculating on the 'ovoid bodies' (the coccoliths) in it for some ten years before his publication. He had long doubted Ehrenberg's conclusion that the 'ovoid bodies' were inorganic in origin. Sorby had discovered sometime earlier that the ovoid bodies were not simply flat discs, as Ehrenberg had described and drawn them, but instead were concave on one side and convex on the other. This, Sorby believed, was a shape entirely unlike anything that could result from inorganic crystallization. To him, the concave–convex shape of the bodies pointed to their having once formed small hollow spheres, rather similar to the shapes of some of the foraminifers Sorby had seen in the chalk.

After Huxley's note appeared in 1858, Sorby requested some of the same deep-sea mud Huxley had worked on. Huxley readily furnished the samples. In the mud, Sorby recognized the same 'ovoid bodies' (coccoliths) that he had seen in the chalk. He later (1860) found, moreover, what he had previously hypothesized: coccoliths occurring on small, hollow spheres (coccospheres) in the chalk. This proved his earlier – and remarkably astute – prediction which had been based solely on the concave–convex shapes of the individual coccoliths.

In October, 1860, Sorby read a short paper describing his findings at a meeting of the Sheffield Literary and Philosophical Society. In this paper, he proposed that the 'Crystalloids' (coccoliths) of the chalk were of organic origin, and not concretions, as believed by Ehrenberg and Huxley. He suggested, moreover, that the spherical shells were in some respects analogous to foraminiferal cells. In his paper of 1861, Wallich alluded to Sorby's discovery of coccospheres in the chalk. This apparently prompted Sorby to prepare a more detailed account of his discovery. This paper also appeared in 1861, although later than Wallich's paper. By this time Sorby did not believe, as Wallich had suggested, that coccospheres were a rudimentary form of foraminifer. His petrographic work had indeed indicated that both foraminifers and coccoliths were apparently composed of calcite. But Sorby held that coccospheres were from some 'independent organism' (although perhaps *related* to foraminifers), a conclusion based in part on the differences in the optical properties of the calcite making up their respective skeletons.

After Sorby's paper in 1861, almost all biologists accepted the fact that coccoliths/coccospheres had some sort of organic origin. But the venerable microscopist Ehrenberg never accepted this opinion. In 1873, in one of his last papers, and just three years before his death he still alluded to them as 'inorganic structures' (Ehrenberg, 1873).

Influenced by Wallich's and Sorby's reports, Huxley re-examined the *Cyclops* samples, this time using a microscope with almost double the magnifying power (×1,200) of the microscope he had used in 1857. In the *Cyclops* samples, which had been preserved in 'spirit' (alcohol) since 1857, Huxley (1868) found 'innumerable lumps of a transparent gelatinous substance'. These 'lumps' were of all sizes, from minute specks to patches visible without need of a microscope. Within these lumps, Huxley could see coccoliths and coccospheres (Fig. 2). He concluded that the gelatinous mass must be protoplasm, and that the coccoliths and coccospheres were skeletal elements within the protoplasm. Together, they constituted a remarkable living slime, a 'moner'. The famous German biologist Ernst Haekel had only shortly before defined the 'monera' as a class of almost formless beings, consisting of jelly-like protoplasm and reproducing by simple spontaneous division. Huxley named this new species of moner *Bathybius haeckelii*. He believed that coccoliths and coccospheres supported the protoplasm, rather as sponges contain spicules to support their soft parts. Huxley (1868) doubted Sorby's view that coccoliths were disaggregated coccospheres. He believed instead that coccospheres formed by the coalescence of coccoliths, and that there were two different types: 'loose' coccospheres, where coalescence of coccoliths was in an early stage (Fig. 2), and 'compact' coccospheres (Fig. 2) in which coalescence was at an advanced stage.

Haeckel (1870) was naturally interested in the moner bearing his name, *B. haeckelii* (Fig. 3). He viewed *Bathybius* as the most primitive form of life, and believed, moreover, that it was from *Bathybius* that all other life forms had arisen. *Bathybius* was believed to have remained in its primordial state because the deep ocean floor was a stable, unchanging environment. In addition, *Bathybius* was considered by Haeckel to be the ultimate food source for all the higher life forms observed in the deep ocean. W. C. Gümbel (1870) and O. Schmidt (1870) added further support to the *Bathybius* concept when they too reported finding this organism on the sea floor. Schmidt (1870) also found club-shaped rods which he named 'rhabdoliths' in the *Bathybius* slime he reported from the floor of the Adriatic Sea.

C. W. Thompson (1874) viewed the relationship of coccoliths to *Bathybius* somewhat differently. He believed coccoliths were the joints of minute one-celled algae which lived at the sea surface. As coccoliths sank, they were engulfed by *Bathybius* and used as food for the protoplasm. As for coccospheres, Thompson (1874) summarized the opinion of many biologists of the time (but certainly not Wallich) when he stated, 'What the coccospheres are, and what relation, if any, they have to the coccoliths we do not know.'

One of the objectives of the HMS *Challenger* expedition (1872–76) was to learn more about the remarkable *Bathybius* and to chart its distribution on the sea floor. John Murray and his co-workers aboard the *Challenger* searched their samples long and carefully for any sign of *Bathybius* during the first two and a half years of the *Challenger* cruise. Every bottom sample was examined under the microscope

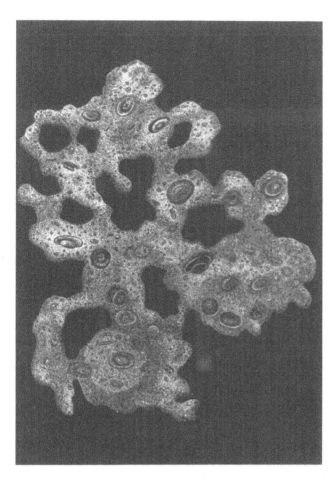

Fig. 3. *Bathybius haeckelii* with its 'skeletal parts' from Haeckel (1870) (reproduced in Thompson, 1874).

'for hours at a time' (Murray, 1876), but no evidence of protoplasmic movement could be found. They did notice that stored specimens preserved with alcohol developed a mobile, jelly-like layer on the surface of the mud in the bottles, whereas specimens preserved without alcohol had no such layer. The *Challenger* scientists experimented on the samples by adding different amounts of alcohol. They found that large quantities of alcohol mixed with sea water created a bulky, gelatinous precipitate that remained jelly-like for years. Analysis showed that the precipitate consisted entirely of sulphate of lime. The staining solutions used by Huxley and Haeckel had given reactions suggesting that *Bathybius* was protoplasm; this reaction Murray attributed to minor amounts of organic matter in the sediment itself.

Even before Murray's (1876) note on the lime sulphate precipitate was published, C. W. Thompson, the leader of the *Challenger* expedition, had advised Huxley that *Bathybius* was disproven. Word spread quickly. G. C. Wallich must have taken some satisfaction in now marshalling the arguments against *Bathybius*, finally and firmly disposing of the concept in a paper in 1875.

In 1877, Wallich published a further paper dealing specifically with coccospheres ('Observations on the Coccosphere'). In this paper he chides Wyville Thompson and others for not crediting him with being the first to understand the connection between coccoliths and coccospheres. Wallich (1877) seemed particularly miffed that they had overlooked a paper he had read before the Royal Microscopical Society in 1865. In that paper he had stated succinctly that he 'had met with coccospheres as free floating organisms in tropical seas' as early as 1857. He stressed, moreover, his long and consistently held opinion that 'free' coccoliths were derived from 'parent' coccospheres. Wallich (1877) proceeded to argue against consideration of single 'coccoliths' as individual cells, a view advocated by some earlier workers. He also dissented from earlier suggestions by Carter (1871) that coccospheres might be algal sporangia belonging to the calcareous alga *Melobesia*. Wallich comments that, if this were the case for coccospheres, 'of the multitudes I have seen, none has ever departed from the sporangial phase, either in those met at the top or bottom of the ocean.' In this same paper, Wallich (1877) noted that he had thus far found coccospheres of only two different species. He illustrated the coccospheres of each species – *Coccosphaera pelagicus* and *Coccosphaera carterii* – (Fig. 4), together with a formal description of each. The following description, taken from Wallich (1877), of the coccosphere cell is an important part of the diagnosis for each of these species:

Genus *Coccosphaera* (Wall.)
1. *Coccosphaera pelagica* (Wall.)

Cell spherical, hyaline, with a distinct membraneous wall. Cell-contents, a perfectly colourless glairy protoplasm. *Coccoliths* generally more or less elliptical, numbering from 16 to 36, arranged side by side, and, in the normal state, not overlapping. Central aperture of *coccolith* single, margin of external disk finely and radially striate. Internal disk plain.

Diameter of *coccosphere* ranging from $\frac{1}{5000}$ to $\frac{1}{830}$ of an inch. Length of *coccoliths* from $\frac{1}{9000}$ to $\frac{1}{1000}$ of an inch.

Habitat. Free-floating, Indian Ocean and North Atlantic; and (dead) in North Atlantic muds. Always most abundant where the *Globigerinae* are in greatest profusion, and the deposit of the purest kind.

2. *Coccosphaera carterii* (Wall.)

Cell oblong. Long diameter about twice that of short diameter. Cell as in *C. pelagica. Coccoliths* varying in number from 16 to 38, more or less oblong, with two central apertures arranged lengthwise, margin finely and radially striate. Internal disk plain. Length of *coccosphere* from $\frac{1}{1000}$ to $\frac{1}{800}$ of an inch. Length of coccolith from $\frac{1}{5000}$ to $\frac{1}{1000}$ of an inch.

Habitat. Free-floating, Indian Ocean, and Mid-Atlantic. (NB I have not observed any intermediate from between the spherical and oblong.)

Fig. 4. Coccospheres and coccoliths of *Coccosphaera pelagica* and *Coccosphaera carterii* from Wallich (1877). *Coccosphaera pelagica* is shown as Figs. 1, 2, 5, 8, 9, 10, and 11. *Coccosphaera carterii* is shown as Figs. 3, 4, 6, and 7. Figure 18 is identified as a sporangium of a 'protophyte' from Bengal.

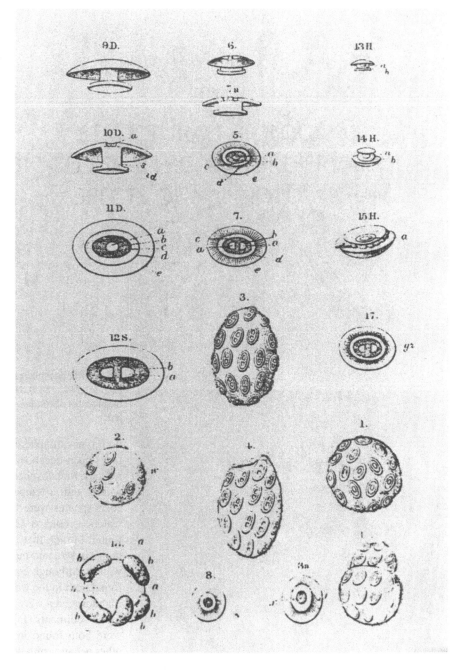

Other late nineteenth century workers observed and briefly described these 'curious organisms' (usually coccoliths; sometimes coccospheres) that they saw in sediments from various parts of the world (see Siesser, 1975, for references to these early papers). Among these, G. M. Dawson (1874) made the first report on coccoliths from the American side of the Atlantic. He noted that they occurred in samples of Cretaceous limestones from both Manitoba and Nebraska, and figured various forms of coccoliths and 'rhabdoliths' from Manitoba, including at least one poorly preserved coccosphere. Another interesting early report was that of N. J.

Sollas (1876), who found coccoliths and coccospheres within granules of glauconite in the Cambridge Greensand.

John Murray and his co-workers aboard HMS *Challenger* firmly established that coccospheres were the skeletons of minute calcareous algae (Tizard *et al.*, 1885; Murray and Renard, 1891). They described the interior of coccospheres as being filled with transparent albuminoid matter in which no nucleus could be detected. They described both coccospheres *sensu stricto* and rhabdospheres in some detail (Fig. 5) and provided the first rudimentary biogeographic information on the distribution of the living algae. They reported

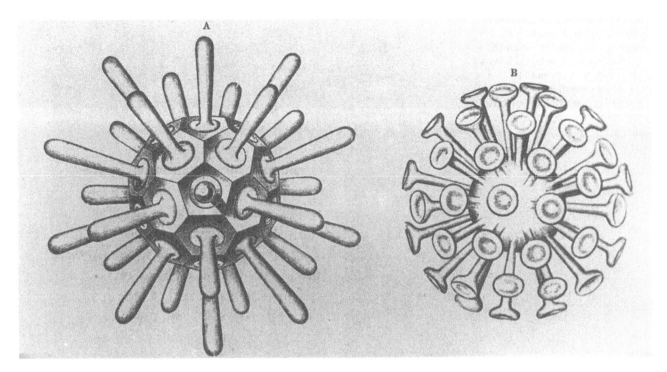

Fig. 5. 'Rhabdospheres' from the HMS *Challenger* reports (Tizard *et al.*, 1885). Note the authors' concept of tight-fitting basal plates of *Rhabdosphaera clavigera* on the specimen at the left.

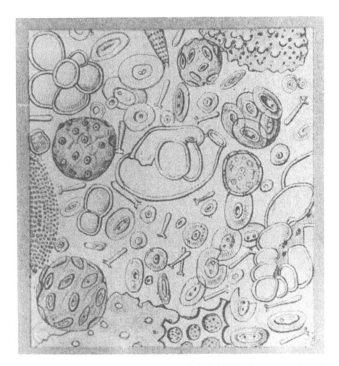

Fig. 6. Coccospheres, coccoliths and rhabdoliths from samples of *Globigerina* ooze, collected from the central Pacific Ocean during the HMS *Challenger* cruise (from Tizard *et al.*, 1885). The *Challenger* scientists found rhabdoliths in sediment samples, but found rhabdospheres only in the water column.

both types of spheres abundant in all surface and near-surface open-ocean waters of the tropical and temperate regions; rhabdospheres were encountered essentially only in waters with a temperature above 65° F (18.3°C); whereas coccospheres were found farther north and south, living in waters as cold as 45° F (7.2°C). The *Challenger* scientists found, in fact, that in colder waters the coccospheres were larger in size and more numerous than they were in warmer waters. Although complete coccospheres *sensu stricto* were found both in the water column and in the sediments, whole rhabdospheres were found only in the water column; never in the sediments (Fig. 6). Coccospheres and rhabdospheres were both found in the stomachs of salps, pteropods and other pelagic animals.

The *Challenger* scientists found that when coccoliths were dissolved with dilute acid, a small gelatinous sphere remained. They reasoned that it was in the outer layer of the gelatin that the coccoliths and rhabdoliths were embedded (Tizard *et al.*, 1885). Murray and his co-workers also illustrated coccospheres of several other genera (now classified as *Helicosphaera*, *Thoracosphaera* and *Scyphosphaera*) in the *Challenger* reports, but did not discuss them separately.

George Murray and V. H. Blackman (1898) collected living coccospheres from Atlantic and Caribbean sea water by pumping the water through 'miller's silk' (Fig. 7). They determined that coccoliths overlapped each other 'in definite order' on the surface of a coccosphere. Murray and

Fig. 7. Plate 15 from Murray and Blackman (1898). Figures 1–7 are of *Coccosphaera leptopora*. Figures 2 and 4 show the central protoplasmic body of the cell, and figs. 6 and 7 illustrate what the authors' called the 'central chromatophore'. The cell in fig. 7 is undergoing fission.

Rhabdosphaera tubifera is shown in figs. 8-10, and *R. clavigera* in Figs. 13, 14, and 15. Figure 11 illustrates a protozoan devouring several rhabdospheres.

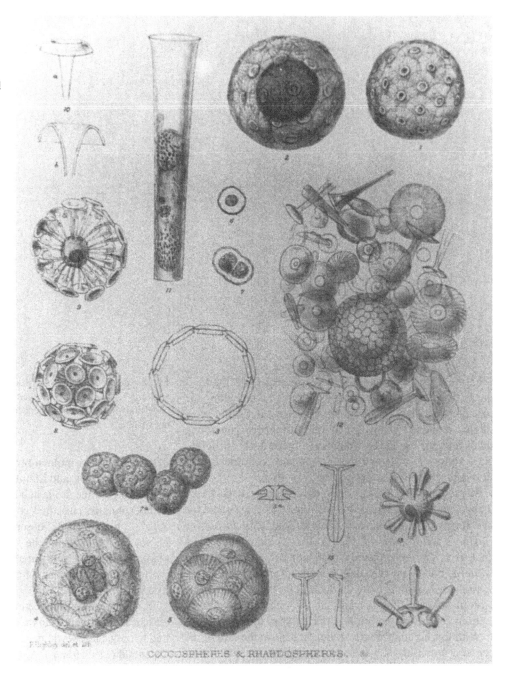

COCCOSPHERES & RHABDOSPHERES.

Blackman were the first to propose a function for coccoliths, supposing them to provide some sort of defensive armor. They also stated that the rhabdosphere illustrations in the HMS *Challenger* reports (e.g., Fig. 5) were misleading, since they showed geometrically arranged, tightly fitting hexagonal plates at the basal ends of the clavate rods of rhabdospheres. Murray and Blackman, conversely, found that these plates were separated from one another (Fig. 7).

Murray and Blackman (1898) described a new species, *Coccosphaera leptopora*, coccospheres of which they found floating in the surface waters of the Atlantic. They gave a detailed description of how the coccoliths of *C. leptopora* overlap one another with the edge of the 'outer limb' (distal shield) of one plate being wedged between the edges of the 'outer and inner limbs' (distal and proximal shields) of a neighboring plate (Fig. 7), and compared the structure of C. leptopora with the structure of the much larger *C. pelagicus* (Fig. 8). They speculated that this system of interlocking plates would be a very effective means of protection for the organism.

In their detailed description of *C. leptopora*, Murray and Blackman (1898) counted the coccoliths occurring on single coccospheres (a range of 20 to 50), and also examined the

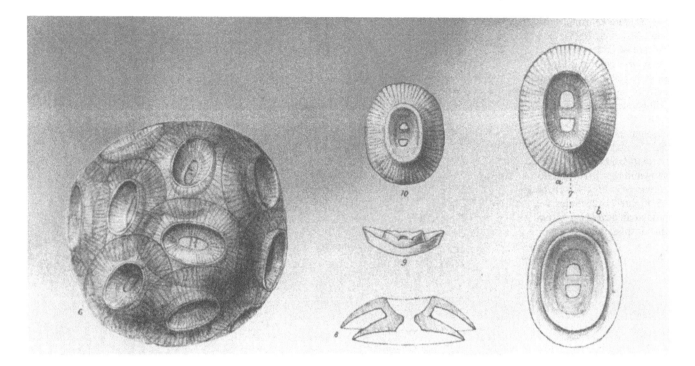

structure of the cell within the coccosphere shell. Figure 7 shows the protoplasmic body they observed lying within the coccosphere. They could not see a nucleus, but could see a single, rounded, yellow-green chromatophore containing small refractive granules which they inferred were oil globules. They also observed the *C. leptopora* cell undergoing simple reproduction by fission (Fig. 7).

Murray and Blackman (1898) described the structure of rhabdospheres in detail for the first time. They separated the two forms common in their samples into *Rhabdosphaera clavigera* and *R. tubifera* (Fig. 7). They could not observe a nucleus or chromatophores within either *R. clavigera* or *R. tubifera*, but noted that both contained protoplasm in the central portion of the body. In *R. tubifera*, they could see the protoplasm extending upward into the trumpet-shaped projections for nearly two-thirds of the length of the projections.

With this amount of information in hand at the turn of the century, a consensus was reached regarding certain fundamental facts about coccospheres. Scientists now agreed – at least for the few living species named – on what they were, on the composition of the living cell, on how the skeletal parts were arranged, and on the few known aspects of their biogeography.

The nucleus of the cell was eventually described by Ostenfeld in 1900. Further discoveries on the biology of the cell resulted from the studies of H. H. Dixon, A. Weber van Bosse and H. Lohmann. Dixon (1900) documented the organic pellicle enclosing coccospheres. Dixon also noted several stages in the growth development of coccoliths before they are extruded to the surface, where they lock together to form a coccosphere. Weber van Bosse (1901)

Fig. 8. Part of Plate 16 from Murray and Blackman (1898) showing *Coccosphaera pelagica*. Figure 6 demonstrates their clear understanding of how coccoliths lock together to form a coccosphere.

described the two golden-brown chromatophores on either side of the nucleus, and Lohmann (1902) first recognized flagella as part of the living cell (Fig. 9). On the basis of the flagella, Lohmann classified coccospheres with chrysomonad algae. Lohmann (1902) described polymorphism in coccospheres for the first time, he also proposed the term 'nannoplankton' for the very small plankton which could pass through the normal mesh of a phytoplankton net (45–60 µm in size).

In the early twentieth century, intensive systematic work on calcareous nannoplankton began, although initially limited to only a few workers. H. Lohmann (1902 *et seq.*), C. H. Ostenfeld (1899 *et seq.*), A. D. Arkhangelsky (1912 *et seq.*), and J. Schiller (1913 *et seq.*) described mostly living, as well as a few fossil, calcareous nannoplankton. Early twentieth century work by these investigators stressed the discovery, description and naming of new species. When coccospheres of a new species were found, they were illlustrated. But in most cases, species names were erected based simply on the morphology of individual coccoliths. Cataloging of species and physiological investigations began in earnest in the late 1920s and the 1930s. These decades were dominated by the work of E. Kamptner in Austria, G. Deflandre in France, and T. Braarud and his colleagues in Norway. Most of the remaining living species were described, as well as many fossil species. During the 1950s and 1960s, the Norwegian

Fig. 9. Plate 4 from Lohmann (1902) showing various species of *Pontosphaera* (figs. 1–20), *Syracosphaera* (figs. 21–25; 31–37), and *Scyphosphaera* (figs. 26–30). Lohmann draws attention to the nucleus, chromatopores and flagella of the living cells on this plate.

group investigated numerous aspects of the cytology, physiology and ecology of living nannoplankton and pioneered attempts at culturing nannoplankton in the laboratory. In England, M. Parke and her colleagues also made many important observations on physiology and ecology, as well as on the growth, calcification, and life cycle of living nannoplankton during this period.

Microscopy techniques were developed during the early 1950s which greatly improved the capability to observe nannoplankton. In 1952 Kamptner used cross-polarized light in his nannoplankton work and also in 1952, T. Braarud and E. Nordli first used phase-contrast illumination and transmission electron microscopy. The transmission electron microscope (TEM) became a tool routinely available to researchers from the mid-1950s onward, and large numbers of TEM photographs of nannoplankton were published to supplement photographs made using transmitted-light optical microscopes.

In 1954, M. N. Bramlette and W. R. Riedel clearly demonstrated for the first time the usefulness of calcareous nannofossils in biostratigraphy. Subsequent initiation (in 1968) of the Deep Sea Drilling Project (DSDP) made the stratigraphic value of calcareous nannofossils apparent to the biostratigraphic community as a whole. Stratigraphic position ('age dates') could be determined quickly and accurately, literally within minutes after a core was brought on deck. Investigation of relatively continuous DSDP cores, which commonly contained abundant, well-preserved nannofossil assemblages, soon led to the development of refined nannofossil biozonations.

The scanning electron microscope (SEM) replaced the TEM as an investigative tool in the late 1960s and early 1970s, and soon became widely used for nannoplankton investigation and illustration. The transmitted-light microscope equipped with cross-polarizers and phase-contrast illumination is, however, still the preferred tool for routine biostratigraphic work.

During the 1970s and 1980s nannoplankton research reached a mature stage. Highly refined biostratigraphic zonations are now available for the Jurassic, Cretaceous and Cenozoic time periods. Nannoplankton have been and are being utilized in detailed taxonomic studies, paleobiogeographic mapping, paleoclimatic/paleoceanographic reconstructions, evolutionary lineage studies, biochronology, diagenesis and stable-isotope investigations. Calcareous nannoplankton investigation has come a long way since Ehrenberg first recognized these tiny 'concretions'.

References

Arkhangelsky, A.D., 1912. Verkhnemelovyya otolozheniga vostoka evropeyskog Rossii (Upper Cretaceous deposits of east European Russia.) *Materialien zur Geologie Russlands*, **25**: 1–631.

Braarud, T. and Nordli, E., 1952. Coccoliths of *Coccolithus huxleyi* seen in an electron microscope. *Nature*, **170**: 361–2.

Bramlette, M.N. and Riedel W. R., 1954. Stratigraphic value of discoasters and some other microfossils related to recent coccolithphores. *J. Paleontol.*, **28**: 385–403.

Carter, H.J., 1871. On *Melobesia unicellularia*, better known as the coccolith. *Ann. Mag. Nat. Hist.*, Ser. 4, **7**: 184–9.

Dawson, G. M., 1874. Note on the occurrence of foraminifera, coccoliths, etc. in the Cretaceous rocks of Manitoba: *Can. Naturalist J. Sci.*, **7**: 252–7.

Dayman, J., 1858. *Deep Sea Soundings in the North Atlantic Ocean Between Ireland and Newfoundland, made in HMS Cyclops*. London, Eyre and Spottiswoode.

Dixon, H.H., 1900. On the structure of coccospheres and the origin of coccoliths. *Proc. Roy. Soc. (London)*, **66**: 305–15.

Ehrenberg, C.G., 1836. Bemerkungen uber feste mikroscopische anorganische Formen in den erdigen und derben Mineralien. *Bericht. Verh. K. Preuss. Akad. Wiss. Berlin*, pp. 84–85.

Ehrenberg, C.G., 1840. Uber die Bildung der Kreidefelsen und Kreidemergels durch unsichtbare Organismen, *Abh. K. Preuss. Akad. Wiss. Berlin*, pp. 59–147.

Ehrenberg, C.G., 1854. *Mikrogeologie. Das Erden und Felsen Schaffende Wirken des Unsichtbar Kleinen Selbstandigen Lebens auf der Erde*. Leipzig. V.I (375 p.); V.II (88 p.).

Ehrenberg, C.G., 1873. Mikrogeologische studien uber das kleinste Leben der Meeres-Tiefgrunde aller Zonen und dessen geologischen Einfluss, *Abh. K. Preuss. Akad. Wiss. Berlin*, pp. 131–398.

Gümbel, W.C., 1870. Coccolithen (Bathybius) in allen Meeresliefen und in den Meeresablagerungen aller Zeiten. *Ausland.*, **43**: 763–4.

Haeckel, E., 1870. Beitrage zur Plastidentheorie. *Jena. Zeitschr. Med. U. Naturw.*, **5**: 492–550.

Huxley, T.H., 1868. On some organisms living at great depths in the North Atlantic Ocean. *Quart. J. Microscopical Sci.*, New Ser., **8**: 203–12.

Kamptner, E., 1952. Das mikroskopische studium des skelettes der Coccolithineen (Kalkflagellaten). *Mikroskopie*, **7**: 232–44; 375–86.

Lohmann, H., 1902. Die Coccolithoporidae. *Arch. Protistenk.* 1: 89–165.

Murray, G. and Blackman, V.H., 1898. On the nature of coccospheres and rhabdospheres. *Phil. Trans. Roy. Soc. (London)*, Ser. B, **190**: 427–41.

Murray, J., 1876. Preliminary Report. *Proc. Roy. Soc. (London)*, **24**: 471–544.

Murray, J. and Renard, A.F. 1891. Report on deep-sea deposits based on the specimens collected during the voyage of HMS Challenger in the years 1872 to 1876. In *Report on the Scientific Results of the Voyage of H.M.S.* Challenger during the Years 1873–76, part 3, *Deep-Sea deposits*. HMSO London.

Ostenfeld, C.H., 1899. Uber Coccosphaera und einige neue Tintinniden im Plankton des nordlichen Atlantischen Oceans. *Zool. Anz.*, **22**: 433–9.

Ostenfeld, C.H., 1900. Uber cocccosphaera. *Zool. Anz.*, **23**: 198–200.

Schiller, J., 1913. Vorlaufige Ergebnisse der Phytoplankton-Utersuchungen auf den Fahrten SMS *Najade* in der Adria 1911–12. Die Coccolithophoriden. *SitzBer. K. Akad. Wiss. (Wien), Math.-Naturw. Kl.*,Abt. 1, **122**: 597–617.

Schmidt, O., 1870. Ueber Coccolithen und Rhabdolithen. *SitzBer. K. Akad. Wiss. (Wien), Math.-Naturw. Kl.* Abt. 1, **62**: 669–82.

Siesser, W.G., 1975. Calcareous nannofossils from the South

African continental margin. *GSO/UCT Marine Geol. Prog. Bull.* 5. Univ. Cape Town.

Siesser, W.G., 1981. Christian Gottfried Ehrenberg: founder of micropaleontology. *Centaurus*, **25**: 166–88.

Sollas, W.J., 1876. On the glauconitic granules of the Cambridge Greensand. *Geol. Mag.* New Ser., **3**: 539–44.

Sorby, H.C., 1861. On the organic origin of the so-called 'crystalloids' of the chalk. *Ann. Mag. Nat. Hist.*, Ser. 3, **8**: 193–200.

Thompson, C.W., 1874. *The Depths of the Sea.* Macmillan & Co., London.

Tizard, T.H., Moseley, H.N., Buchahan, J.V., & Murray, J., 1885. Narrative of the Cruise of HMS *Challenger*, In: *Report of the Scientific Results of the Voyage of the HMS.* Challenger *during the Years 1873–76*, vol. 1, Part 2, Neill and Co. Edinburgh. pp. 511–1110.

Wallich, G.C., 1861. Remarks on some novel phases of organic life, and on the boring powers of minute annelids, at great depths in the sea. *Ann. Mag. Nat. Hist.*, Ser. 3, **8**: 52–58.

Wallich, G.C., 1875. On the true nature of the so-called *Bathybius* and its alleged function in the nutrition of the Protozoa. *Ann. Mag. Nat. Hist.*, Ser. 4, **16**: 322–39.

Wallich, G.C., 1877. Observations on the coccosphere. *Ann. Mag. Nat. Hist.*, Ser. 4, **19**: 342–50.

Weber Van Bosse, A., 1901. Etudes sur les algues de l'archipel Malaisien. III. Note preliminaire sur les resultats algologiques de l'expedition du Siboga. *Ann. Jard. Bot. Buitenz.*, 17: Ser. 2, **2**: 126–41.

2 Ultrastructure and calcification of coccolithophores

RICHARD N. PIENAAR

Introduction

The purpose of this chapter is to provide information on the ultrastructure and calcification of coccolithophores. It is also hoped that this chapter will bring the micropaleontologist and the neontologist closer together and foster an appreciation of each other's work on this fascinating group of organisms.

If we are ever to develop a real understanding of the biology of coccolithophores it is important that we recognize the need for simultaneous studies to be undertaken using the light, transmission, and scanning electron microscopes. Ideally such studies should be done on living material using species that have been isolated and maintained as unialgal cultures. Priority areas for future research on coccolithophores should include perfection of isolation techniques, and the development of culture media which would enable us to isolate and maintain unialgal cultures for experimental purposes. The majority of coccolithophores that have thus far successively been isolated and maintained in culture are those that could be regarded as the more 'weedy species', e.g., *Pleurochrysis* Pringsheim and *Emiliania* Hay and Mohler.

Cytology

Coccolithophores belong to the Class Prymnesiophyceae (Hibberd, 1976). Figure 1 summarizes the basic organization of the prymnesiophycean cell.

The majority of living coccolithophores are unicellular motile or non-motile coccoid forms. In addition to these two morphological forms there are reports of groups of non-motile colonies as well as filamentous forms.

Periplast boundary

Generally coccolithophore cells are characterized by possessing a cell covering, often referred to as a 'periplast', which is made up of one or several layers of scales. Some scales are entirely organic (see Fig. 3A–D) whereas others are composed of an inorganic component associated with an organic matrix (see Figs. 4A–B, 5D). These latter forms are

the coccoliths which are found as microfossils in sedimentary rocks from the Triassic to the present time.

Because of the extremely small size of scales and coccoliths, it is essential that the periplast-scale boundary be studied using the transmission or scanning electron microscope. Most useful detailed information on the structure of scales and coccoliths is obtained with the transmission electron microscope using direct preparations, heavy-metal shadowed and surface-replica preparations. The scanning electron microscope (TEM) is useful for studying the structure of coccoliths but is of very limited use for studying organic-scale structure.

Organic scales are usually too small and too thin to be resolved with even the most sophisticated light microscope. To observe these structures one needs to use direct preparations together with preparations that have been shadowed, under vacuum, with a heavy metal at an angle of 20–30° (Fig. 3A–D, 8C). The commonly used heavy metals are gold/palladium, platinum, and chromium (see Pienaar, 1976b; Inouye and Pienaar, 1983 for details of these techniques).

Calcified scales (coccoliths) are often larger than organic scales and because of the difference in the refractive index of $CaCO_3$ and the water mountant, one is usually able to distinguish the presence or absence of coccoliths with the aid of the light microscope (Figs. 2A–C, 5A–B, 7B–C). In some instances one is even able to identify coccoliths of different shape and structure (coccolith dimorphism) with the light microscope (Fig. 8D–F).

Organic Scales

Organic scales, as the name implies, lack the presence of an inorganic component. The majority of coccolithophores that have been studied possess one or more layers of organic scales immediately external to the plasma membrane (Fig. 5D). Notable exceptions are certain stages in the life cycle of *Emiliania huxleyi* (Lohmann) Hay and Mohler (Klaveness, 1972a) (Fig. 9).

Organic scales are usually characterized by possessing distinct proximal and distal surface patterns. The proximal surface has radiating microfibrils (Fig. 12). These circular, unmineralized scales frequently have a distinct microfibril-

Fig. 1. Diagrammatic representation to illustrate the basic organization of the Prymnesiophyceae. Abbreviations: F = flagellum; H = haptonema; Tr = transitional zone; FB = flagellar bases; HB = haptonema base; R = microtubular root; Sc = scales; PER = peripheral endoplasmic reticulum; CE = chloroplast envelope; PN = periplastidial network; Chl = chloroplast; Pyr = pyrenoid; ch.ER = chloroplast endoplasmic reticulum; G = Golgi body; Pd = peculiar dilations; V = 'hairy' vesicles; M = mitochondrion. Modified from Hibberd (1976).

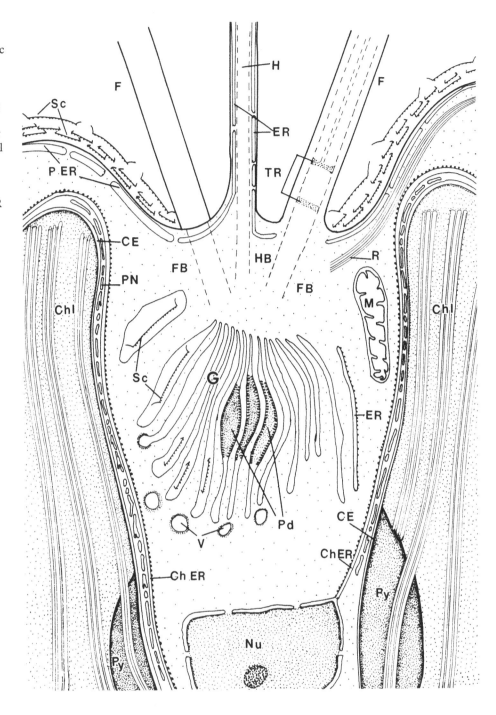

lar rim, for example, *Pleurochrysis* (Figs 3A, B) but in some coccolithophores, for example, *Syracosphaera pulchra* Inouye and Pienaar, (1988) such a rim is absent (Fig. 8C). These organic scales make up several layers of scales immediately external to the plasma membrane and have been shown to be present in coccolithophores that possess holococcoliths, for example, *Calyptrosphaera* Lohmann (Fig. 7D) and heterococcoliths, for example, *Pleurochrysis* (Pienaar, 1969a; Klaveness, 1973) and Hymenomonas (Pienaar, 1976b) (Figs 5D–E, 6D). In *Pleurochrysis* a smaller unmineralized scale possessing an ultrastructure similar to the circular organic scale has been reported (Pienaar, 1969a; Manton and Peterfi, 1969). This scale is restricted to the surface of the reduced bulbous haptonema (Fig. 8C).

In those coccolithophores that have been shown to possess a life cycle stage composed of small, branched, filamentous colonies, the area identified as a cell wall with the light microscope can be demonstrated with the TEM to be composed of many organic-scale profiles that have become

Fig. 2. *Pleurochrysis carterae*

(A)–(C) Through focus of a motile coccolith bearing cell. Large arrowhead = bulbous haptonema; small arrowhead = coccoliths. Note the two well defined chloroplasts with bulging pyrenoids.

(D) Two motile coccolith bearing cells that have been decalcified. Note the new intracellular coccolith (arrowheads) near the flagellar poles.

(E) Non-motile coccolith-bearing colonial stage.

(F) Haploid benthic 'Apistonema' stage. Note the well defined cell wall (arrowhead) made up of scale profiles.

(G) Low power view of benthic filamentous stage.

(H) Ultrathin section of the boundary between four cells (C_1–C_4) note the cell wall made up of scales (Sc) and the cementing substance (arrowhead).

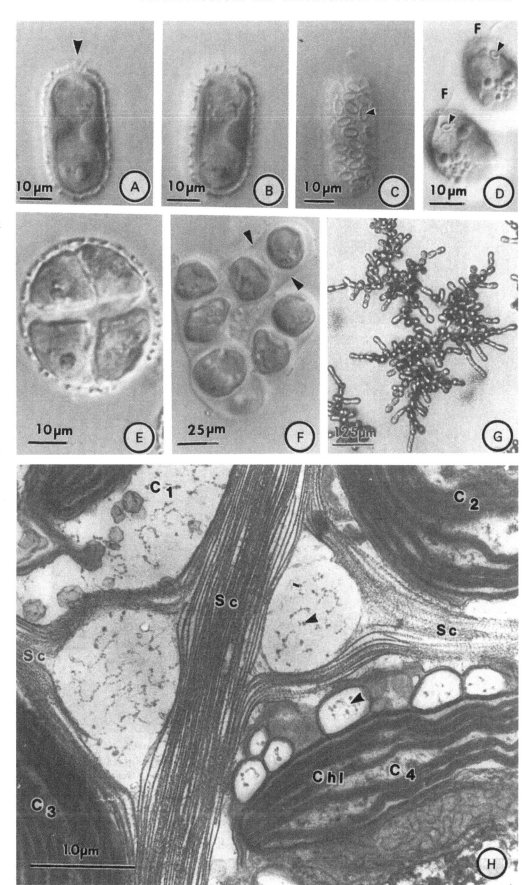

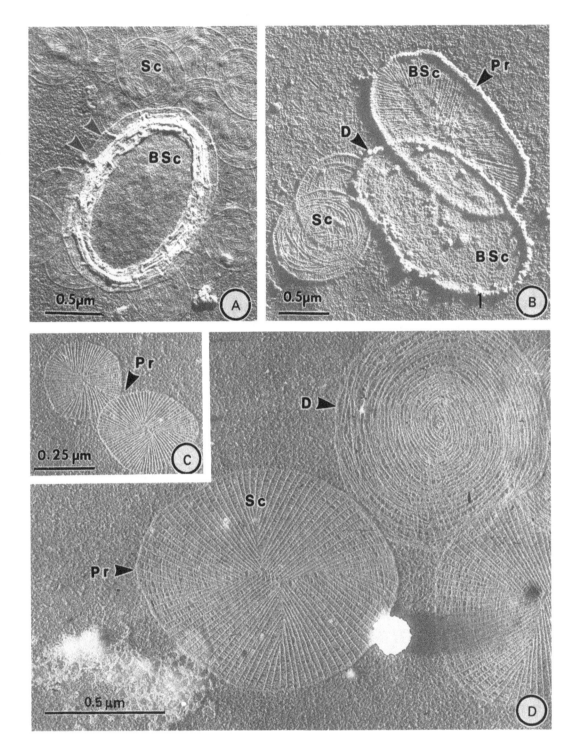

Fig. 3. Heavy metal shadowed (AuPd) preparations of the scale and coccolith component of *Pleurochrysis*.

(A) View of circular organic body scales (Sc) and a base-plate scale (BSc) seen in distal view. Note the organic matrix (arrowheads). Preparation from *P. carterae*.

(B) Organic body scales (Sc) and the base-plate scale (BSc) seen in proximal (Pr) and distal (D) view. Preparation from *P. carterae*.

(C) View of small organic haptonemal scales seen in distal view. Preparation from *P. scherffelli*.

(D) Circular organic body scales of (Sc) of *P. scherffelii* seen in proximal (P) and distal (D) view to illustrate the two distinct surface patterns. Note the absence of rim.

16

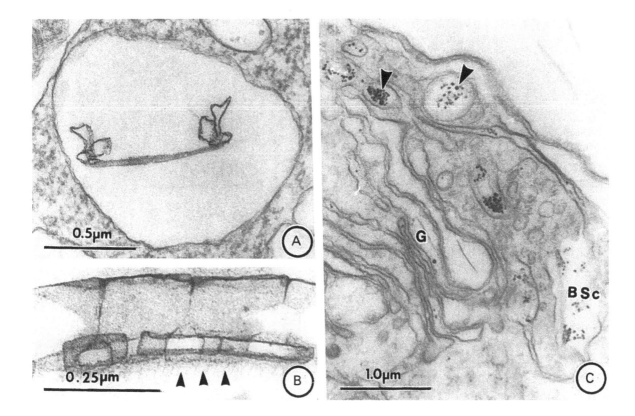

Fig. 4. Coccolithogenesis *P. carterae*

(A) A coccolith seen in sectional view inside a Golgi vesicle. The organic matrix surrounding $CaCO_3$ elements is clearly seen as the $CaCO_3$ has been dissolved away.

(B) A mature coccolith from the exterior of a cell. The $CaCO_3$ has been dissolved during preparation. Note the proximal fibers of the base-plate scale (arrowheads).

(C) Portion of a Golgi body (G) showing vesicles containing coccolithosomes (arrowheads) some of which are associated with the base-plate scale (BSc).

arranged as a well defined cell wall (Figs 2F–H). Individual scales are held together by what may be the equivalent of hemicellulose in plant cell walls (Fig. 2H).

There are coccolithophores which lack the presence of distinct organic scales. The most well known example is *E. huxleyi* (Klaveness, 1972a; Green *et al.*, 1990). In this species the dominant coccolith-bearing stage lacks the presence of organic scales whereas the flagellate motile stage possesses unmineralized scales (Klaveness, 1972b). The chemical composition of unmineralized scales has been investigated by several research groups (for recent reviews see Klaveness and Paasche, 1979; Romanovicz, 1981). The organic scales of *Pleurochrysis scherffelii* Pringsheim were shown to contain a cellulose-like polymer, as indicated by the glucose residues produced when isolated scales were subjected to strong acid hydrolysis. Using more refined techniques it was demonstrated that in addition to the cellulose-like polymer, the scales also possess pectin-like sulphated and carboxylated polysaccharides. More detailed analysis revealed that the spiral microfibrils yielded 37% protein and 63% carbohydrate of which the principal monosaccharide was glucose. Other fractions, such as the coating of the spiral microfibrils, the radial microfibrils and an amorphous polysaccharide fraction, yielded much lower proportions of protein (6–9%) as well as a variety of monosaccharide residues such as galactose, fucose, arabinose, mannose, and glucose (Romanovicz, 1981). Both types of unmineralized scales have been shown to be produced in the dictyosomal system.

Some coccolithophore genera possess organic scales as well as coccoliths, for example, *Pleurochrysis* (Pienaar, 1969a; Gayral and Fresnel, 1983), *Hymenomonas* (Pienaar, 1976b) and *Syracosphaera* Lohman (Inouye and Pienaar, 1988) whereas some lack the presence of organic scales, for example, *Emiliania* (Klaveness, 1972a) and *Umbilicosphaera* Lohman (Inouye and Pienaar, 1984).

CALCIFIED SCALES (COCCOLITHS)

One of the most distinctive features of the coccolithophore cell is the presence of highly characteristic calcified scales called coccoliths. These calcified scales usually occur as a single layer immediately external to the organic scales (if present), for example, *Hymenomonas lacuna* (Figs 5D-E). In some stages of the life cycle the coccolith is the only type of scale that covers the cell, for example, *Emiliania huxleyi*

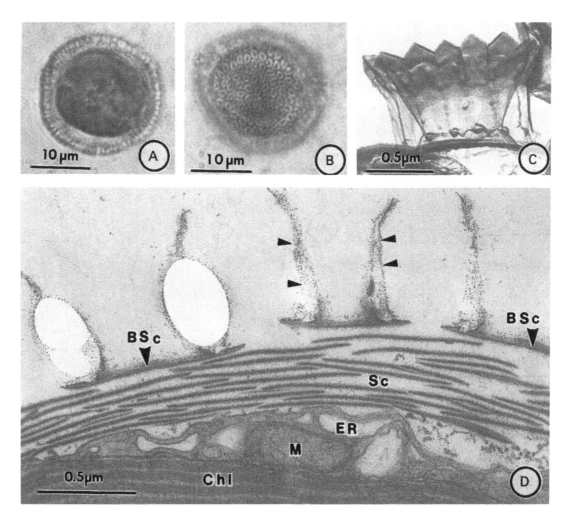

Fig. 5. *Hymenomonas lacuna*

(A), (B) Light micrographs of a through focus of a non-motile cell. Note the well defined coccolith casing.

(C) Carbon/platinum replica of a single coronate coccolith.

(D) Section of a periphery of a cell. Note (i) the single layer of coccolith made up of a base-plate scale (BSc) and the organic matrix (arrow heads). The $CaCO_3$ has been dissolved during prepartion. (ii) the well defined region of numerous circular organic body scales (Sc). The position of the peripheral endoplasmic reticulum beneath the plasma membrane can be clearly seen (ER).

(Klaveness, 1972a). In other coccolithophores the deposition of the inorganic components takes place on a preformed organic base-plate scale, for example, *Pleurochrysis*. The base-plate scale is usually oval to circular in shape. When studied after heavy-metal shadowing the proximal surface is seen to possess radiating fibrils and the distal surface to be composed of an amorphous material devoid of any specific ornamentation (Figs 3A–B). The base-plate scale has been shown to be coated with a polysaccharide.

The inorganic component of the coccolith is $CaCO_3$ and is deposited in the form of calcite. Individual $CaCO_3$ elements are deposited in characteristic patterns. The final form of the coccolith is used in the taxonomic subdivision of coccolithophores. The importance of coccolith morphology in systematics must always be used with circumspection as many coccolithophores exhibit coccolith dimorphism and polymorphism, for example, *Syracosphaera* (Inouye and Pienaar, 1988)(see also Chapter 4), whereas others have different types of coccoliths in different life cycle stages, for example, *Coccolithus pelagicus* (Wallich) Schiller (Parke and Adams, 1960).

In *Syracosphaera pulchra* Lohmann there are three distinct types of coccoliths. These are proximal (made up of ordinary and stomatal coccoliths) and distal coccoliths (Figs 8A–F). Proximal coccoliths are of two types – the one referred to as 'ordinary coccoliths' (which cover the entire cell surface except at the flagellar pole) (Fig. 8F) and the 'stomatal coccoliths' which are only found at the flagellar pole (Fig. 8D).

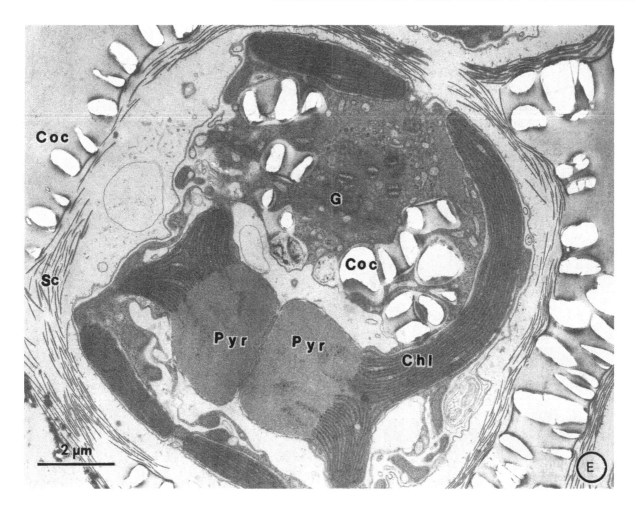

Fig. 5. *Cont.*
(E) Section of a cell of *Hymenomonas lacuna* to show the coccolith (Coc) and scale (Sc) boundary, the production of numerous intracellular coccoliths (Coc) and the two chloroplasts (Chl) with conspicuous bulging pyrenoids (Pyr).

The ordinary coccolith has an elliptical base-plate scale and three kinds of calcified elements, two of which build up the raised ridge and the third element which is lamellate and arranged like radiating spokes on the base-plate scale. The stomatal coccoliths are identical to the ordinary coccolith except they possess a central rod (Fig. 8D). The distal coccolith is monomorphic, dome shaped, and has no associated organic base-plate scale. The distal coccolith is composed of four different $CaCO_3$ elements.

In most coccolithophores the coccoliths form a single layer external to organic scales, if present. Some species have more than a single layer of coccoliths, for example, *Syracosphaera pulchra* (Inouye and Pienaar, 1988) which exhibits a double-layered coccolith case.

Coccoliths and scales are always found in precise positions relative to the plasma membrane. It has been suggested that to explain their precise position on the cell surface there may be a rhythmic production of the periplast components (Pienaar, 1976a). The production of scales and coccoliths could also be linked to the cell division cycle. Once the scales and coccoliths have been released to the exterior of the cell they must be held in position to produce the distinc-

tive species specific periplasts. In some coccolithophores there is evidence that individual overlapping coccoliths become firmly attached to each other by some extracellular calcification. In others, where the coccoliths do not overlap, there is evidence that a mucilage-like material may be involved in maintaining the integrity of the periplast region (Pienaar, 1969a). In scale-bearing algae like *Pyramimonas* Schmarda (belonging to the class Prasinophyceae) it has been suggested that the diverse and complex scale types that cover the cell and flagellar surfaces may be held in their precise positions by electrical charges (Aken, 1985).

In *Calyptrosphaera sphaeroidea* (which produces holococcoliths and organic scales) the coccoliths and scales are found external to the plasma membrane but are often found covered by a 'skin-like' layer which may play some role in maintaining the position of the periplast components (Fig.

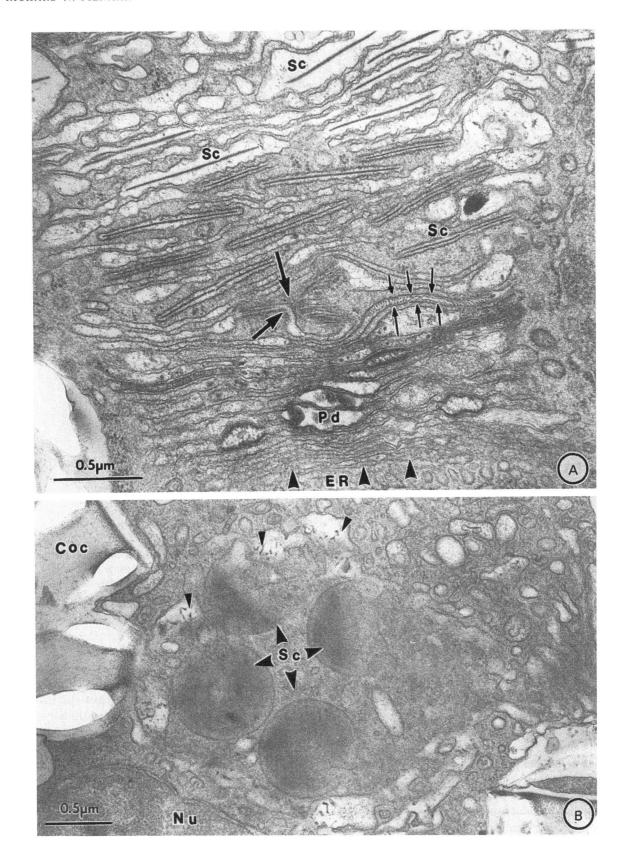

Fig. 6. Scale production in *Hymenomonas lacuna*

(A) Section of an entire Golgi body to show the sequential production of scales (Sc). Note the peculiar dilations (Pd) with the remains of an electron dense metabolite. The small arrows point to a vesicle in which the two surfaces of the scales are being laid down.

Another view of scale production can be seen (large arrows).

The forming face of the Golgi body and associated ER (ER) can be seen towards the base of the micrograph.

(B) Horizontal section through a single Golgi cisternum to show the simultaneous production of four circular organic scales (Sc) and the tubular material (arrowheads) which is eventually found external to the plasma membrane. Intracellular coccoliths (Coc) are also evident.

(C) Section of the Golgi body of *Hymenomonas lacuna* to show its large size and the sequential production of scales (Sc). The peculiar dilation (Pd) and the earliest stage of scale formation (arrows) is clearly visible.

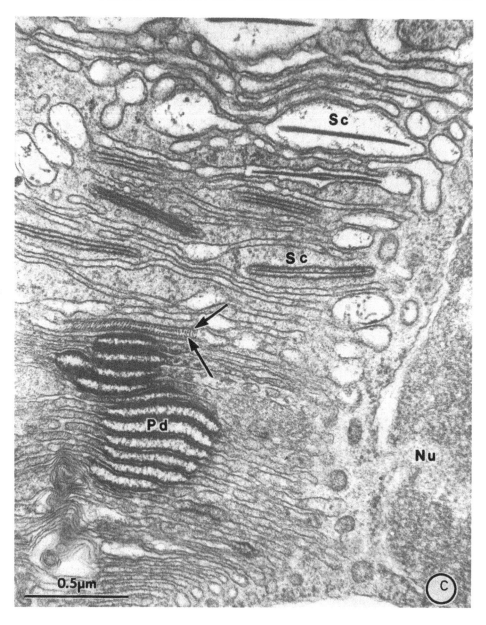

7D). In some coccolithophores a layer of columnar/tubular-like material has been found immediately external to the plasma membrane, for example, *C. sphaeroidea* and *Pleurochrysis carterae*, *Coccolithus pelagicus* and *Hymenomonas lacuna* (Fig. 6D). These tubules produce a dense mat and have been implicated in the positioning of the scales on the cell surface (Manton and Leedale, 1969).

Flagellar Apparatus

Many coccolithophores are motile or produce motile cells as a part of their life cycle, for example, *Coccolithus pelagicus* (Parke and Adams, 1960) and *Syracosphaera pulchra* (Inouye and Pienaar, 1988). Those coccolithophores that possess flagella usually have two subequal to unequal fla-gella which lack any hair-like appendages. The flagellar action is usually heterodynamic and the flagellar surfaces are devoid of any trace of scale ornamentation.

The flagellar apparatus has been demonstrated to be of importance in that it is considered to be an evolutionarily conservative feature and as such is of phylogenetic importance. Only recently have detailed studies on the ultrastructure of the flagellar apparatus of chlorophyll *c* containing algae been undertaken. Members of the class Prymnesiophyceae that have been investigated include the non-coccolithophores *Diacronema vlkianum* Prauser (Green and Hibberd, 1977), *Pavlova pinguis* Green (Green, 1980), *Imantonia rotunda* Reynolds (Green and Hori, 1986), *Chrysochromulina apheles* Moestrup and Thomsen (Moestrup and Thomsen, 1986), and the coccolithophores

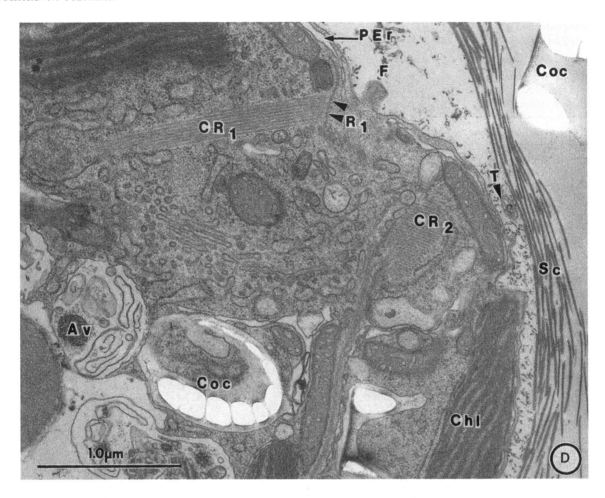

Umbilicosphaera sibogae var. *foliosa* (Kamptner) Okada and McIntyre (Inouye and Pienaar, 1984), *Pleurochrysis carterae* (Braarud and Fagerland) Christensen (Beech and Wetherbee, 1988), and an unnamed species of *Pleurochrysis* (Inouye and Pienaar, 1985).

Terminology relating to the flagellar apparatus of coccolithophores has been established. The flagella and haptonema usually emerge from a shallow apical depression. Beech and Wetherbee (1988) have shown that the longer flagellum is associated with root 1(R1) and hence corresponds to the left flagellum; the right flagellum is therefore the shorter. These workers have demonstrated that a peculiar helical band in the basal region of the axoneme and the transitional region consists of a distinctive axosome and at least five tiers of transitional rings (Fig. 10). All motile coccolithophores possess the usual '9 + 2' arrangement of microtubules in the axoneme. The species that have been studied possess two complex microtublar roots (R1 and R2) both attached to the left flagellum and each comprised of two portions. The principle portion of each root is composed of a sheet of closely aligned microtubules (R1 and R2) from which a crystalline bundle of microtubules arises at right angles (CR1 and CR2). The principle sheet of the R1 root is much broader

Fig. 6. *Cont.*

(D) Section through the anterior of cell to illustrate certain aspects of the microtubular flagellar roots (R1, CR1 and CR2). The peripheral endoplasmic reticulum (PEr), scale boundary (Sc), coccolith (Coc) and tubular region (T) is clearly visible. Note the small autophagic vacuole (lysozyme) (Av).

(20–25 microtubules) than that of the R2 root (± 7 microtubules) (Inouye and Pienaar, 1984, 1985; Beech and Wetherbee, 1988). The third, and least conspicuous microtubular root (R3), is associated with the right basal body (Fig. 11). In addition to the microtubular roots (R1, R2 and R3) there is a non-striated fibrous root (F) which emerges from the left basal body and merges with the microtubular sheet portion of the R1 root. An interesting observation that Beech and Wetherbee (1988) made was that the fibrous root (F) and the principle sheet of the R1 root together constitute a composite root that forms the core of the peculiar cytoplasmic tongue (CT). The fibrous root tapers off with distance away from the basal bodies and gives rise to fibres that interconnect the microtubules of the cytoplasmic tongue (CT). Beech and Wetherbee (1988) have also indicated that the

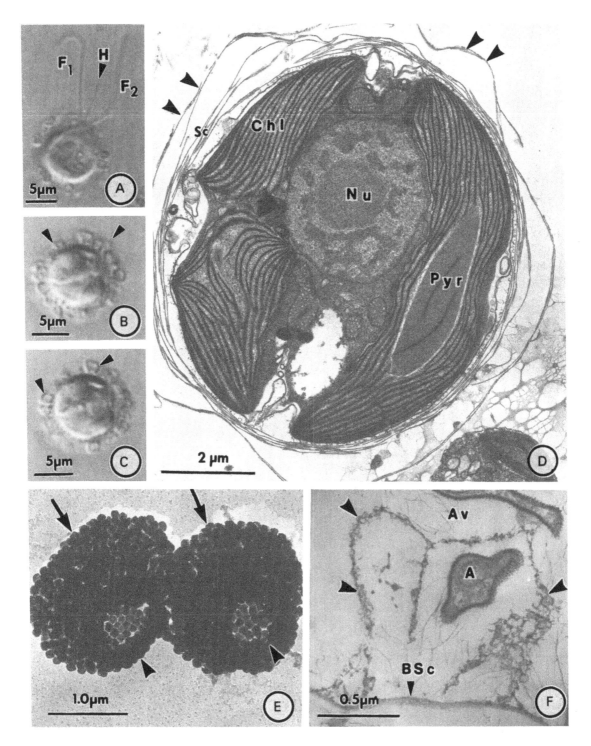

Fig. 7. *Calyptrosphaera sphaeroidea*

(A) Light micrograph (microflash) to show the non-coiling haptonema (H) and the two flagella of a motile coccolith bearing cell.

(B), (C) Through focus to show the well defined hat-shaped calyptroliths (arrows).

(D) Section of a cell. Note the scale boundary (Sc) covered by a 'skin-like' layer (arrow heads) the calyptrolith were dissolved during preparation.

(E) Direct preparation to show two coccoliths (calyptroliths) with their distinct proximal region (arrowhead/and the numerous rhombic crystals of $CaCO_3$ (arrows).

(F) View of a single coccolith which had been injested by an amoeba (A) and is contained in a food vacuole (Av). The organic matrix is visible (arrowheads) as is the base-plate scale (BSc).

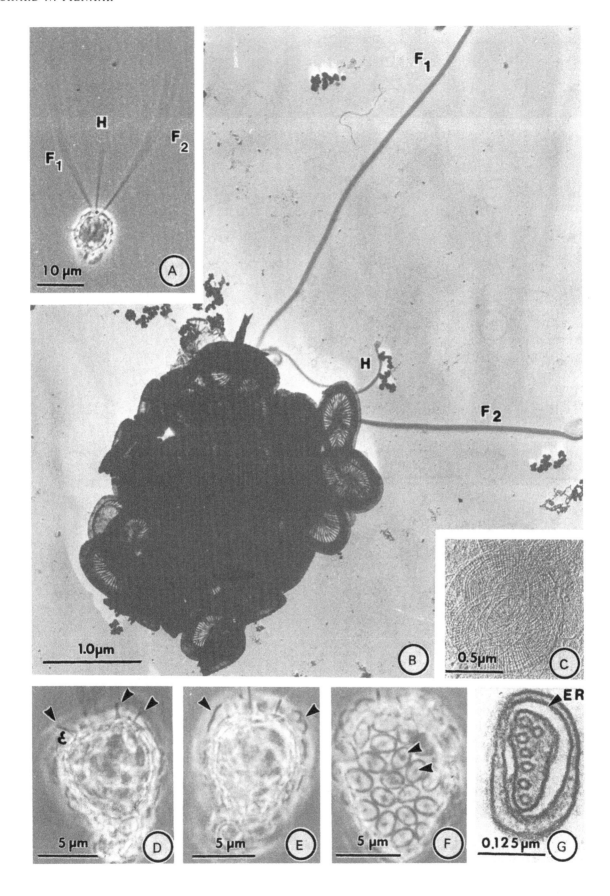

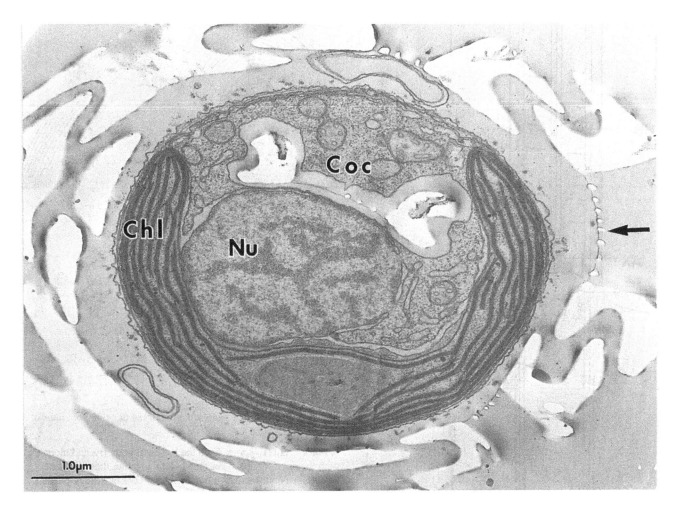

Fig. 9. *Emiliania huxleyi*
Section of a cell show the production of a single intracellular and many extra cellular coccoliths (Coc). Note the lack of organic body scales.

Fig. 8. *Syracosphaera pulchra*

(A) Motile coccolith-bearing cell. Note the extended haptonema (H) and the two flagella (F_1 and F_2).

(B) Heavy metal (Au/Pd) preparation of a motile coccolith bearing cell. The haptonema (H) is much thinner than the flagella (F_1 and F_2).

(C) Circular organic body scale – note the lack of a rim.

(D)–(F) Through focus of a cell to show the different types of coccoliths. (D) arrowhead = stomatal coccolith (proximal coccolith); (E) arrowhead = distal coccolith; (F) arrowhead = ordinary coccolith (proximal coccolith).

(G) Transverse section through the haptonema. Note the layer of endoplasmic reticulum (ER) and the seven single microtubules.

broad sheet component of the R1 is associated proximally with the forming face of the Golgi body and then becomes partially enveloped by the peripheral endoplasmic reticulum (PER) as it extends for the entire length of one side of the cell. This whole structure has been described as the cytoplasmic tongue.

In addition to the microtubular and fibrous roots there are a number of electron-dense bands linking the flagellar bases and the haptonema base. These are:

1. AB1 – a striated accessory band connecting the right flagellar base with the haptonema base.
2. AB2 – a striated accessory band connecting the left flagellar base to the microtubular R1 root.
3. AB3 – a non-striated accessory band connecting the base of the left flagellum to the base of the haptonema.

25

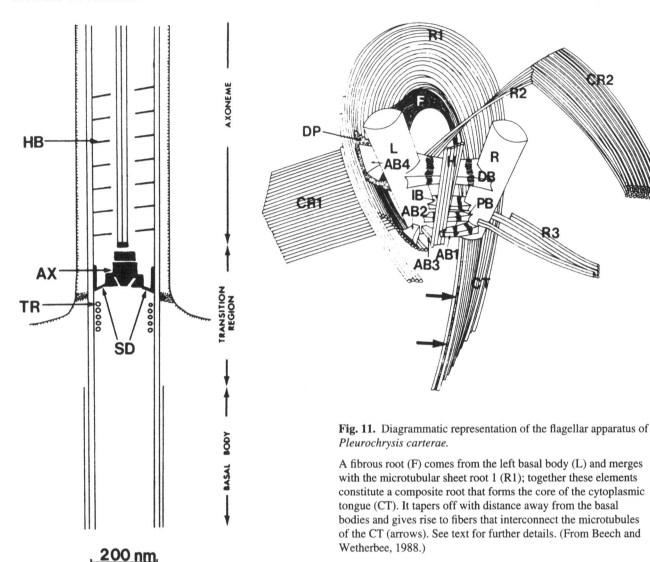

Fig. 10. Diagrammatic representation of a median longitudinal section through the axoneme, transition region and basal body of a flagellum of *Pleurochrysis carterae.*

The helical band (HB) is illustrated as it would appear in a tangential section of a flagellum.

AX = Axosome; TR = Transitional ring; SD = Septal disc.

(From Beech and Wetherbee, 1988.)

Fig. 11. Diagrammatic representation of the flagellar apparatus of *Pleurochrysis carterae.*

A fibrous root (F) comes from the left basal body (L) and merges with the microtubular sheet root 1 (R1); together these elements constitute a composite root that forms the core of the cytoplasmic tongue (CT). It tapers off with distance away from the basal bodies and gives rise to fibers that interconnect the microtubules of the CT (arrows). See text for further details. (From Beech and Wetherbee, 1988.)

4. AB4 – a non-striated accessory band connecting the distal connecting fibrous band to the left flagellar basal body.
5. PB – a striated proximal band connecting the lower regions of the basal bodies of the left and right flagella.
6. DB – a large striated distal band connecting the upper region of the basal bodies of the two flagella.
7. IB – an intermediate striated fibrous band connecting the mid region of the two flagellar basal bodies.

8. DP - a distal connecting fibrous plate associated with the R1 root.

All coccolithophores studied have a similar flagellar apparatus and share one common character, namely, two of the three microtubular roots (root 1, CR1 and root 2, CR2) are of the compound type. The compound flagellar root has been observed in many other coccolithophores even though the detailed structure of their flagellar apparatus is presently unknown (see Inouye and Pienaar, 1985). This suggests that the secondary bundle of microtubules is common to most coccolithophores. The one known exception is *Syracosphaera pulchra* (Inouye and Pienaar, 1988). In this species the secondary bundles of microtubules (CR1 and CR2) are absent suggesting that *S. pulchra* is a very unusual coccolithophore with respect to its flagellar roots. It seems to more closely resemble the Prymnesiales – a point borne out by its haptonemal structure.

Haptonema

One of the characteristics of the Prymnesiophyceae is the presence of a haptonema in those cells that have a motile stage. The haptonema was first observed with the light microscope and referred to as a third flagellum-like appendage even though earlier workers were able to observe that it did not beat like a true flagellum. Manton (1964b, 1967) recognized the uniqueness of the haptonema. Its ultrastructure was first elucidated in members of the order Prymnesiales, for example, *Chrysochromulina* and *Prymnesium*. The haptonema is a highly variable organelle. In the genus *Chrysochromulina* it can be long and capable of coiling and uncoiling. In the genus *Prymnesium* it is short and non coiling. In coccolithophores the haptonema is also variable. In the much studied genus *Pleurochrysis* the haptonema is bulbous and can vary in length from 2–4 μm (Pienaar 1969a,b; Inouye and Pienaar 1985) and be covered with one or more layers of small elliptical organic scales. In *Syracosphaera pulchra* (Figs. 8A–B) and the motile (*Crystallolithus*) stage of *Coccolithus pelagicus* the haptonema is long and capable of coiling (Manton and Leedale, 1963; Inouye and Pienaar, 1988). In *Calyptrosphaera sphaeroidea* the haptonema is straight and does not coil (Fig. 7A) although it bends once the cell dies (Klaveness, 1973; Pienaar, pers. obs.).

The ultrastructure of the haptonema is not well studied in coccolithophores although the number of microtubules organized in the emergent part and haptonema base is known in several species. In *Hymenomonas lacuna* Pienaar (Pienaar, 1976b) there are three microtubules in the haptonema base, four in *Pleurochrysis carterae* (Braarud) Gayral and Fresnel, five in *Ochrosphaera neapolitana* Schussing (Gayral and Fresnel-Morange, 1971) and *Jomonlithus littoralis* Inouye and Chihara (Inouye and Chihara, 1983), and eight in *Calyptrosphaera sphaeroidea* Schiller (Klaveness, 1973), *H. globosa* (Magne) Gayral and Fresnel (Gayral and Fresnel, 1976), *Pleurochrysis* sp. (Inouye and Pienaar, 1985) and *S. pulchra* Lohmann (Inouye and Pienaar, 1988).

The emergent part of the haptonema is less variable and contains six microtubules in the motile phase of *Coccolithus pelagicus* (Manton and Leedale, 1963) and *Calyptrasphaera sphaeroidea* (Klaveness, 1973). In *S. pulchra* there are seven microtubules in the emergent part of the haptonema (Fig. 8G). In many members of the Prymnesiales the emergent part of the haptonema possesses seven and the haptonema base eight microtubules. In some Prymnesiales the number of microtubules in the haptonema base increases to nine.

Many coccolithophores have vestigial or reduced haptonemata with regard to length and/or the number of microtubules. *S. pulchra* is currently the only known coccolithophore possessing a haptonema of the more common type suggesting a close affinity between *S. pulchra* and representatives of the Prymnesiales. This is further supported by the similarity in structure of the flagellar apparatus of *S. pulchra* and members of the Prymnesiales (Inouye and Pienaar, 1988).

The function of the haptonema is still largely unknown. It has been shown to be responsible for the attachment of some cells of the Prymnesiales to a substratum, for example, *Prymnesium* and *Chrysochromulina* (Leadbeater, 1971a). It has also been implicated as a sensory and tactile organelle. Some Prymnesiophyceae have been demonstrated to be able to ingest particulate matter, for example, other algal cells, bacteria, etc. This has been observed in the genus *Chrysochromulina* (Pienaar and Norris, 1979) and may also occur in certain coccolithophore cells, for example, *Pleurochrysis* (Manton and Peterfi, 1969). More recently Kawachi *et al.* (1991) using ultra high speed video techniques were able to demonstrate the possible capturing of particulate matter by a coiling haptonema in a species of *Chrysochromulina*. Particulate matter attaches to the haptonemal tip and moves to a position on the haptonemal shaft, where the particulate matter collects. Once a certain amount has collected, the haptonema bends and deposits the particulate matter at the posterior end of the cell where it is taken in by phagocytosis and digested. A great deal more work on haptonemal function and ultrastructure clearly needs to be done before we fully understand this unusual organelle.

Cell Ultrastructure

A common feature of all prymnesiophycean cells is the layer of peripheral endoplasmic recticulum which is located just internal to the plasma membrane (Figs. 5D, 6D). This layer of endoplasmic reticulum is the same one which is found to occur in the haptonemal shaft. Beech and Wetherbee (1988) have demonstrated how the R1 flagellar root consisting of a broad sheet of microtubules becomes partially enveloped by the peripheral endoplasmic reticulum as it extends for the entire length of one side of the cell. They referred to this area as the cytoplasmic tongue. This cytoplasmic tongue has been observed in *Pleurochrysis carterae* (Beech and Wetherbee, 1988) and *P. pseudoroscoffensis* Gayral and Fresnel (Gayral and Fresnel, 1983).

The peripheral endoplasmic reticulum is very difficult to preserve during the fixation process and often results in 'blebbing' of the cell which in turn results in displacement of the scales and coccoliths during the fixation and dehydration process. The exact function of this peripheral endoplasmic reticulum layer is as yet unknown.

Plastids

Cells usually possess one or two plastids surrounded by a double unit-membrane chloroplast envelope. Internal to the chloroplast envelope the plastid possesses transversing bands of two or three closely adpressed thylakoid lamellae (Figs. 5D–E). The stroma contains 70S ribosomes and regions of DNA nucleoids. All coccolithophores that have been studied possess pyrenoids. These are localized electron-dense areas which may be lens shaped and immersed in the chloroplast (fusiform pyrenoids), or they occur as bulges on the inner face of the plastid towards the center of the cell, for example, *Hymenomonas* (Pienaar, 1976b) and *Pleurochrysis* (Pienaar, 1969b). The pyrenoid matrix is transversed by one or a few pairs of parallel thylakoids (Figs. 5E, 7D). Chloroplasts of all members of the Prymnesiophyceae do not possess girdle lamellae (Hibberd, 1976).

Plastids of the Prymnesiophyceae contain chlorophylls a, c_1 and c_2 together with a number of carotenoids, the most commonly occurring being β carotene, diatoxanthin, diadinoxanthin and fucoxanthin. *Emiliania huxleyi* is unusual in having the fucoxanthin derivative 19'-hexanoyloxy fucoxanthin which occurs in some dinoflagellates (Berger et al., 1977).

Surrounding the plastids and associated pyrenoids is a layer of endoplasmic reticulum which is confluent with the nuclear envelope (Figs. 7D, 9). The outermost membrane surface of this chloroplast endoplasmic reticulum possesses attached ribosomes whereas the membrane surface nearest the chloroplast is devoid of ribosomes (Fig. 1). In some coccolithophores, for example, *P. carterae*, there is often a complicated network of membranous profiles between the chloroplast endoplasmic reticulum and the chloroplast envelope. This periplastidal network may be involved in the transportation of synthesized metabolites from the plastid to other regions in the cell.

The position of the nucleus is variable and probably to some extent depends on the cell division cycle but usually it occupies a median to posterior position and occurs between the two plastids where its nuclear envelope often buds off and is confluent with the chloroplast endoplasmic reticulum (Fig. 9).

Eyespots have not been observed in coccolithophores or in the Prymensiophyceae in general. One exception is within the order Pavlovales where *Pavlova* Butcher has been shown to possess a well defined intra-plastidial eyespot (Green, 1980).

Dictyosome

A very conspicuous organelle in the Prymnesiophyceae is the single dictyosome which makes up the Golgi body. The Golgi body dominates the anterior region of the cell and is situated immediately beneath the flagellar basal region. The Golgi body is highly distinctive in that it is large and composed of many cisternae. A well defined *trans* and *cis* face can be observed. A distinctive feature of the Golgi body is its characteristic fan shape with the one edge being polarized towards the flagellar bases and the opposite end fanning out towards the center of the cell. Midway up the cisternal stack are well defined cisternal dilations located towards the middle of the cisternae (Figs. 6A, C). These were first described by Manton (1967) and referred to as 'peculiar dilations'. These dilations may be filled with an electron-dense substance or may be devoid of any densely stained material depending upon the stage in the cell division cycle when the cells were fixed for electron microscopy. The contents of these dilations have been implicated in scale formation (Parke *et al.*, 1959; Romanovicz, 1981). The single dictyosome is responsible for the production of organic body, haptonemal and base-plate scales and ultimately with the calcification phase to produce the coccolith. In some coccolithophore genera the cisternum in which the coccoliths are produced is too large to be accommodated in the normal array of Golgi cisternae and the vesicle in which coccoliths are produced occurs as a 'T'-shaped extension (Manton and Leedale, 1969). The role of the Golgi body in scale and coccolith production is discussed later in this chapter.

Mitochondria and Other Organelles

Mitochondrial profiles are observed in most sections but are more conspicuous towards the anterior of the cell. There is evidence to suggest that each cell possesses a single branched and often fenestrated mitochondrion, for example, *Pleurochrysis carterae* (Beech and Wetherbee, 1984).

In addition to the normal complement of organelles that have been previously described, some coccolithophore cells possess other membrane-bound organelles which may only be observed under certain culture conditions or as the cells age. One of these is an autophagic vacuole which has been shown to be rich in acid hydrolases (Pienaar, 1971) and probably has a similar function to a lysozyme (Fig. 6D). These organelles are bound by a single unit membrane and sometimes possess the remains of membranes, scales, and even coccolith profiles. There may be several autophagic vacuoles per cell and they are normally closely associated with the main posteriorly located vacuole. These organelles may originate from the dictyosomal system.

Other unusual unit membrane-bound organelles have been detected in certain coccolithophore cells and have been implicated in coccolith production. These are a special vacuolar system not directly connected to the Golgi body consisting of a coccolith vesicle and a reticular body as observed in *Emiliania huxleyi* (Westbroek *et al.*, 1989). In addition to the various membrane-bound organelles there are other structures such as lipid droplets which are often located adjacent to the plastid. These have been reported to become more abundant as the cells age (Pienaar, 1969b).

The storage metabolite of coccolithophores is reported to be leucosin (chrysolaminarin) which is a β 1-3 glucan. Strangely, this has never been confirmed but is reported to occur in vacuolar-like regions towards the posterior of the cell. In addition to the storage metabolite leucosin, *E. huxleyi* possesses lipids and sterols. Marlowe *et al.*, (1984) have reported the following unusual lipids and sterols that have not been reported elsewhere in the algae. These are polyunsaturated methyl and ethyl ketones and esters. *Emiliania huxleyi* together with the non-coccolithophore genera *Chrysotila* Anand and *Isochrysis* Parke also have particularly high concentrations of the sterol 24-methycholesta-5, 22 E-dien- 3β-o1 which has also been detected in some members of the order Coccolithophorales. Under certain conditions both in the natural state and under culture conditions, cells may become infected with viral-like particles (Pienaar, 1976c). These are usually located in the cytoplasm.

It is important to note the position of the two flagella relative to the haptonema. These three appendages are not found in one plane which means that a single longitudinal section will never show all three organelles in median longitudinal section. The flagella and haptonema are usually found emerging from a slight apical depression (Fig. 1). The endoplasmic reticulum associated with the haptonema is an extension of the PER. The highly polarized Golgi body occurs just beneath the flagellar and haptonema bases and the flagellar roots can be clearly observed.

The anterior flagellar pole (Figs 2A–C, 6D) is the only region of the cell where the plasma membrane is not occluded by the plastids and associated membranous boundaries. This is of importance when considering scale and coccolith production and their subsequent release as well as the ingestion of particulate matter. In some instances there may be a region at the opposite posterior region that may not be occluded by the plastid and peripheral endoplasmic reticulum.

Scale and Coccolith Production

A distinctive feature of the Class Prymnesiophyceae is the presence of scales external to the plasmalemma. As previously mentioned these may be of several types. The two main types are: (1) purely organic scales which when present form one or more layers immediately external to the plasmalemma and (2) coccoliths having an inorganic component of $CaCO_3$ elements in association with an organic matrix but with or without an organic base-plate scale.

Because of the importance of the architecture and structure of scales and coccoliths in the identification of many prymnesiophycean taxa, a great deal of emphasis has been placed on scale and coccolith structure and formation.

Particular attention has been given to the process of calcification during coccolith production as many workers believe that coccolithophores may be useful unicellular organisms in understanding the process of $CaCO_3$ deposition. Some workers have used these unicellular organisms as single-cell models of bone formation (Isenberg *et al.*, 1966). For convenience, the production of scales and coccoliths will be discussed separately.

Scale production

Hibberd (1980) and Romanovicz (1981) have given comprehensive reviews of scale production and readers are referred to these two excellent articles for detailed accounts. Since the early studies using electron microscopy to examine scale-bearing flagellates, scale profiles have been detected in the Golgi cisternae and vesicles closely associated with the Golgi body. The organelle is therefore implicated in the synthesis and subsequent deposition of material to produce scales. In the large Golgi body found in certain coccolithophore genera, for example, *Hymenomanas lacuna* (Pienaar, 1976a) (Figs 6A, C) and *Pleurochrysis carterae* (Pienaar 1969b; Outka and Williams, 1971) sequential stages of scale production can be observed.

Scale profiles have never been detected near the forming face of the Golgi body. Evidence for scale synthesis can be observed in the cisterna just posterior to the peculiar dilations (Figs 6A, C). Here one is able to detect the synthesis of the two distinct structural surfaces of organic and base-plate scales. The immature scales are frequently observed towards the middle of the Golgi body and the mature scales are found towards the mature face where they are often seen in more dilated Golgi-derived vesicles (Figs 6A, C). In some genera, as many as four circular unmineralized organic scales can be found associated with a single Golgi cisternum (Pienaar, 1976b) (Fig. 6B). Tubules within the dictyosomal cisternae may also be involved in scale synthesis and this may account for the highly distinctive pattern of deposition of the radial microfibrils (Pienaar, 1969b). In *Pleurochrysis scherffelii* there is evidence that the radial fibrils are laid down before the spiral fibrils and that the amorphous polysaccharide coating material is added concomitantly with the spiral fibrils.

Numerous authors have reported that *Emiliania huxleyi* in its coccolith-bearing phase does not possess organic body scales. In a more detailed study, workers have detected 'flake-like' structures immediately external to the cell membrane and between the coccolith covering and the cell membrane (van der Wal *et al.*, 1983B). These flake-like structures have only been observed when cells have been subjected to specific staining for polysaccharides using the method described by Thiery (1967). Similar structures have also been observed in the Golgi body which suggests that they are homologous to the organic scales found in other coccolithophores.

Within the Prymnesiophyceae as a whole, the orientation of the scale within the Golgi vesicle may vary from species to species and depends to some extent on the size and morphology of the scales being synthesized. In the case of coccolithophores where the shape, size and ornamentation of the organic scale is simple and relatively small, the scales are produced within the normal array of Golgi cisternae. Pienaar (1976a) and Ariovich (1980) have indicated that there may be a rhythmic production of organic body scales in *P. carterae* and that the synthesis of organic scales is probably controlled by an endogenous circadian rhythm. Once the complete scales have been synthesized, they migrate to the cell surface, usually towards the region of the flagellar insertion, where they are released via a process of reversed pinocytosis.

It is evident from the literature on scale production in the Prymnesiophyceae that we know very little about the underlying mechanisms of scale production and that much work still needs to be done before this complex process is fully understood. For example, what controls the precise shape and size of the scale being produced? This question is even more pertinent when one considers the very complicated, diverse, and species specific scales that are characteristic of other species in the Prymnesiophyceae, for example, *Chrysochromulina*, and in particular those species which produce spine scales whose length is orders of magnitude larger that the cell in which they are produced.

Coccolith production

In the majority of coccolithophores that have been studied in detail it has been shown that coccoliths are associated with a structure referred to as a base-plate scale (Pienaar 1969a, 1969b). This base-plate scale often has a distinctive shape and two distinctive surfaces. It is on this organic base-plate scale that $CaCO_3$ elements are deposited. Although not all coccolithophores have been demonstrated to possess organic base-plate scales, all have been shown to possess an organic matrix which plays an important role in determining the form and shape of the individual $CaCO_3$ elements that are deposited. In the case of *Emiliania huxleyi* the coccolith-bearing phase does not possess base-plate scales or organic scales but certain alternate stages of its life cycle are motile and possess organic scales.

Syracosphaera pulchra is even more unusual in that it produces proximal and distal coccoliths of which the proximal coccoliths are of two distinctive types. The proximal coccoliths have organic base plates but the distal coccoliths not only have a different arrangement of the $CaCO_3$ elements but lack any trace of a base-plate scale (Inouye and Pienaar, 1988). *Umbilicosphaera sibogae* var. *foliosa* is also unusual in that it does not possess any unmineralized scales but does possess an unusual base-plate scale which is devoid of any pattern and does not have the usual radiating pattern

of microfibrils on the proximal surface (Inouye and Pienaar, 1984). In *Coccolithus pelagicus* no patterned base-plate scale has been reported when specimens are studied using heavy-metal-shadowed preparations although ultrathin-sectioned material reveals a base-plate with a well defined pattern of radiating fibrils (Manton and Leedale, 1969).

In summary, four combinations of scales, base-plates and coccoliths exist:

1. Those coccolithophores that do not possess any base-plate scale associated with the mature coccolith, for example, *E. huxleyi* (Klaveness, 1972a, 1972b).
2. Those that possess a base-plate scale which appears patternless when heavy metal shadowed preparations are viewed. When the same species is viewed with ultra-thin sections the base plate has a distinct radiating pattern of microfibrils, for example, *U. sibogae* var. *foliosa* (Inouye and Pienaar, 1984) suggesting that the patternless amorphous material originally covered the radiating fibrils but during preparation for electron microscopy the amorphous material was removed – possibly during dehydration. An alternate explanation could be that the amorphous material was not fixed and was lost during dehydration.
3. Those that possess two layers of coccoliths (proximal and distal coccoliths) in which the proximal coccoliths possess a base-plate scale but the distal coccoliths are devoid of a base-plate scale, for example, *S. pulchra* (Inouye and Pienaar, 1988).
4. Those that possess well defined base-plate scales with an amorphous distal surface devoid of ornamentation and a proximal surface with well defined radiating fibrils, e.g., *Hymenomonas lacuna* (Pienaar, 1976b) and *Pleurochrysis carterae* (Pienaar, 1969b).

Because of the limited number of species available in culture, our understanding of coccolith formation comes from experimental work that has centered around two genera each bearing heterococcoliths. These are *Pleurochrysis* (Manton and Leedale, 1969; Manton and Peterfi, 1969; Outka and Williams, 1971; Pienaar, 1971, 1976a, 1976b; van der Wal, *et al.*, 1983, 1983a, 1987; Westbroek *et al.*, 1986) and *Emiliania* (Klaveness, 1972a; Klaveness and Paasche 1979; Westbroek *et al.*, 1984, 1986, 1989). More recently some work has been done on the motile (Crystallolithus) stage of *Coccolithus pelagicus* which produces holococcoliths. In this species the calcification of the base-plate scale takes place extracellularly (Rowson *et al.*, 1986). These different coccolithophores show different processes of coccolith production.

COCCOLITH FORMATION IN *PLEUROCHRYSIS*

In *Pleurochrysis* the coccolith consists of an oval ring of calcified elements arranged peripherally on the amorphous distal surface of the base-plate scale (Pienaar 1969a, 1971; Outka and Williams, 1971; van der Wal *et al.*, 1983, 1983a)

(Figs 3A–B, 4A–B). The base-plate scale is produced in the Golgi body. Once the base-plate scale has been produced it is transferred to a more distal region. Vesicles containing electron-dense particles called coccolithosomes (Outka and Williams, 1971) are also produced in the Golgi body and also migrate distally (Fig. 4C). The coccolithosomes have been shown to contain polysaccharide but are also rich in calcium (van der Wal et al., 1983a). At this stage the vesicle containing the base-plate scale (which will eventually become the coccolith) takes on a shape that resembles the form of the mature coccolith. This is particularly evident when these vesicles and scales are seen in sectional view (Fig. 9).

The vesicles containing coccolithosomes fuse with the vesicle possessing the base-plate scale. The coccolithosomes then disintegrate as the calcified rim of the coccolith develops. It appears that calcification and the formation of a polysaccharide coating, which is particularly evident over the rim of the coccolith, are simultaneous processes, the coating material and the calcium being transported simultaneously to the coccolith (van der Wal et al., 1983a). This coating material is often referred to as the 'organic matrix' (Figs 3A, 4A–B, 5D).

Irradiance level has been implicated in the calcification process. Early work implied that light was essential for calcification. Ariovich and Pienaar (1979) presented evidence to suggest that calcium uptake (Ca^{2+}) was a light-dependent process and that calcification could take place in the light and in the dark but that dark calcification was dependant upon a reservoir of Ca^{2+} having been built up during the previous light period. Van der Wal et al., (1983a), however, found no evidence for such Ca^{2+} accumulation. They demonstrated that Ca^{2+} ions were channelled quickly through the cell to the calcification site. In their experiments some Ca^{2+} uptake took place in darkness and was stimulated by previous decalcification (van der Wal et al., 1987).

Coccolith production, that is, base-plate scale production and its subsequent calcification, is a fairly rapid process. This can be demonstrated by taking exponentially growing cells and decalcifying the cells by dissolving the external coccoliths. Lowering the pH of the external medium results in dissolution, which can easily be done by gently bubbling CO_2 through the medium. The resultant drop in pH causes the external coccoliths to lose their $CaCO_3$ elements. In the light microscope the cells appear 'naked' but still possess the base- plate scales and the organic body scales (Fig. 2D). When these cells are placed into fresh medium with the normal pH of 8–8.3 and exposed to light, new intracellular coccoliths are produced and within three hours the first newly produced coccolith can be seen in the region of the Golgi body (Fig. 2D) and shortly thereafter the coccolith can be observed emerging from the region of the flagellar pole. This is the region closest to the mature face of the Golgi body and the only region of the cell where the plasmalemma is not obscured by the chloroplasts and the peripheral endoplasmic reticulum. Within six hours half the cell is covered by newly produced coccoliths and after 12 hours a completely new set of coccoliths cover the entire cell (Ariovich, 1980).

Light and electron microscope observations have revealed that several coccoliths may be produced at any one time in the Golgi body but usually only one base-plate scale per cisterna has been detected. There have also been reports that the formation of coccoliths in Pleurochrysis carterae is a rhythmic process where the base-plate scale and its subsequent calcification takes place within the first eight hours of light when the cells are subjected to a 16L:8D cycle (Pienaar, 1976a; Ariovich, 1980). Coccolithophores that have small cells relative to their large coccoliths, for example, Umbilicosphaera (Inouye and Pienaar, 1984) and Emiliania huxleyi have been shown to produce only one coccolith at a time (Fig. 9). Further production of new coccoliths will not commence until the mature coccolith has been released.

COCCOLITH PRODUCTION IN *EMILIANIA HUXLEYI*

Coccolith formation in E. huxleyi is also an intracellular process, although in this organism the role of the Golgi body in coccolithogenesis is not as obvious as in Pleurochrysis. In Emiliania huxleyi at least three distinct cell types can be found in laboratory cultures (Wilbur and Watabe, 1963; Klaveness, 1972a; Klaveness and Paasche, 1979). These are:

1. The common coccolith-bearing cells, each cell usually has about 15 coccoliths, (C-cells).
2. Naked cells lacking conspicuous extracellular scales or coccoliths (N-cells).
3. Flagellate cells that possess organic body scales but do not produce coccoliths (S-cells). These S-cells may produce smaller crystalline bodies, probably of $CaCO_3$, within a vacuole. This is the X-body described by Wilbur & Watabe (1963).

In addition to these three major cell types a fourth cell type which is amoeboid, is rarely found. The various cell types are probably life cycle stages.

Each coccolith of a C-cell of E. huxleyi consists of a radial array of unit elements. Each element processes four distinct parts: two parts extend up from the center of the element. The inner of these is a block, the outer bends over and has a hammer-shaped end. Below is a region called a lath and directed towards the center of the coccolith is a flake (Westbroek et al., 1989). Each unit element of the coccolith is a single crystal (Watabe, 1967; Mann and Sparks, 1988; also see chapter 4).

Westbroek et al., (1989) have illustrated successive stages of coccolith production starting off with deposition of an oval ring of rhombohedra. Each rhobohedron represents a future element. The four parts of the element grows out from these rhombohedral roots. The transmission electron microscope has enabled workers to demonstrate the sequential phases in coccolith production in *E. huxleyi*. The process occurs intracellularly in a special membrane-bounded coccolith-production compartment (CPC). This CPC compartment consists of a highly structured 'coccolith room', which in the initial stages is closely associated with the nuclear envelope. It surrounds the growing coccoliths as a 'glove'. Distally there is a reticular body, which is interconnected with the coccolith room.

Purified coccoliths have revealed an acidic coccolith polysaccharide (CP). The primary structure of the CP has been partially elucidated (Fichtinger-Schepman, *et al.*, 1981). Immunocytochemical techniques (Van Emburg, *et al.*, 1986) have demonstrated that the CP is associated with the mineral phase of isolated coccoliths. The CP is capable of binding calcium ions and it strongly inhibits the *in vitro* crystallization of $CaCO_3$ (Borman *et al.*, 1982, 1987). Immunocytochemistry of ultra-thin sections through lowicryl-embedded cells showed the presence of the coccolith polysaccharide throughout the coccolith production compartment, the Golgi apparatus and extra cellularly in the space between the cell and the coccosphere (Van Emberg *et al.*, 1986). From these results these workers deduced that the CP plays a regulatory role in the biomineralization process and probably also acts as a 'glue' by which the adhesion of the coccoliths in the coccosphere is secured.

Also associated with the coccoliths is a thin organic plate (the base-plate), which is subtended between the calcite rhombohedra at the onset of calcification (van der Wal *et al.*, 1983b). This base-plate is found in the coccolith room prior to commencement of the calcification process and it may serve as a substrate for crystal nucleation. There is evidence that a protein fraction, which is present in minute quantities, in the coccoliths may be derived from the base-plate (unpublished results of Lingschooten cited in Westbroek *et al.*, 1989).

Cells homogenized with 10% trichloroacetic acid (TCA) have produced intracellular polysaccharide. A small protein is associated with these polysaccharide precursors and it is believed to act as an anchor by which the polysaccharide is fixed to the membrane surrounding the CPC so that the polysaccharide protrudes into the lumen of the vesicle.

In *E. huxleyi*, coccolith production is a light-dependent process and in actively growing cultures one coccolith is produced every two hours (Westbroek *et al.*, 1989). Westbroek *et al.*, (1989) have used the above observations to postulate a working hypothesis for coccolithogenesis in *E. huxleyi*. Their hypothesis is as follows: small polysaccharide-containing vesicles, probably also carrying calcium and carbonate, are pinched off from the Golgi apparatus and are transported to the nuclear surface to form the incipient ICP. Differentiation between the coccolith room and the reticular body, and the formation of the base-plate follows soon afterwards. The precipitation of $CaCO_3$ is inhibited at this stage by the polysaccharide lining the CPC membrane and pervading the lumenal space.

As more Golgi-derived vesicles are produced and fuse with the CPC, the CPC expands. The coccolith room extends in a precisely programmed fashion, gradually assuming the final shape of the coccolith. First a hollow space is created along its periphery. The polysaccharide is pulled away from its interior so that the initial calcite rhombohedra can form. From the rim of the base-plate a hypothetical nucleation factor must extend into the hollow space giving the initial coccolith elements their correct crystallographical orientation. While new membrane material and other components of the coccolith room are continuously supplied from the reticular body, the coccolith room further dilates and the crystals grow out into the extending lumen until they reach the polysaccharide lining of the coccolith room membrane. Eventually, the expansion of the coccolith room is terminated and crystal growth is arrested by the polysaccharide. The polysaccharide is then detached from its anchor, so that it remains fixed only to the crystal surface. The reticular body disappears and the coccolith is extruded to join the shell of coccoliths, which make up the coccosphere.

Cell division

Cell division within the algae has been of considerable interest to phycologists because of its diversity in so many algal groups. (See Pickett-Heaps, 1976; Hori, 1979, for reviews.) Most of the ultrastructural work that has been done on coccolithophores has concentrated on the interphase cell. There has, however, been some work on the ultrastructural changes that have been observed during mitosis and cytokinesis. Species that have been studied include *Cricosphaera roscoffensis* (Hori and Inouye, 1981), *Pleurochrysis carterae* (Stacey and Pienaar, 1980), '*Apistonema*' stage of *Pleurochrysis* (Mesquita and Santos, 1983) and *Emiliania huxleyi* (Hori and Green, 1985).

There is a general pattern among these studied taxa which seems to be similar with impending cell division being heralded by the duplication of major cell organelles. These include the flagellar bases (kinetosomes) and the haptonema, chloroplasts and Golgi body (Manton, 1964a; Hori and Inouye, 1981; Stacey and Pienaar, 1980; Mesquita and Santos, 1983; Klaveness, 1972b). The chloroplasts always commence division on the inner side of the chloroplasts at the mid point of the chloroplast. By late prophase to early metaphase the major organelles have been duplicated.

All coccolithophores that have been studied with respect to their cell division cycle have been shown to possess an open spindle with the complete disassociation of the nuclear envelope. Groups of spindle microtubules are found to traverse through the chromatin mass; this can most easily be seen during metaphase.

Emiliania huxleyi (Hori and Green, 1985) has been shown to exhibit certain dissimilarities when compared with other coccolithophores. Hori and Green (1985) demonstrated for the first time the presence of flagellar bases in the non-motile coccolith-bearing phase of *E. huxleyi*. The flagellar bases are relatively short which may explain why they have never been observed before. They are also unusual in that these workers could not detect any flagellar roots. The most conspicuous features of the mitotic events in *E. huxleyi* are concerned with the metaphase plate and the behavior of the nuclear envelope during nuclear division, particularly during metaphase and anaphase.

Other members of the Prymnesiophyceae exhibit the following stages in the disorganization and reconstitution of the nuclear envelope.

1. Complete breakdown of the nuclear envelope by late prophase.
2. Re-organization of the new nuclear envelope by coalescence and fusion of many vesicles, commencing on the poleward surface of condensed chromatin at or after metaphase.
3. Further development of the nuclear envelope during metaphase and anaphase.
4. Re-establishment of physical continuity of the new daughter nuclear envelope with the sac of chloroplast endoplasmic reticulum after late anaphase (Mesquita and Santos, 1983) or during interphase (Manton, 1964a).

Studies on *E. huxleyi* show that some larger fragments of the old nuclear envelope remain in the cytoplasm and maintain continuity with the chloroplast endoplasmic reticulum. During metaphase these fragments seem to behave as rough surfaced endoplasmic reticulum and they delimit the region in which mitosis takes place. The re-organization of the new nuclear envelope begins early in metaphase. Many vesicles aggregate and eventually fuse along the poleward surface of the metaphase plate resulting in the formation of larger flattened vesicles which are the precursors of the new daughter nuclear envelopes.

When the metaphase cell begins to separate into two daughter masses of chromatin, the endoplasmic reticulum extensions of the chloroplast endoplasmic reticulum expand again towards the metaphase plate and are then incorporated as parts of the new daughter nuclear envelope. This means that the continuity of the new nuclear envelope with the chloroplast endoplasmic reticulum is established at this early stage. There are small (3–5) and large (>10) bundles of spindle microtubules. The main bundles of spindle microtubules may play a role as spindle axis, not contributing directly to the separation of the daughter chromatin plates.

Life cycle

Despite an increasing base of knowledge of coccolithophores from many parts of the world (estimated by some to be in excess of 200 living species), we know surprisingly little about their life cycles (Klaveness and Paasche, 1979). What is clear is that they exhibit considerable diversity with respect to their vegetative and sexual reproduction. Possible reasons as to why so little is known about the coccolithophore life cycle may be because:

1. We still know very little about the precise culture and nutritional requirements of the majority of species, especially the truly planktonic forms making study of life cycles in the laboratory very difficult.
2. The majority of culture isolates have been obtained by single-cell isolation. In cases where a particular species was heterothallic, requiring two strains to complete the life cycle, it would be impossible to demonstrate a complete life cycle under laboratory conditions. Those species that are homothallic would not pose such a problem.

Currently there is information available on the life cycle stages of the inshore euryhaline genus *Pleurochrysis,* such as *P. carterae* (Leadbeater, 1970), *P. scherffelii* (Leadbeater, 1971b; Merrick and Leadbeater, 1979), *P. pseudoroscoffensis* (Gayral and Fresnel, 1983), and two planktonic species *Emiliania huxleyi* (Klaveness, 1972b) and *Coccolithus pelagicus* (Parke and Adams, 1960). In all the elucidated life cycles there is always a non-motile stage which alternates with one or more motile stages. The actual nature of the motile and non-motile stages differs significantly among species.

In some *Pleurochrysis* species the non-motile benthic stage may be in the form of a colony of several cells contained in an investiture. This resembles a cell wall but is in fact made up of many circular unmineralized scales 'cemented' together to produce what at the light microscope level looks like a normal cell wall. This has been found in *P. scheffelii* (Leadbeater, 1971b). In *P. carterae* and *P. pseudoroscoffensis,* the non-motile benthic stage is filamentous and is usually referred to as the 'Apistonema' stage. Here, the cells are surrounded by a complex layer of scales forming a distinct cell wall (Figs. 2F–G). The regular formation of the cell walls of the non-coccolith-bearing benthic phase has been shown to be produced by the rotation of the protoplast within the scale layer (Fig. 2H). This results in the scales being deposited on the cell surface in a regular manner (Brown, 1969).

The life cycle of *Pleurochrysis* is summarized in Fig. 12. It is important to note that the motile coccolith-bearing phase is capable of perpetuating itself by simple binary fission and occasionally it will produce two- or four-celled colonies which possess a coccolith boundary (Fig. 2E).

In the case of *Pleurochrysis*, it is considered that the motile coccolith-bearing phase is diploid and the benthic non-coccolith-bearing phase is haploid (Rayns, 1962).

In addition to the motile coccolith-bearing phase, two

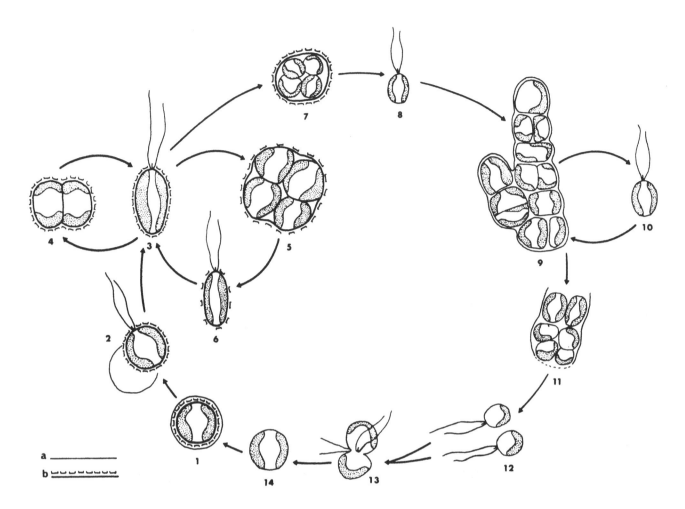

Fig 12. Schematic representation of the life cycle of *Pleurochrysis pseudoroscoffensis* (from Gayral and Fresnel, 1983).

1 = zygote (2n); 2 = release of zygote from the envelope; 3 = motile coccolith bearing cell; 3–6 = various stages of coccolith bearing cells; 7 = meiosis producing four haploid cells still within a coccolith casing; 8 = meiospore; 9 = pseudofilamentous phase (haploid0; 10 = zoospore; 11 = formation of gametes; 12 = motile gametes; 13 = plasogamy; 14 = karyogamy resulting in the formation of a diploid zygote.

a——— = boundary of cells that are haploid; b <u>uuu</u> = boundary of cells that are diploid – these possess coccoliths.

motile non-coccolith-bearing cells (swarmers and gametes) have been observed. One is produced by the motile coccolith-bearing cell and is thought to be produced as a result of meiosis. This haploid swarmer produces the non-motile haploid benthic phase. Under certain conditions it can produce motile cells which can function either as isogametes or as vegetative swarmers that will settle and reform the benthic haploid phase. Should sexual fusion (a process that has rarely been observed and more usually inferred) take place, a diploid zygote is produced which will eventually release a single motile coccolith-bearing diploid cell (Figs. 2A–C). *Emiliania huxleyi* (Klaveness, 1972b) has also been shown to possess a non-motile coccolith-bearing phase which alternates with a motile non-coccolith scale bearing stage.

In the planktonic non-motile species *Coccolithus pelagicus*, the cell is covered by highly characteristic placoliths (heterococcoliths) and the cells are passively transported by ocean currents. Under certain conditions (as yet poorly understood), the non-motile placolith-bearing stage gives rise to a motile stage with a well defined coiling haptonema. This stage now possesses crystalloliths (holococcoliths) and is considered to be the haploid stage of the life cycle. The observation of Parke and Adams (1960) that a single cocco-

lithophore species can have two life-cycle stages that produce two completely different types of coccoliths has serious implications with respect to the systematics of the group. At present coccolith type and morphology have been regarded as being useful in the taxonomy of the group as it was considered that the coccoliths were species specific. Thus many of the planktonic species that have been described may in fact not be separate species but life cycle stages of coccolithophores that await elucidation as our knowledge of culture and nutritional requirements improves.

This problem has recently been highlighted by Thomsen *et al.*, (1991).

A great deal of work needs to be done in an attempt to understand exactly what environmental and culture conditions need to be applied to initiate various life cycle stages. There is some evidence that in *Pleurochrysis* the transition from the benthic non-coccolith-bearing phase to the coccolith-bearing phase may be brought about by the experimental manipulation of the nitrogen supply (Brown and Romanovicz, 1976).

Hussain and Boney (1971) demonstrated that a species of *Cricosphaera* was capable of producing certain plant-growth substances and that these substances appeared to have an effect on the motility of the cells. In *C. elongata* both the motile and non-motile phases synthesized gibberellin-like substances which on extraction showed properties similar to those of GA_3 and GA_7. These substances were released into the medium during exponential growth. Plant growth substances were not produced by *P. carterae*. In addition to the gibberellin-like substances both *C. elongata* and *P. carterae* produce compounds that exhibit auxin-like activity. Hussain and Boney (1971) have also identified plant-growth inhibitors in extracts of motile cells and in the spent medium but they were absent from extracts of non-motile phases. There is also evidence that some interaction occurs between gibberellins and the inhibitors. It is therefore possible that these substances play a role in inducing the non-motile stage in Cricosphaera.

Acknowledgements

This work was supported by the Foundation for Research Development of the C.S.I.R. I am grateful to colleagues and various publishers for allowing me to reproduce certain figures and to Miss C. Sam for typing the manuscript.

References

Aken, M.E., 1985. A study of the marine phytoflagellate *Pyramimonas pseudoparkeae* Pienaar et Aken (Prasinophyceae). Unpub. Ph.D. Thesis, Univ. of Natal, Pietermaritzburg, vol. I and vol. II.

Ariovich, D., 1980. An investigation into the role of light in the intracellular formation of coccoliths in *Hymenomonas carterae* (Braarud and Fagerland) Manton and Peterfi. Unpub. Ph.D. Thesis, Univ. of the Witwatersrand, Johannesburg.

Ariovich, D. and Pienaar, R.N., 1979. The role of light in the incorporation and utilization of Ca^{2+} ions by *Hymenomonas carterae* (Braarud et Fagerland) Braarud (Prymnesiophyceae). *Br. Phycol. J.* **14**: 17–24.

Beech, P.L. and Wetherbee, R., 1984. Serial reconstruction of the mitochondrial reticulum in the coccolithophorid, *Pleurochrysis carterae* (Prymnesiophyceae). *Protoplasma*, **123**: 226–9.

Beech, P.L. and Wetherbee, R., 1988. Observations on the flagellar apparatus and peripheral endoplasmic reticulum of the coccolithophorid, *Pleurochrysis carterae* (Prymnesiophyceae). *Phycologia*, **27**(1): 142–58.

Berger, R., Liaaen-Jensen, S., McAllister, V. and Guillard, R.R.L., 1977. Carotenoids of the Prymnesiophyceae (Haptophyceae). *Biochem. Syst. Ecol.*, **5**: 1–75.

Borman, A.H., De Jong, E.W., Huizinga, M., Kok, D., Westbroek, P., and Bosch, L., 1982. The role in $CaCO_3$ crystallization of an acid Ca^{2+}-bing polysaccharide associated with coccoliths of *E. huxleyi*. *Eur. J. Biochem.*, **129**: 179–182.

Borman, A.H., De Jong, E.W., Thierry, R., Westbroek, P., Bosch, L., Gruter, M., and Kamerling, J.P., 1987. Coccolith-associated polysaccharides from cells of *Emiliana huxleyi* (Haptophyceae). *J. Phycol.*, **23**: 118–123.

Brown, R.M., 1969. Observations on the relationship of the Golgi apparatus to wall formation in the marine chrysophycean algae, *Pleurochrysis scherffelii* Pringsheim. *J. Cell Biol.*, **41**: 109–23.

Brown, R.M. and Romanovicz, D.K., 1976. Biogenesis and structure of Golgi-derived cellulosic scales in Pleurochrysis. I. Role of the endomembrane systems in scale assembly and exocytosis. *Applied Polymer Symposium*, **28**: 537–85.

Fichtinger-Schepman, A.M.I., Karmerling, J.P., Versluis, C., and Vliegenthart, J.F.G., 1981. Structural studies of the methylated acidic polysaccharide associated with coccoliths of *Emiliania huxleyi* (Lohman) Kamptner. *Carbohydrate Res.*, **93**: 105–23.

Gayral, P. and Fresnel, J., 1976. Nouvelles observations sur deux Coccolithophoracees marines: *Cricosphaera roscoffensis* (P. Dangeard) comb. nov. et *Hymenomonas globosa* (F. Magne) comb. nov. *Phycologia*, **15**: 399–55.

Gayral, P. and Fresnel, J., 1983. Description, sexualite et cycle de developpement d'une nouvelle Coccolithophoraceae (Prymnesiophyceae): *Leurochrysis pseudoroscoffensis* sp. nov. *Protistologia*, **19**: 245–61.

Gayral, P. and Fresnel-Morange, J., 1971. Resultats preliminaires sur la structure et la biologie de la coccolithacee *Ochrosphaera neopolitana* Schussing. *C.R. Hebd. Seanc. Acad. Sci. Paris*, **273**: 1683–786.

Green, J.C., 1980. The fine structure of *Pavlova pinquis* Green and a preliminary suvey of the order Pavlovales (Prymnesiophyceae). *Br. Phycol. J.*, **15**: 151–91.

Green, J.C., Perch-Nielsen, K. and Westbroek, P., 1990. Phylum Prymnesiophyta. In *Handbook of Protoctista*, ed. L. Margulis, J.O. Corliss, p. 293–317. Jones and Bartlett Publishers, Boston.

Green, J.C. and Hibberd, D.J., 1977. The ultrastructure and taxonomy of *Diacronema vlkianum* (Prymnesiophyceae) with special reference to the haptonema and flagellar apparatus. *J. Mar. Biol. Assn. U.K.*, **57**: 1125–36

Green, J.C. and Hori, T., 1986. The ultrasturucture of the flagellar root system of *Imantonia rotunda* (Prymnesiophyceae). *Br. Phycol. J.*, **21**: 5–18.

Hibberd, D.J., 1976. The ultrastructure and taxonomy of the Chrysophyceae and Prymnesiophyceae (Haptophyceae): survey with some new observations on the ultrastructure of the Chrysophyceae. *Bot. J. Linn. Soc.*, **72**: 55–80.

Hibberd, D.J., 1980. Prymnesiophytes (= Haptophytes) In *Phytoflagellates* ed. E.R. Cox, Elsevier North Holland, New York. pp. 273–318..

Hori, T., 1979. Ultrastructure of cell division in the eucaryotic algae exclusive of green algae. *Jap. J. Phycol.*, **27**: 217–29.

Hori, T. and Green T.C., 1985. An ultrastructural study of mitosis in non-motile coccolith-bearing cells of *Emiliania huxleyi* (Lohm.) Hay and Mohler (Prymnesiophyceae). *Protistoligica*, **21**(1): 107–20.

Hori, T. and Inouye, I., 1981. The ultrastructure of mitosis in *Cricosphaera roscoffensis* var. *haptonemifera* (Prymnesiophyceae). *Protoplasma*, **106**: 121–35.

Hussain, A. and Boney, A.D., 1971. Plant growth substances associated with the motile and non-motile phases of two *Cricosphaera species* (Order Prymnesiales, Class Haptophyceae). *Botanica Mar.*, **14**: 17–21.

Inouye, I. and Chihara, M., 1983. Ultrastructure and taxonomy of *Jomonlithus littoralis* gen. et sp. nov. (Class Prymnesiophyceae) a Coccolithophorid from the northwest Pacific. *Bot. Mag. Tokyo*, **96**: 365–76.

Inouye, I. and Pienaar, R.N., 1983. Observations on the life cycle and microanatomy of *Thoracosphaera heimii* (Dinophyceae) with special reference to its systematic position. *S. Afr. J. Bot.*, **2**: 63–75.

Inouye, I. and Pienaar, R.N., 1984. New observations on the coccolithophorid *Umbilicosphaera sibogae* var *foliosa* (Prymnesiphyceae) with reference to cell covering, cell structure and flagellar apparatus. *Br. Phycol. J.*, **19**: 357–69.

Inouye, I. and Pienaar, R.N., 1985. Ultrastructure of the flagellar apparatus in *Pleurochrysis* (Class Prymnesiophyceae). *Protoplasma*, **125**: 24–35.

Inouye, I. and Pienaar, R.N., 1988. Light and electron microscope observations of the type species of *Syracosphaera*, *S. pulchra* (Prymesiophyceae). *Br. Phycol. J.*, **23**: 205–17.

Isenberg, H.D., Lavine, L.S., Spicer, S.S., and Weissfellner, H., 1966. A protozoan model of hard tissue formation. *Ann. N.Y. Acad. Sci.*, **136**: 155–90.

Kawachi, M., Inouye, I., Maeda, O. and Chihara, M., 1991. The haptonema as a food capturing device; observations on *Chrysochromulina hirta* (Prymnesiophyceae). *Phycologia*, **30**(6): 563–73.

Klaveness, D., 1972a. *Coccolithus huxleyi* (Lohman) Kamptner. I. Morphological investigations on the vegetative cell and the process of coccolith formation. *Protistologica*, **8**: 335–346.

Klaveness, D., 1972b. *Coccolithus huxleyi* (Lohm.) Kamptn. II. The flagellate cell, aberrant cell types, vegetative propagation and life cycles. *Br. Phycol. J.*, **7**: 309–18.

Klaveness, D., 1973. The microanatomy of *Calyptrosphaera sphaeroidea*, with some supplementary observations on the motile stages of *Coccolithus pelagicus*. *Norw. J. Bot.*, **20**: 151–62.

Klaveness, D. and Paasche, E., 1979. Physiology of coccolithophorids. In *Biochemistry and Physiology of Protozoa*, ed. M. Levandowsky, M. and S.H. Hunter, vol. I, pp. 191-213. New York, Academic Press. second edn.

Leadbeater, B.S.C., 1970. Preliminary observations on differences of scale morphology at various stages in the life-cycle of 'Apistonema-Syracosphaetra' sensu von Stosch. *Br. Phycol. J.*, **5**: 57–69.

Leadbeater, B.S.C., 1971a. Observations by means of cine photography on the behaviour of the haptonema in plankton flagellates of the class Haptophyceae. *J. Mar. Biol. Assn.*, *U.K.* **51**: 207–17.

Leadbeater, B.S.C., 1971b. Observations on the life-history of the haptophycean alga *Pleurochrysis scherffelii* with special reference to the micro-anatomy of the different types of motile cell. *Ann. Botany*, **35**: 429–39.

Mann, S. and Sparks, N.H.C., 1988. Single crystalline nature of coccolith elements of the marine alga *Emiliania huxleyi* as determined by electron diffraction and high resolution transmission electro-microscopy. *Proc. Roy. Soc. Lond. B.*, **234**: 441–53.

Manton, I., 1964a. Observations with the electron microscope on the division cycle of the flagellate *Prymnesium parvum* Carter. *J. Roy. Microsc. Soc.*, **83**: 317–25.

Manton, I., 1964b. Further observations on the fine structure of the haptonema of *Prymnesioum parvum. Archiv. Mikrobiol.*, **49**: 315–30.

Manton, I., 1967. Further observations on the fine structure of *Chrysochromulina chiton* with special reference to the haptonema, peculiar Golgi structure and scale production. *J. Cell Sci.*, **2**: 265–72.

Manton, I. and Leedale, G.F., 1963. Observation on the microanatomy of *Crystallolithus hyalinus* Gaarder and Markali. *Arch. Protistenk*, **47**: 115–36.

Manton, I. and Leedale, G.F., 1969. Observations on the microanatomy of *Coccolithus pelagicus* and *Cricosphaera carterae* with special reference to the origin and nature of coccoliths and scales. *J. Mar. Biol. Assn. U.K.*, **49**: 1–16.

Manton, I. and Peterfi, L.S., 1969. Observations on the fine structure of coccoliths, scales and the protoplast of a freshwater coccolithophorid *Hymenomonas roseola* Stein, with supplementary observations on the protoplast of *Cricosphaera carterae*. *Proc. Roy. Soc. Series B.*, **172**: 1–15

Marlowe, I., Green, J.C., Neal, A., Brassell, S., Eglington, G. and Course, P.A., 1984. Long-chain (n-C_{37}-C_{39}) alkenones in the Prymnesiophyceae. Distribution of akenones and other lipids and their taxonomic significance. *Br. Phycol. J.*, **19**: 203–16.

Merrick, P.J. and Leadbeater, B.S.C., 1979. Release and settlement of swarmers in *Pleurochrysis scherffelii* Pringsheim. *Br. Phycol. J.*, **14**: 339–47.

Mesquita, J.F. and Santos, M.F., 1983. Cytological studies in golden algae (Chrysophyceae). III. Fine Structure of mitosis and cytolinesis in the 'Apistonema stages' of a Coccolithophoraceae. *J. Submicrosc. Cytol.*, **15**: 751–65.

Moestrup, O. and Thomsen, H.A., 1986. Ultrastructure and reconstruction of the flagellar appartus in *Chrysochromulina apheles* sp. nov. (Prymnesiophyceae = Haptophyceae). *Can. J. Bot.*, **64**: 593–610.

Outka, D.E. and Williams, D.C., 1971. Sequential coccolith morphogenesis in *Hymenomonas carterae*. *J. Protozool.*, **18**: 285–97.

Parke, M. and Adams, I., 1960. The motile (*Crystallolithus hyalinus* Gaarder and Markali) and non-motile phases in the life history of *Coccolithus pelagicus* (Wallich) Schuller. *J. Mar. Biol. Assn. U.K.*, **39**: 263–74.

Parke, M. Manton, T. and Clarke, B., 1959. Studies on marine flagellates. V. Morphology and microanatomy of *Chrysochromulina strobilus* sp. nov. *J. Mar. Biol. Assn. U.K.*, **38**: 69–188.

Pickett-Heaps, J.D., 1976. Cell division in eucaryotic algae. *BioScience*, **26**: 445–50.

Pienaar, R.N., 1969a. The fine structure of *Cricosphaera carterae*. I. External morphology. *J. Cell Sci.*, **4**: 56167.

Pienaar, R.N., 1969b. The fine structure of *Hymenomonas* (*Cricosphaera*) *carterae*. II. Observations on scale and coccolith production. *J. Phycol.*, **5**: 321–31.

Pienaar, R.N., 1971. Coccolith production in *Hymenomonas carterae*. *Protoplasma*, **73**: 217–14.

Pienaar, R.N., 1976a. The rhythmic production of body covering components in the Haptophycean flagellate *Hymenomonas carterae*. *Mechanisms of Mineralization in the Invertebrates and Plants*, In ed. N. Watanabe, and K.J. Willow, Univ. of South Carolina Press, Columbia.

Pienaar, R.N., 1976b. The microanatomy of *Hymenomonas lacuna* sp. nov. (Haptophyceae). *J. Mar. Biol. Assn. U.K.*, **56**: 1–11.

Pienaar, R.N., 1976c. Viral-like infections in three species of phytoplankton from San Juan Island, Washington. *Phycologia*, **15**: 2–6.

Pienaar, R.N. and Norris, R.E., 1979. The ultrastructure of the flagellate *Chrysochromulina spinifera* (Fournier) comb. nov. (Prymnesiophyceae) with special reference to scale production. *Phycologia*, **17**: 41–51.

Rayns, D.G., 1962. Alternation of generations in a coccolithophorid, *Cricosphaera carterae* (Braarad and Fagerl.) Braarud. *J. Mar. Biol. Assn. U.K.*, **42**: 481–4.

Romanovicz, D.K., 1981. Scale formation in flagellates. In *Cytomorphogenesis in Plants, Cell Biology Monographs*, vol. 8, pp. 27–62. O. Kiermayer, Springer-Verlag. Berlin.

Rowson, J.D., Leadbeater, B.S.C., and Green, J.C., 1986. Calcium carbonate deposition in the motile (*Crystallolithus*) phase of *Coccolithus pelagicus* (Prymnesiophyceae). *Br. Phycol. J.*, **21**: 359–70.

Stacey, V.J. and Pienaar, R.N., 1980. Cell division in *Hymenomonas carterae* (Braarud et Fagerland) Braarud (Prymnesiophyceae). *Br. Phycol. J.*, **15**: 365–76.

Thiery, P.J., 1967. Mise en evidence des polysaccharides sur coupes fines en microscopie electronique. *J. Microscopie*, **6**: 987–1018.

Thomsen, H.A., Ostergaard, J.B., and Hansen, L.E., 1991. Heteromorphic life histories in Arctic coccolithophorids (Prymnesiophyceae). *J. Phycol.*, **27**: 634–42.

Van Emburg, P.R., De Jong, E.W., and Daems, W.Th., 1986. Immunochemical localization of a polysaccharide from biomineral structures (coccoliths) of *Emiliania huxleyi*. *J. Ultrast. Mol. Structure Res.*, **94**: 246–59.

Wal, P. van der, De Bruijn, W.C., and Westbroek, P., 1985. Cytochemical and X-ray microanalysis studies of intracellular calcium pools in scale-bearing cells of the coccolithophorid *Emiliania huxleyi*. *Protoplasma*, **124**: 1–9.

Wal, P. van der, De Jong, E.W., Westbroek, P., and De Bruijn, W.C., 1983. Calcification in the coccolithophorid alga *Hymenomonas carterae*. *Ecol. Bull.*, **35**: 251–8.

Wal, P. van der, De Jong, E.W., Westbroek, P., De Bruijn, W.C. and Molder-Stapel, A.A., 1983a. Polysaccharide localization, coccolith formation and Golgi dynamics in the coccolithphorid *Hymenomonas carterae*. *J. Ultrastruct. Res.*, **85**: 139–58.

Wal, P. van der, De Jong, E.W., Westbroek, P. De Bruijn, W.C., and Mulder-Stapel, A.A., 1983b. Ultrastructural polysaccharide localization in calcifying and naked cells of the coccolithophorid *Emiliania huxleyi*. *Protoplasma*, **118**: 157–168.

Wal, P. van der, De Vrind, P.M., De Vrind-De Jong, E.W., and Borman, A.H., 1987. Incompleteness of the coccosphere as a possible stimulus for coccolith formations in *Pleurochrysis carterae* (Prymnesiophyceae). *J. Phycol.*, **23**: 218–22.

Wal, P. van der, Leunissen-Bijvelt, J.J.M. and Verkley, A.J., 1985. Ultrastructure of the membranous layer enveloping the cell of the coccolithophorid *Emiliania huxleyi*, *J. Ultrastr. Res.*, **91**: 24–9.

Watabe, N., 1967. Crystallographic analysis of the coccolith of *Coccolithus huxleyi*. *Calc. Tiss. Res.*, **1**: 114–21.

Westbroek, P., De Jong, E.W., Van Der Wal, P., Borman, A.H., De Vrind, J.P.M., Kok, D., De Bruijn, W.C., and Parker, S.B., 1984. Mechanism of calcification in the marine alga *Emiliania huxleyi*. *Phil. Trans. Roy. Soc. Ser. B.*, **304**: 435–44.

Westbroek, P., Wal, P. van der, Van Emburg, P.R., De Vrind-de Jong, E.W., and De Bruijn, W.C. 1986. Calcification in the coccolithophorids *Emiliania huxleyi* and *Pleurochrysis carterae*. I. Ultrastructural aspects. In *Biomineralization in Lower Plants and Animals*, ed. B.S.C. Leadbeater, and R. Riding, Systematics Assn. Spec. vol. 30, pp. 188–203. Clarendon Press, Oxford.

Westbroek, P., Young, J.R., and Linschooten, K., 1989. Coccolith production (Biomineralization) in the marine alga *Emiliania huxleyi*. *J. Protozool.*, **34**(4): 368–73.

Wilbur, K.M. and Watabe, N., 1963 Experimental studies on calcifications in molluscs and the alga *Coccolithus huxleyi*. *Ann. N.Y. Acad. Sci.*, **109**: 82–112.

3 Physiological ecology of marine coccolithophores

LARRY E. BRAND

Introduction

Ultimately an ecologist would like to be able to explain or predict the spatial and temporal distribution of a species based on information on the basic biology of the species and its environment. This chapter examines how well our knowledge of coccolithophores can help explain their distribution. The basic premise is that biogeography, mesoscale distribution, patchiness, long-term changes on both geological and historical time scales, seasonality, and bloom dynamics are ultimately the result of the biochemistry, physiology, morphology, behavior, and life history characteristics of the species in interaction with environmental factors and physical-transport mechanisms. The interaction between the biological characteristics and ecological distributions must be envisioned as a two-way process. Not only do the various biological adaptations result in species being able to live in a certain range of environmental conditions on the time scale of ecological events, but many of the adaptations will evolve to 'fit' the ecological circumstances to which the species is exposed on an evolutionary time scale.

Hutchinson (1957) envisioned a 'niche hyperspace' in which biological characteristics such as reproduction rate and other components of fitness could be plotted as a function of various environmental factors, each environmental factor on a different axis. This generates what is known as the n-dimensional niche hyperspace, a geometric representation of the capabilities of a species throughout a range of environmental situations. Coupled with information on the distribution of environmental factors and a hydrodynamic model that describes the physical transport of the species, this conceptual model should theoretically be capable of explaining or predicting the spatial and temporal distribution of a planktonic species.

In reality of course, we have enough data to fill in only a small portion of the niche hyperspace for only a few of the coccolithophore species that live in the ocean. This forces us to make broader generalizations and to take a more comparative approach. This chapter will compare the biology of a number of coccolithophore species that have been studied in laboratory culture with their ecological distributions.

An underlying assumption in this approach is that organisms in laboratory cultures respond to environmental factors in the same way as when they are in the ocean. Despite all of our efforts, the laboratory environment cannot recreate the natural environment perfectly. This is obvious from the fact that we still cannot culture many oceanic species for reasons upon which we can only speculate (Brand, 1986). Therefore, we must always consider our laboratory models of the niche hyperspace to be imperfect. Furthermore, our knowledge of the physiological characteristics of coccolithophores is limited to those species that have been successfully cultured. This introduces a strong bias in our data to the extent that the cultures are a non-random sample of the species in the ocean. A comparison of coccolithophore species that have and have not been cultured with their distribution reveals this bias. Many species that increase in abundance in more nutrient-rich waters are in culture but most species that do not increase in abundance in response to any natural enrichment processes are not in culture (Brand, 1986). Clearly we still have a very poor understanding of the biology of coccolithophores in oligotrophic waters.

Most laboratory research on coccolithophores has been conducted on the coastal species *Pleurochrysis carterae*, subpolar species *Coccolithus pelagicus*, cosmopolitan species *Emiliania huxleyi*, eutrophic oceanic species *Gephyrocapsa oceanica*, and oligotrophic oceanic species *Calcidiscus leptoporus* and *Umbilicosphaera sibogae*. A major goal of this chapter is to examine how well our information on these species acquired in the laboratory explains their distribution in the ocean and how the physiological ecology of coccolithophores differs from other marine phytoplankton groups.

Most of these studies are conducted with single clones of the species. Extrapolating the niche hyperspace of a clone to an entire species assumes that there is no genetic variability within populations, genetic differentiation between populations or genetic change over time, an assumption known to be false (Brand, 1991a). This is discussed in detail later in the chapter.

Ecological distribution of coccolithophores

The spatial and temporal distribution of marine coccolithophores is discussed in more detail by Winter *et al.*

(chapter 8), but a brief overview which emphasizes the species that have been studied in culture is given here.

The overall distribution of coccolithophores as a phylogenetic group has changed substantially over geological time. In the Cretaceous, coccolithophores were quite cosmopolitan and diverse (Tappan, 1980). They were very abundant in both coastal and oceanic waters and in both polar and equatorial waters. The vast majority of species in the diverse coccolithophore community became extinct at the end of the Cretaceous. The diatoms, a phytoplankton group that had been rather sparse in the Mesozoic, were little affected by the terminal Cretaceous extinction event and took over much of the habitat previously occupied by coccolithophores. Today diatoms are much more abundant than coccolithophores in most coastal and polar waters. Coccolithophores have regained their dominance in temperate and tropical oceanic waters although they are not as diverse as they were in the Mesozoic. Many of these changes are suspected to be the result of competition with other phytoplankton, primarily the newly evolved diatoms. It appears that competitive interaction was greatly shifted by the sudden event at the end of the Cretaceous. Most coccolithophore species today live in warm, stratified, nutrient-poor offshore waters.

The biogeographic distributions of individual species also change substantially through geological time as a result of changes in the environment. McIntyre (1967) has demonstrated that the subpolar species *Coccolithus pelagicus* expanded its range further toward the equator during the last glacial peak while the tropical species *Calcidiscus leptoporus*, *Umbilicosphaera irregularis*, *U. mirabilis*, *U. tenuis*, *Helicosphaera carteri*, *Rhabdosphaera stylifera*, and *Syracosphaera pulchra* all had reduced biogeographic ranges withdrawn closer to the equator during the last glacial peak. The relative abundance of the subpolar species *Coccolithus pelagicus* and the tropical species *Calcidiscus leptoporus* reflects the glacial cycles during the Quaternary quite well (Geitzenauer, 1969). These data show that the biogeographic ranges of coccolithophore species change over time, in this case in response to temperature and the other environmental factors correlated with temperature and the associated water masses.

In addition to biogeographic distributions being changed by environmental changes, they can change as a result of genetic alterations within a species, leading it to occupy a new habitat. For example, although *Coccolithus pelagicus* is a subpolar species today, it evolved in the tropics in the early Cenozoic and migrated toward the poles during the mid-Cenozoic (Haq and Lohmann, 1976). This demonstrates that species distributions do not simply respond to environmental factors over time, but that genetic changes within species can drive changes in biogeographic distributions.

The highest diversity of coccolithophores today is in the subtropical oceanic gyres (Hulburt, 1963, 1964). Diversity is lower in temperate oceanic waters and only a few species live in the subpolar oceanic waters, although they can be quite abundant there (Hulburt, 1963). Similarly, diversity is low in coastal waters. Unlike subpolar oceanic waters where coccolithophores can comprise a high fraction of the phytoplankton community, coccolithophores in coastal waters are usually only a small fraction of the community. Species of the genera *Pleurochrysis* and *Hymenomonas* can be found in estuaries and lagoons, even tidal pools, but are rarely a large fraction of the phytoplankton community.

McIntyre and Bé (1967) were able to distinguish five different assemblages of coccolithophores in the Atlantic: tropical, subtropical, transitional, subarctic and subantarctic. Okada and Honjo (1973) identified six biogeographic provinces in the Pacific, with patterns similar to those in the Atlantic.

Although coccolithophore diversity is much lower in polar waters than in the tropics, a number of species do live there. *Coccolithus pelagicus* is the most abundant species found in polar waters and its range extends into subpolar waters (McIntyre and Bé, 1967). Other less abundant coccolithophore species have also been found in the waters of the Arctic and Antarctic (Manton *et al.*, 1977; Thomsen, 1981). *Emiliania huxleyi* is very abundant in subpolar waters of both the Arctic and Antarctic (McIntyre and Bé, 1967; Hasle, 1969; Okada and Honjo, 1973).

Most coccolithophore species live in the stratified waters of the temperate and tropical regions of the world. They are present during the entire year in the tropics where the water column is permanently stratified. In temperate regions where the water is stratified only in the summer, coccolithophores are abundant only in the summer. *Emiliania huxleyi* becomes abundant in the spring when the water column first stratifies. *Emiliania huxleyi* and *Gephyrocapsa oceanica* both also tend to become very abundant in nutrient-rich waters along the edges of the subtropical central gyres, in equatorial upwelling regions and other mature upwelling areas, and on the outer continental shelves (Berge, 1962; Smayda, 1966; Hulburt, 1967, 1976, 1979, 1983; Hulburt and Mackenzie, 1971; Marshall, 1976; Marshall and Cohn, 1983; Hallegraeff, 1984; Mitchell-Innes and Winter, 1987; Verbeek, 1989; Winter, 1985). *Emiliania huxleyi* tends to be more abundant in the colder waters and *G. oceanica* is more abundant in the warmer waters of these regions. In the well stratified waters of the central gyres these two species are usually present but in much lower abundance and a large number of other coccolithophore species dominate the community. Although *E. huxleyi* and *G. oceanica* do increase in abundance in response to both natural and artificial nutrient enrichments, most of the tropical oceanic species remain in low abundance (Hulburt, 1967, 1979, 1983; Hulburt and Corwin, 1969).

Seasonal distribution

Seasonal patterns reflect the geographic distribution quite well. In the winter during times of more vertical mixing in the open ocean, cosmopolitan species such as *Emiliania huxleyi* flourish while the tropical species become quite rare (Okada and McIntyre, 1979). During summer stratification, the diverse tropical community becomes dominant and *E. huxleyi* becomes rather rare. In the more temperate regions of the central gyres where the stratification is broken down in the winter, *E. huxleyi* and *Gephyrocapsa oceanica* are major components of the spring bloom as stratification develops (Hulburt *et al.*, 1960). In coastal waters on the other hand, *E. huxleyi* is most abundant during summer stratification (Marshall, 1978).

Vertical Stratification

In stratified waters, the vertical distribution of the species can be quite distinct. Venrick (1982) has demonstrated that some species are restricted primarily to the upper photic zone while others live only in the lower photic zone. Some species of coccolithophores such as *Anthosphaera* sp., *Thorosphaera flabellata* and *Florisphaera profunda* live almost exclusively between 100 and 200 m, below what is generally regarded as the photic zone (Honjo and Okada, 1974; Venrick, 1982). It is not known if they can photosynthesize with extremely low levels of light or if they are heterotrophic.

These are the basic ecological distributional patterns we want to explain. The following biological characteristics and environmental factors are thought to be important in determining these distributions.

Overall structure and morphology of coccolithophores

Coccolithophores are relatively small (3 to 40 μm) and most are coccoid in shape. The small size provides an advantage in stratified, oligotrophic waters because the resulting high surface to volume ratio reduces diffusion limitation of nutrient uptake at low nutrient concentrations. They are capable of motility using two flagella, and their phototactic behavior (Mjaaland, 1956) may allow them to maintain the distinct vertical distributions observed, presumably for optimum combinations of light intensity and nutrient concentration.

One characteristic that distinguishes coccolithophores (and other prymnesiophytes) from other groups of phytoplankton is the possession of a haptonema. Details about this structure are given by Pienaar (chapter 2). It can bend, rotate, coil or attach to substrates very quickly (Hibbard, 1980). Given the wide range of sizes and behaviors of this structure found in different species, it probably serves different functions, depending on the ecology of the species in question. Some species use it to attach to surfaces (Hibbard, 1980) and others use it to capture particles (Inouye, 1990). As other groups of microalgae have methods to accomplish the same tasks, there is no evidence at present that the haptonema provides coccolithophores with a unique ecological niche.

The calcified plates covering the cell – the coccoliths – distinguish coccolithophores from other prymnesiophytes as well as from other phytoplankton. These coccoliths clearly have a significant effect on sinking rate. Smayda and Bienfang (1983) found that coccolithophores have higher sinking rates than most other groups of phytoplankton for their size, presumably because of their coccoliths. Eppley *et al.* (1967) found that cells of *Emiliania huxleyi* without coccoliths sank at rates of 0.28 m/day whereas cells with coccoliths sank 1.3 m/day.

A number of hypotheses have been proposed for the advantage of calcifying the organic plates on the outside surface of the cell and these are discussed in detail by Young (chapter 5). Most unicellular algae have some type of cell covering (Sournia, 1982) and it seems likely that the cover has some protective function. The fact that the entire surface area of coccolithophores, rather than just a portion, is covered with coccoliths suggests that reducing access to the cell surface by potential parasites and predators may be an important function of the coccoliths. Young (chapter 5) points out just how well coccolith architecture and placement provide a protective barrier around the cell. The fact that the coccoliths are rearranged to cover the cell after cell division in the dark when no new coccoliths are being produced (Paasche, 1967) also suggests a protective function. Van der Wal *et al.* (1987) found that coccolith production in *Pleurochrysis carterae* is stimulated by an incomplete cell cover and inhibited by a complete cell cover of coccoliths. The control mechanism appears to be an adaptation to insure that the entire surface of the cell is covered by coccoliths at all times. Although Sikes and Wilbur (1982) found that calcified and non-calcified cells of *Emiliania huxleyi* and *Pleurochrysis carterae* were eaten by copepods at the same rate, suggesting that coccoliths are not very effective protective devices against large predators, coccoliths could perhaps prevent the attachment and attack of viruses, bacteria or protozoa. This may have been the initial function of organic prymnesiophyte scales. Viruses have recently been hypothesized to have a strong impact on phytoplankton populations (Bratbak *et al.*, 1990; Proctor and Fuhrman, 1990; Suttle *et al.*, 1990) and it seems likely that bacteria and protozoa also have significant impacts on algal populations (Sieburth, 1979; Sherr and Sherr, 1988). Hard plates covering the cell could deter their attack.

Whatever the initial function of the cell covering, one

must also question why the organic prymnesiophyte scales are calcified and what advantage $CaCO_3$ has over SiO_2 or cellulose or other organic polymers used by other groups of algae. Organic polymers are energetically costly because carbon dioxide must be reduced using NADPH derived from photosynthesis. Inorganic materials are expected to be less costly to deposit because there is no chemical reduction involved. Silica is at low limiting concentrations in most photic zone waters, thus restricting its use, whereas $CaCO_3$ is actually at supersaturating concentrations in temperate and tropical photic zones (Broecker and Peng, 1982) where most coccolithophores reside. The SiO_2 frustules of diatoms also generate an architectural constraint on the diel timing of cell division because of the necessary temporal linkage between frustule formation and cell division in diatoms and the energy costs of SiO_2 deposition (Nelson and Brand, 1979). The architecture of individual coccoliths does not force a temporal linkage between the formation of coccoliths and cell division in coccolithophores.

Although calcification does alleviate the accumulation of hydroxyl ions generated by photosynthesis, and coccolith production is largely dependent on light and occurs almost exclusively during the day, there is no evidence that coccolith production is necessary for photosynthesis (Klaveness and Paasche, 1979). The relationship is most likely simply based upon the energetic needs of calcification.

Another advantage of coccoliths over diatom frustules and dinoflagellate thecae is that they are detachable and thus can be used as ballast. The resulting higher specific density could be used to enhance their sinking rate to deeper, nutrient-rich water when nutrient-depleted water is encountered. The higher sinking rate also can reduce diffusion limitation of nutrient uptake. Most phytoplankton sink faster when nutrient limited than when nutrient replete (Smayda, 1970), a response that would more quickly lead phytoplankton to nutrient-rich deep waters. It is reasonable to expect most coccolithophores to have the same response to nutrients. Indeed, calcification has been shown to be higher under lower nutrient conditions in *Hymenomonas* sp. (Baumann *et al.*, 1978) and in *E. huxleyi* (Wilbur and Watabe, 1963; Linschooten *et al.*, 1991). An additional response of coccolithophores is to lose their coccoliths (thus reducing their specific density) in the presence of high nutrient concentrations (Klaveness and Paasche, 1979).

Although most coccoliths are produced during the day (Paasche, 1964, 1966, 1969; Linschooten *et al.*, 1991), sinking rate is highest during the latter part of the night and lowest near the end of the day in *Pleurochrysis carterae* (Bienfang, 1981). Therefore coccolith cover cannot explain diel variation in sinking rates. Furthermore, it seems unlikely that ion exchange could account for the reduced sinking rate during the day. Although large diatoms may use ion exchange to counteract the weight of the SiO_2 frustule, it appears unlikely that coccolithophores can rely upon the

same mechanism because of the much larger cell wall to vacuole ratio in the smaller cells (Smayda, 1970). Overall then, coccoliths can explain sinking rate responses to nutrient concentrations but not to diel light variation.

Light intensity

Coccolithophores have chlorophylls *a* and *c*, beta-carotene, fucoxanthin, diatoxanthin, and diadinoxanthin, similar to diatoms and chrysophytes, and their overall absorption spectra do not appear to be significantly different from diatoms, chrysophytes, and dinoflagellates (Prezelin and Boczar, 1986; Dring, 1990). It has been hypothesized that the coccoliths may focus light onto the chloroplasts (Gartner and Bukry, 1969) or reflect light (Braarud *et al.*, 1952) from the cell but there are no direct data to support these hypotheses. Paasche and Klaveness (1970) found that calcified and non-calcified cells did not differ significantly in their ability to utilize different light intensities. The levels of light needed to keep coccolithophores alive and to saturate growth rates and photosynthetic rates also do not appear to be significantly different from other eukaryotic phytoplankton (Brand and Guillard, 1981). An exception may be coccolithophore species such as *Florisphaera profunda* and *Thorosphaera flabellata* that are usually found between 100 and 200 m (Okada and Honjo, 1973; Honjo and Okada, 1974). It is suspected that they can grow at much lower light intensities than most phytoplankton, but so far no one has cultured these species and tested their abilities.

The cosmopolitan species *Emiliania huxleyi* and the coastal species *Pleurochrysis carterae* have their growth rates light saturated at around 0.1 ly/min whereas the more oligotrophic oceanic species *Gephyrocapsa oceanica* and *Calcidiscus leptoporus* are saturated at around 0.023 ly/min (Brand and Guillard, 1981). These saturation levels are similar to those of the diatoms and dinoflagellates also investigated. Species living in more eutrophic waters appear to need more light than species living in oligotrophic waters. This may be because species in the stratified oligotrophic waters must live deeper in the photic zone in order to obtain nutrients from below.

Production of the coccoliths is strongly, but not completely, light dependent (Paasche, 1964, 1966, 1969; Dorigan and Wilbur, 1973; Klaveness and Paasche, 1979; van der Wal *et al.*, 1983). The calcification process saturates at a lower light intensity than does photosynthesis (Paasche, 1964, 1969).

Photoperiod

Photoperiod is particularly important to oceanic coccolithophores. Brand and Guillard (1981) demonstrated that many oceanic phytoplankton, including coccolithophores, are harmed or even killed by continuous light. They hypothesized that the diel partitioning of various physiological processes in oceanic phytoplankton is so strong that a light–dark cycle is necessary for the proper metabolism of the cells. *Calcidiscus leptoporus* was completely inhibited by continuous light whereas the more cosmopolitan species *Emiliania huxleyi* and *Gephyrocapsa oceanica* were only slightly inhibited. The coastal coccolithophore *Pleurochrysis carterae* was not inhibited by continuous light. Brand and Guillard (1981) hypothesized that diel partitioning is not as strong or advantageous in coastal species because of the larger light fluctuations on time scales less than a day resulting from vertical mixing and the greater light attenuation with depth. The overall pattern appears to be greater sensitivity to continuous light (and thus stronger diel partitioning of metabolic activity) in tropical oceanic waters and less sensitivity in polar, temperate, and coastal waters.

Increasing photoperiods do lead to higher growth rates up to some point simply because of energetic constraints. The growth rate of *E. huxleyi* increases with increasing photoperiod up to about 16 hours of light a day (Paasche, 1967). Brand (unpub. data) has found that coastal species and *E. huxleyi* and *G. oceanica* grow faster in longer photoperiods but oceanic species such as *Calcidiscus leptoporus* do not. Oceanic species appear to 'saturate' at shorter photoperiods.

Diel periodicity

The fact that many oceanic coccolithophores cannot grow in continuous light indicates that diel partitioning of various processes is important. Most biological characteristics that have been examined in coccolithophores has been found to exhibit diel periodicity, indicating that coccolithophores are well adapted to the highly predictable light–dark cycle.

Emiliania huxleyi and *Pleurochrysis carterae* both subdivide primarily at night (Paasche, 1967; Eppley *et al.*, 1967; Nelson and Brand, 1979). It is thought that cell division at night avoids the disruption of the photosynthetic process that can only occur during the day. Photosynthetic capacity in *E. huxleyi* is highest during the day (Eppley *et al.*, 1971) and an indicator of photosynthetic capacity has been demonstrated to be under the control of a biological clock (Brand, 1982b). Eppley *et al.* (1967) found that nitrite reductase activity was highest late in the day and early night and nitrate reductase and glutamate dehydrogenase activities were highest toward the end of the night.

Emiliania huxleyi and *Coccolithus pelagicus* produce coccoliths almost exclusively during the day (Paasche, 1964, 1966, 1969; Linschooten *et al.*, 1991), as calcification is strongly light dependent. As discussed earlier, sinking rates are highest toward the end of the night in *P. carterae* (Bienfang, 1981), as seen in many phytoplankton species.

Many of these characteristics that exhibit diel periodicity are probably not just responding to the light–dark cycle but are also influenced by a biological clock. Brand (1982b) found evidence of a biological clock in *E. huxleyi*, *Gephyrocapsa oceanica, and P. carterae*.

Nutrients

Most coccolithophores that have been cultured are autotrophic and utilize inorganic nutrients as do other phytoplankton. They can use nitrate or nitrite as a nitrogen source but use ammonium preferentially. *Emiliania huxleyi* can use urea (Antia *et al.*, 1975), but amino acids only poorly (Wheeler *et al.*, 1974), and uric acid not at all (Pintner and Provasoli, 1963) as nitrogen sources. They can use phosphate or various forms of organic phosphates. Organic phosphates can be utilized as phosphate sources because they are hydrolyzed by phosphatases at the cell surface (Kuenzler and Perras, 1965). As with many algae, some coccolithophores require certain vitamins in addition to inorganic nutrients. *Emiliania huxleyi* has been shown to require thiamine (Carlucci and Bowes 1970).

Of the 16 species examined by Eppley *et al.* (1969), *E. huxleyi* had the lowest half-saturation constant for nitrate and ammonium uptake, indicating its adaptation to the low nutrient conditions of the open ocean. The oceanic coccolithophores *E. huxleyi*, *Gephyrocapsa oceanica*, *Calcidiscus leptoporus*, *Umbilicosphaera hulburtiania*, and *U. sibogae* are all able to grow at much lower concentrations of iron, zinc, and manganese than the coastal coccolithophore *Pleurochrysis carterae* or other coastal phytoplankton species (Brand *et al.*, 1983). All these data indicate that nutrient limitation by a number of different nutrients is an important selective force on phytoplankton. Oceanic coccolithophores have been able to evolve mechanisms that allow them to survive the low nutrient conditions found in subtropical central gyres. In the case of iron, it appears that oceanic coccolithophores achieve this ability by reducing the amount of iron per cell they need (Brand, 1991b).

Although *E. huxleyi* and *G. oceanica* respond to nutrient enrichment by increasing their population size both in the ocean as well as in culture, most of the exclusively subtropical oceanic species do not (Hulburt and Corwin, 1969). Species such as *Discosphaera tubifera, Calyptrosphaera oblonga, Oolithotus fragilis, Helicosphaera carteri, Rhabdosphaera clavigera*, and *Umbellosphaera irregularis* have never been cultured with any type of nutrient-enriched culture media and they do not increase in abundance in upwelling areas or other regions of the ocean that have elevated nutrient concentrations. This may be an indication of extreme K-selection in the oligotrophic conditions. They may grow well at extremely low nutrient concentrations and simply do not grow any faster at higher nutrient concentrations.

Alternatively, some coccolithophores may be heterotrophic. Some may be capable of utilizing the relatively high concentrations of dissolved organic molecules present in the photic zone of the ocean. Some species have been demonstrated to use organic compounds in laboratory studies, but it is not clear how relevant these data are because of the unrealistically high concentrations used. Littoral species such as *Pleurochrysis* appear to be able to utilize some organic compounds and this may indicate their exposure to relatively high organic concentrations in littoral waters (Paasche, 1968).

Another possibility is that some coccolithophores are phagotrophic, consuming bacteria, other phytoplankton species or other organic particulate matter. This has been observed by Pienaar and Norris (1979) and Parke and Adams (1960). Parke and Adams (1960) demonstrated that the subpolar species *Coccolithus pelagicus* can ingest bacteria and small phytoplankton cells. This ability may allow it to survive the dark winters of the polar regions. Inouye (1990) has shown how the non-coccolithophore prymnesiophyte *Chrysochromulina* can use its haptonema to capture particles, coiling the haptonema in as little as 1/100 of a second. Capture of food particles could be the function of the haptonema, as demonstrated in a few cases. This is in agreement with the observation that clearly autotrophic species such as *E. huxleyi* and *G. oceanica* have no haptonema (Klaveness, 1972). There may be a whole range of nutritional strategies ranging from complete autotrophy to photoheterotrophy to complete heterotrophy, with presence and length of a haptonema an indication of the nutritional strategy of a species.

Toxic metals

Besides nutrients that can stimulate the growth of coccolithophores, some elements in the ocean can inhibit growth. Sunda and Huntsman (1983) have hypothesized that upwelled copper could be toxic to phytoplankton. In support of this idea is the fact that the coccolithophore most often found in upwelling areas, *Emiliania huxleyi*, is adapted to tolerate much higher concentrations of copper than most other species (Brand *et al.*, 1986). Cadmium is known to interfere with calcification and it has been found that coccolithophores are more sensitive to cadmium toxicity than other phytoplankton (Brand *et al.*, 1986).

Temperature

Although temperature affects all biological processes, there is no evidence that coccolithophores are affected any differently than other phytoplankton. Of the 10 species whose distributions were examined by McIntyre *et al.* (1970),

Emiliania huxleyi had the broadest temperature range, from 1 to 31 °C. While present in the tropics, it is most abundant in subpolar waters (Okada and Honjo, 1973). In the Pacific, *Gephyrocapsa oceanica* lives throughout the tropics and subtropics in temperatures from 19 to 31 °C, but is most abundant in the equatorial upwelling region (McIntyre *et al.*, 1970). *Calcidiscus leptoporus* is primarily a tropical species in waters ranging from 20 to 30 °C, but a cold-water form of the species is found down to 6 °C (McIntyre *et al.*, 1970). *Discosphaera tubifera*, *Rhabdosphaera stylifera*, *Umbilicosphaera irregularis*, and *U. tenuis* are all tropical species that live primarily in the oligotrophic central gyres with a narrow temperature range primarily from 20 to 30 °C (McIntyre *et al.*, 1970).

Laboratory studies (Brand, unpub. data) show the same trend in temperature tolerance. The coastal species *Pleurochrysis carterae* has the widest temperature tolerance; then the cosmopolitan species *E. huxleyi*; then the tropical species *G. oceanica*, *C. leptoporus*, *U. sibogae*, *Syracosphaera pulchra*, and *Crenalithus sessilis;* and the tropical oceanic species *Scyphosphaera apsteinii* with the narrowest temperature tolerance.

Watabe and Wilbur (1966) and Fisher and Honjo (1991) found the growth rate in culture of *E. huxleyi* clones isolated from the Sargasso Sea to be highest between 18 and 24 °C and drop off precipitously at 7 and 27 °C. These data are in agreement with the early data of Paasche (1967). Individual temperature tolerance, however, does not explain the complete temperature range. It is clear that there are genetically different populations of *E. huxleyi* with different temperature responses. Coccoliths of species from cold water are morphologically different from those from warm water (McIntyre *et al.*, 1970). Brand (1982a) demonstrated that clones from cold water in the Gulf of Maine are genetically adapted to lower temperatures than those from warm water in the Sargasso Sea and coastal waters south of Cape Cod. Fisher and Honjo (1991) found that a clone of *E. huxleyi* from coastal waters tolerated high temperatures better than an oceanic clone from the Sargasso Sea.

Salinity

Coccolithophores also do not appear to respond to salinity in any way different from other phytoplankton. As one would expect, oceanic species tolerate only a narrow range of salinities and estuarine species can tolerate a much wider range (Brand, 1984). A few species of coccolithophores live in freshwater or low salinity tidal pools but most are marine (Paasche, 1968; Tappan, 1980). Others, such as *Coccolithus pelagicus*, have been found in the Dead Sea at salinities as high as 250 ppt (Tappan, 1980).

Emiliania huxleyi is found in salinities as high as 41 ppt in the Red Sea (Winter *et al.*, 1979) and as low as 18 ppt in Oslo

Fjord (Braarud, 1962) and 11 ppt in the Black Sea (Bukry, 1974). *Emiliania huxleyi* apparently did not invade the Black Sea until about 3000 years ago, presumably indicating that the salinity was not high enough until that time (Bukry, 1974). *Emiliania huxleyi* in culture can grow in salinities ranging from 15 to 45 ppt with highest growth rates between 20 and 35 ppt (Pintner and Provasoli, 1963; Brand, 1984). Although coastal genotypes of *E. huxleyi* can grow in 15 ppt, some oceanic genotypes from the Sargasso Sea cannot (Brand, 1984). Fisher and Honjo (1991) also found that an oceanic genotype of *E. huxleyi* tolerated high salinities better than a coastal genotype. *Gephyrocapsa oceanica* also cannot grow in 15 ppt and *Calcidiscus leptoporus* cannot grow in even 25 ppt (Brand, 1984). The coastal cocolithophore *Pleurochrysis carterae*, on the other hand, is able to grow in salinities from 15 to 45 ppt quite well (Brand, 1984).

Life cycles

The primary mode of reproduction of coccolithophores is asexual binary fission. Under conditions of light and nutrient saturation, *Emiliania huxleyi* and *Pleurochrysis carterae* can divide 2.5 times a day, *Gephyrocapsa oceanica* twice a day, and *Calcidiscus leptoporus* only once a day (Brand and Guillard, 1981). This reflects r-selection for high growth rates in species from coastal and eutrophic waters and K-selection for low growth rates in species from oligotrophic tropical waters. In general, asexual reproduction rates in coccolithophores are higher than in cyanobacteria and dinoflagellates but lower than in diatoms. Mitosis is similar to that found in most eukaryotic cells and each daughter cell gets approximately half the coccoliths. Sexual reproduction is known to occur in some species, but it is not well studied. If the sexual–asexual switching in coccolithophores is similar to that in many other microbes, one would expect sexual reproduction to occur primarily during times of poor growing conditions such as low nutrient concentrations or light intensities.

Many species are also known to have several life phases (see chapter 2). *Emiliania huxleyi* has at least three different life phases (Paasche and Klaveness, 1970; Klaveness, 1972). From the few species that have been studied, it appears that motile life stages alternate with non-motile stages. In coastal species such as *P. carterae*, the non-motile stage appears to be benthic, whereas the motile stage is planktonic (Rayns, 1962; Paasche, 1968; Klaveness and Paasche, 1979). In the case of *Coccolithus pelagicus*, both phases are planktonic but only one is motile (Parke and Adams, 1960). The two phases produce different types of coccoliths. One suspects that the shift between stages in temperate and polar coastal species occurs on a seasonal basis. The fact that only the planktonic stage possesses coccoliths suggests that they play an important role in the planktonic mode of existence.

Too few species have been studied, but it is likely that most coastal species have part of their life cycle as a benthic stage for 'hibernating' during times of the year when growing conditions are poor, whereas oceanic species are most likely planktonic in all life phases.

Genetic variation

Attempts to explain the distribution of a species usually make assumptions about the concept of the species and its genetic constitution. In the case of coccolithophores, species identification is by the morphology of the coccoliths. To what extent coccolith morphology reflects genetic relatedness is not well studied. Genetic variability within populations and genetic differentiation between populations have been demonstrated to exist in the niche hyperspace of coccolithophores. Genetic variability within populations of *E. huxleyi*, *Gephyrocapsa oceanica* and *Calcidiscus leptoporus* in growth rates ranges from 4.1 to 13.4% coefficient of variation (Brand, 1981).

Brand (1982a) found populations of *E. huxleyi* in the Gulf of Maine to be genetically different from those in the Sargasso Sea and coastal waters south of Cape Cod and populations of *Gephyrocapsa oceanica* in coastal waters to be genetically different from those in the Sargasso Sea. The genetic differentiation between coastal waters and the Sargasso Sea found in *G. oceanica* but not *E. huxleyi* indicates that different species have different spatial patterns of genetic differentiation.

Geographically separate populations of some species also sometimes differ from each other morphologically, although in many cases it is not known if this is caused by genetic or environmental differences. Young and Westbroek (1991) have documented several genetically different morphological forms of *E. huxleyi* that correlate only partially with temperature in the ocean. It seems likely that the morphological variation observed in other species such as *Calcidiscus leptoporus* (McIntyre *et al.*, 1970) also has a genetic basis and is related to adaptational differences. It is also known that different life stages of one species can have totally different forms of coccoliths, which could easily lead to them being identified as two separate species.

Ecological strategies

How well can we explain the spatial and temporal distribution of coccolithophores from what we have learned in laboratory studies? Coastal species such as *Pleurochrysis carterae* and cosmopolitan species such as *Emiliania huxleyi* are subjected to a wider range of environmental conditions than species living in the subtropical central gyres and indeed they do show a wider tolerance to factors such as temperature and salinity than do the oceanic species such as *Calcidiscus leptoporus* and *Umbilicosphaera sibogae*. Such environmental fluctuations lead to large variations in the potential for active growth by phytoplankton, resulting quite often in coastal and polar waters with nutrients remaining at relatively high concentrations in the photic zone. By contrast, in the permanently stratified waters of the subtropical central gyres, phytoplankton have depleted the photic zone of nutrients to extremely low concentrations. Obtaining nutrients is probably the most important ecological challenge for these oceanic species and indeed laboratory studies do show that they are better adapted for growing with low concentrations of nutrients than are coastal and polar species. Because the flux of nutrients into the oceanic photic zone is low, species living there have relatively low maximum growth rates, as a high potential growth rate could not be realized as a result of nutrient limitation. Species that live in coastal or polar waters appear to be adapted to more variable but nutrient-rich environments by having high maximum growth rates which can generate large blooms when environmental conditions are optimal. Most oceanic species appear to be adapted to low nutrient, non-variable environmental conditions in such a way that they do not respond significantly to nutrient-enrichment events. *Emiliania huxleyi* and *Gephyrocapsa oceanica* are in between these two endpoints of ecological strategies, often labelled r- and K-selected. *Emiliania huxleyi* and *Gephyrocapsa oceanica* are found in low abundance in nutrient-poor oceanic waters but become more abundant along the edges of the gyres, in equatorial upwelling areas and along the outer continental shelves where more nutrients are available. *Emiliania huxleyi* is most abundant in nutrient-rich subpolar waters while *G. oceanica* is most abundant in enriched tropical waters. Both species have higher maximum growth rates than the oceanic species that do not respond to enrichment. Overall, it appears that coccolithophore species are distributed along nutrient gradients from the eutrophic coastal and polar waters to oligotrophic subtropical central gyres and their biological characteristics reflect r- and K-selection respectively.

Effects of coccolithophores on the ecosystem

Coccolithophores not only respond to their environment but modify it as well. They are one of the major producers of calcareous sediments in the open ocean and the details of this are discussed by Steinmetz (chapter 9). Approximately 80% of the carbon buried in marine sediments each year is in the form of $CaCO_3$, most of this biologically produced (Broecker and Peng, 1982). Honjo *et al.* (1982) estimate that 20 to 40% of this $CaCO_3$ is formed by coccolithophores. It was the evolution of the pelagic coccolithophores and foraminifera in the Mesozoic that led to the shift of $CaCO_3$ deposition from occurring primarily in shallow coastal waters to more open-ocean areas, thus leading to a large change in the sedimentary record (Kennett, 1982).

Coccolithophores have an inordinate effect on light scattering in the ocean because of their coccoliths. As a result, high concentrations of coccolithophores can cause very inaccurate estimates of phytoplankton abundance based on remote-sensing measurements (Balch *et al.*, 1989). The light-scattering properties of coccolithophores, however do allow one to document the spatial and temporal extent of surface blooms using satellite imagery (Holligan *et al.*, 1983; Groom and Holligan, 1987; Balch *et al.*, 1991). Surface coccolithophore blooms reduce light levels below them. Balch *et al.* (1991) have speculated that blooms may reduce subsurface phytoplankton populations through light limitation enough to allow more nutrients to be transported up to the coccolithophores at the surface. They can also increase light reflectance back out into space (albedo) five to ten-fold (Balch *et al.*, 1991).

Coccolithophores, along with other prymnesiophytes and dinoflagellates, tend to produce more dimethylsulfoniopropionate than most other groups of phytoplankton (Keller *et al.*, 1989). This compound is apparently used for osmoregulation and leads to the excretion of dimethylsulfide (Vairavamurthy *et al.*, 1985). Charlson *et al.* (1987) have pointed out that this gas is oxidized to sulfate aerosols in the atmosphere and these aerosol particles are among the most important nuclei in the atmosphere over the ocean for promoting cloud condensation. Because increasing cloud cover increases the Earth's albedo, it is thought that coccolithophores may be affecting the Earth's climate because of their production of dimethyl-sulfoniopropionate and dimethylsulfide. Correlations between biological productivity, atmospheric dimethylsulfide, and sulfate aerosols have been demonstrated (Ayers *et al.*, 1991; Prospero *et al.*, 1991).

Long chain alkenones have been found to be specific to certain coccolithophores and other prymnesiophytes in the order Isochrysidales (Marlowe *et al.*, 1984, 1990) and to constitute around 8% of their organic biomass (Prahl *et al.*, 1988). The fact that the degree of unsaturation in these alkenones is a function of the temperature at which the algae live (Prahl and Wakeham, 1987) makes these compounds potentially useful as paleooceanographic indicators of past temperatures and prymnesiophyte abundances (Prahl *et al.*, 1989).

Summary

Physiologically and ecologically, coccolithophores appear to be similar to other groups of microalgae, the most striking difference being that most coccolithophore species tend to be more K-selected. As a result, coccolithophore species tend to be more important components of the phytoplankton community in warm, stratified, nutrient-poor waters and less important in nutrient-rich waters. Although most coccolithophore species are K-selected, many of the species that have been cultured and studied in the laboratory are more r-selected. Overall, there is good correspondence between the laboratory physiological data and species distributions in the ocean. The data show a spectrum from r-selection to K-selection as follows: *Pleurochrysis carterae – Emiliania huxleyi – Gephyrocapsa oceanica – Calcidiscus leptoporus – Umbilicosphaera sibogae*. The more r-selected species have wider temperature tolerances and are more easily limited by low nutrient concentrations. Although laboratory data cannot make precise quantitative predictions of coccolithophore abundance in the ocean, they do predict the relative distributions of species in time and space quite well, with more r-selected species in coastal and polar waters and more K-selected species in the oligotrophic subtropical central gyres.

It appears that coccolithophores have a significant impact on their environment as well. They are one of the major producers of calcareous sediments in the ocean and therefore influence global biogeochemical cycles. Coccolithophores apparently increase the albedo of the Earth by reflecting light with their coccoliths and by producing dimethylsulfide, a gas which ends up enhancing cloud formation in the atmosphere by increasing sulfate aerosol nuclei.

References

Antia, N. J., Berland, B. R., Bonin, D. J. and Maestrini, S. Y., 1975. Comparative evaluation of certain organic and inorganic sources of nitrogen for phototrophic growth of marine algae. *J. Mar. Biol. Assn. U.K.*, 519–39.

Ayers, G. P., Ivey, J. P., and Gillett, R. W., 1991. Coherence between seasonal cycles of dimethyl sulphide, methanesulphonate and sulphate in marine air. Nature, 349: 404–6.

Balch, W. M., Eppley, R. W., Abbott, M. R., and Reid, F. M. H., 1989. Bias in satellite-derived pigment measurements due to coccolithophores and dinoflagellates. *J. Plankt. Res.*, 11: 575–81.

Balch, W. M., Holligan, P. M., Ackleson, S. G., and Voss, K. J., 1991. Biological and optical properties of mesoscale coccolithophore blooms in the Gulf of Maine. *Limnol. Oceanogr.*, 36: 629–43.

Baumann, F. G., Isenberg, H. D., and Gennaro, J., 1978. The inverse relationship between nutrient nitrogen concentration and coccolith calcification in cultures of the coccolithophorid *Hymenomonas* sp. *J. Protozool.*, 25: 253–6.

Berge, G., 1962. Discoloration of the sea due to a *Coccolithus huxleyi* 'bloom'. Sarsia, 6: 27–40.

Bienfang, P. K., 1981. Sinking rate dynamics of *Cricosphaera carterae* Braarud. I. Effects of growth rate, limiting substrate, and diurnal variation in steady-state populations. *J. Exp. Mar. Biol. Ecol.*, 49: 217–33.

Braarud, T., 1962. Species distribution in marine phytoplankton. *J. Oceanogr. Soc. Japan*, 20: 628–49.

Braarud, T., Gaarder, K. R., Markali, J., and Nordli, E., 1952. Coccolithophorids studied in the electron microscope. *Nytt Mag. Bot.*, 1: 129–34.

Brand, L. E., 1981. Genetic variability in reproduction rates in marine phytoplankton populations. *Evolution*, 35: 1117–27.

Brand, L. E., 1982a. Genetic variability and spatial patterns of genetic differentiation in the reproductive rates of the marine coccolithophores *Emiliania huxleyi* and *Gephyrocapsa oceanica*. *Limnol. Oceanogr.*, 27: 236–45.

Brand, L. E., 1982b. Persistent diel rhythms in the chlorophyll fluorescence of marine phytoplankton species. *Mar. Biol.*, 69: 253–62.

Brand, L. E., 1984. The salinity tolerance of forty-six marine phytoplankton isolates. *Est. Coast. Shelf Sci.*, 18: 543–56.

Brand, L. E., 1986. Nutrition and culture of autotrophic ultraplankton and picoplankton. *Can. Bull. Fish. Aquat. Sci.*, 214: 205–33.

Brand, L. E., 1991a. Review of genetic variation in marine phytoplankton species and the ecological implications. *Biol. Oceanogr.*, 6: 397–409.

Brand, L. E., 1991b. Minimum iron requirements of marine phytoplankton and the implications for the biogeochemical control of new production. *Limnol. Oceanogr.*, 36: 1756–71.

Brand, L. E. and Guillard, R. R. L., 1981. The effects of continuous light and light intensity on the reproduction rates of twenty-two species of marine phytoplankton. *J. Exp. Mar. Biol. Ecol.*, 50: 119–32.

Brand, L. E., Sunda, W. G., and Guillard, R. R. L., 1983. Limitation of marine phytoplankton reproductive rates by zinc, manganese, and iron. *Limnol. Oceanogr.*, 28: 1182–98.

Brand, L. E., Sunda, W. G., and Guillard, R. R. L., 1986. Reduction of marine phytoplankton reproduction rates by copper and cadmium. *J. Exp. Mar. Biol. Ecol.*, 96: 225–50.

Bratbak, G., Heldal, M., Norland, S., and Thingstad, T. F., 1990. Viruses as partners in spring bloom microbial trophodynamics. *Appl. Env. Microbiol.*, 56: 1400–5.

Broecker, W. S. and Peng, T.-H., 1982. *Tracers in the Sea*. Lamont-Doherty Geological Observatory.

Bukry, D., 1974. Coccoliths as paleosalinity indicators –evidence from Black Sea. *Mem. Amer. Assn. Petrol. Geol.*, 20: 353–63.

Carlucci, A. F. and Bowes, P. M., 1970. Vitamin production and utilization by phytoplankton in mixed culture. *J. Phycol.*, 6: 393–400.

Charlson, R. J., Lovelock, J. E., Andreae, M. O., and Warren, S. G., 1987. Oceanic phytoplankton, atmospheric sulphur, cloud albedo and climate. *Nature*, 326: 655–61.

Dorigan, J. L. and Wilbur, K. M., 1973. Calcification and its inhibition in coccolithophorids. *J. Phycol.*, 9: 450–6.

Dring, M. J., 1990. Light harvesting and pigment composition in marine phytoplankton and macroalgae. In *Light and Life in the Sea*, ed. P. J. Herring, A. K. Campbell, M. Whitfield, and L. Maddock, pp. 89–103. Cambridge University Press, Cambridge.

Eppley, R. W., Holmes, R. W., and Strickland, J. D. H., 1967. Sinking rates of marine phytoplankton measured with a fluorometer. *J. Exp. Mar. Biol. Ecol.*, 1: 191–208.

Eppley, R. W., Rogers, J. N., and McCarthy, J. J., 1969. Half-saturation constants for uptake of nitrate and ammonium by marine phytoplankton. *Limnol. Oceanogr.*, 14: 912–20.

Eppley, R. W., Rogers, J. N., McCarthy, J. J. and Sournia, A., 1971. Light/dark periodicity in nitrogen assimilation of the marine phytoplankters *Skeletonema costatum* and *Coccolithus huxleyi* in N-limited chemostat culture. *J. Phycol.*, 7: 150–4.

Fisher, N. S. and Honjo, S. 1991. Intraspecific differences in temperature and salinity responses in the coccolithophore *Emiliania huxleyi*. *Biol. Oceanogr.* **6**: 355–61.

Gartner, S. and Bukry, D., 1969. Tertiary holococcoliths. *J. Paleontology*, **43**: 1213–21.

Geitzenauer, K. R., 1969. Coccoliths as late Quaternary palaeoclimatic indicators in the subantarctic Pacific Ocean. *Nature*, **223**: 170–2.

Groom, S. B. and Holligan, P. M., 1987. Remote sensing of coccolithophore blooms. *Adv. Space Res.*, **7**: 73–8.

Hallegraeff, G. M., 1984. Coccolithophorids (calcareous nanoplankton) from Australian waters. *Bot. Mar.* **27**: 229–47.

Haq, B. U. and Lohmann, G. P., 1976. Early Cenozoic calcareous nannoplankton biogeography of the Atlantic Ocean. *Mar. Micropaleontol.*, **1**: 119–20.

Hasle, G. R., 1969. Analysis of the phytoplankton of the Pacific Southern ocean. *Hvalrad. Skr.*, **52**: 1–168.

Hibberd, D. J., 1980. Prymnesiophytes (=Haptophytes). In *Phytoflagellates*, ed. E. R. Cox, pp. 273–317.

Holligan, P. M., Viollier, M., Harbour, D. S., Camus, P., and Champagne-Philippe, M., 1983. Satellite and ship studies of coccolithophore production along a continental shelf edge. *Nature*, **304**: 339–42.

Honjo, S. and Okada, H., 1974. Community structure of coccolithophores in the photic layer of the mid-Pacific. *Micropaleontol.*, **20**: 209–30.

Honjo, S., Manganini, S. J., and Cole, J. J., 1982. Sedimentation of biogenic matter in the deep ocean. *Deep-Sea Res.*, **29**: 609–25.

Hulburt, E. M., 1963. The diversity of phytoplanktonic populations in oceanic, coastal, and estuarine regions. *J. Mar. Res.*, **21**: 81–93.

Hulburt, E. M., 1964. Succession and diversity in the plankton flora of the western North Atlantic. *Bull. Mar. Sci. Gulf Carib.*, **14**: 33–44.

Hulburt, E. M., 1967. A note on regional differences in phytoplankton during a crossing of the southern North Atlantic Ocean in January, 1967. *Deep-Sea Res.*, **14**: 685–90.

Hulburt, E. M., 1976. Limitation of phytoplankton species in the ocean off western Africa. *Limnol. Oceanogr.*, **21**: 193–211.

Hulburt, E. M., 1979. An asymmetric formulation of the distribution characteristics of phytoplankton species: An investigation in interpretation. *Mar. Sci. Comm.*, **5**: 245–68.

Hulburt, E. M., 1983. Quasi K-selected species, equivalence, and the oceanic coccolithophorid plankton. *Bull. Mar. Sci.*, **33**: 197–212.

Hulburt, E. M. and Corwin, N., 1969. Influence of the Amazon River outflow on the ecology of the western tropical Atlantic. III. The planktonic flora between the Amazon River and the Windward Islands. *J. Mar. Res.*, **27**: 55–72.

Hulburt, E. M. and Mackenzie, R. S., 1971. Distribution of phytoplankton species at the western margin of the North Atlantic Ocean. *Bull. Mar. Sci.*, **21**: 603–12.

Hulburt, E. M., Ryther, J. H., and Guillard, R. R. L., 1960. The phytoplankton of the Sargasso Sea off Bermuda. *J. Cons. Int. Explor. Mer.*, **25**: 115–28.

Hutchinson, G. E., 1957. Concluding remarks. *Cold Spr. Harb. Symp. Quant. Biol.*, **22**: 415–27.

Inouye, I., 1990. High speed video analysis on the flagellar beat and swimming pattern of algae. Paper delivered at ISEP 8 meeting, University of Maryland, June, 1990.

Keller, M. D., Bellows, W. K., and Guillard, R. R. L., 1989. Dimethyl sulfide production in marine phytoplankton. *Amer. Chem. Soc. Symp., Ser.* **393**: 167–82.

Kennett, J. P., 1982. *Marine Geology*. Prentice-Hall, Englewood Cliffs, New Jersey.

Klaveness, D., 1972. *Coccolithus huxleyi* (Lohm.) Kamptn. II. The flagellate cell, aberrant cell types, vegetative propagation and life cycles. *Br. Phycol. J.*, **7**: 309–18.

Klaveness, D. and Paasche, E., 1979. Physiology of coccolithophores. In *Biochemistry and Physiology of Protozoa*, ed. M. Levandowsky, and S.H. Hutner, vol. 1, pp. 191–213.

Kuenzler, E. J. and Perras, J. P., 1965. Phosphatases of marine algae. *Biol. Bull.*, **128**: 271–84.

Linschooten, C., van Bleijswijk, J. D. L., van Emburg, P. R., de Vrind, J. P. M., Kempers, E. S., Westbroek, P. and de Vrind-de Jong, E. W., 1991. Role of the light-dark cycle and medium composition on the production of coccoliths by *Emiliania huxleyi* (Haptophyceae). *J. Phycol.*, **27**: 82–6.

Manton, I., Sutherland, J., and Oates, K., 1977. Arctic coccolithophorids: *Wigwamma arctica* gen. et sp. nov. from Greenland and arctic Canada, *W. annulifera* sp. nov. from South Africa and S. Alaska and *Calciarcus alaskensis* gen. et sp. nov. from S. Alaska. *Proc. Roy. Soc. London*, Ser. B 197: 145–68.

Marlowe, I. T., Brassell, S. C., Eglinton, G., and Green, J. C., 1990. Long-chain alkenones and alkyl alkenoates and the fossil coccolith record of marine sediments. *Chem. Geol.*, **88**: 349–75.

Marlowe, I. T., Green, J. C., Neal, A. C., Brassell, S. C., Eglinton, G., and Course, P. A., 1984. Long chain (n-C37-C39) alkenones in the Prymnesiophyceae. Distribution of alkenones and other lipids and their taxonomic significance. *Br. Phycol. J.*, **19**: 203–16.

Marshall, H. G., 1976. Phytoplankton distribution along the eastern coast of the USA. I. Phytoplankton composition. *Mar. Biol.*, **38**: 81–9.

Marshall, H. G., 1978. Phytoplankton distribution along the eastern coast of the USA. Part II. Seasonal assemblages north of Cape Hatteras, North Carolina. *Mar. Biol.*, **45**: 203–8.

Marshall, H. G. and Cohn, M. S., 1983. Distribution and composition of phytoplankton in northeastern coastal waters of the United States. *Est. Coast. Shelf Sci.*, **17**: 119–31.

McIntyre, A., 1967. Coccoliths as paleoclimatic indicators of Pleistocene glaciation. *Science*, **158**: 1314–17.

McIntyre, A. and Bé, A. W. H., 1967. Modern coccolithophorids of the Atlantic Ocean. I. Placoliths and cyrtoliths. *Deep-Sea Res.*, **14**: 561–97.

McIntyre, A., Bé, A. W. H., and Roche, M. B., 1970. Modern Pacific coccolithophorida: a paleontological thermometer. *N.Y. Acad. Sci. Trans.*, **32**: 720–31.

Mitchell-Innes, B. A. and Winter, A., 1987. Coccolithophores: a major phytoplankton component in mature upwelled waters off the Cape Peninsula, South Africa in March, 1983. *Mar. Biol.*, **95**: 25–30.

Mjaaland, G., 1956. Some laboratory experiments on the coccolithophorid *Coccolithus huxleyi*. *Oikos*, **7**: 251–5.

Nelson, D. M. and Brand, L. E., 1979. Cell division periodicity in 13 species of marine phytoplankton on a light:dark cycle. *J. Phycol.*, **15**: 67–75.

Okada, H. and Honjo, S., 1973. The distribution of oceanic coccolithophorids in the Pacific. *Deep-Sea Res.*, **20**: 355–74.

Okada, H. and McIntyre, A., 1979. Seasonal distribution of modern coccolithophores in the western North Atlantic Ocean. *Mar. Biol.*, **54**: 319–28.

Paasche, E., 1964. A tracer study of the inorganic carbon uptake during coccolith formation and photosynthesis in the coccolithophorid *Coccolithus huxleyi*. *Physiol. Plant. Suppl.*, **3**: 1–82.

Paasche, E., 1966. Adjustment to light and dark rates of coccolith formation. *Physiol. Plant.*, **19**: 271–8.

Paasche, E., 1967. Marine plankton algae grown with light-dark cycles. I. *Coccolithus huxleyi. Physiol. Plant.*, **20**: 946–56.

Paasche, E., 1968. Biology and physiology of coccolithophorids. *Ann. Rev. Microbiol.*, **22**: 71–86.

Paasche, E., 1969. Light-dependent coccolith formation in the two forms of *Coccolithus pelagicus. Arch. Mikrobiol.*, **67**: 199–208.

Paasche, E. and Klaveness, D., 1970. A physiological comparison of coccolith-forming and naked cells of *Coccolithus huxleyi. Arch. Mikrobiol.*, **73**: 143–52.

Parke, M. and Adams, I., 1960. The motile (*Crystallolithus hyelinum* Gaarder and Markali) and non-motile phases in the life history of *Coccolithus pelagicus* (Wallich) Schuller. *J. Mar. Biol. Assn. U.K.*, **39**: 263–74.

Pienaar, R. N. and Norris, R. E., 1979. The ultrastructure of the flagellate *Chrysochromulina spinifera* (Fournier) comb. nov. (Prymnesiophyceae) with special reference to scale production. *Phycologia*, **17**: 41–51.

Pintner, I. J. and Provasoli, L., 1963. Nutritional characteristics of some Chrysomonads. In: *Symposium on Marine Microbiology*, ed. C.H. Oppenheimer, pp. 114–155.

Prahl, F. G. and Wakeham, S. G., 1987. Calibration of unsaturation patterns in long-chain ketone compositions for palaeotemperature assessment. *Nature*, **330**: 367–69.

Prahl, F. G., Muehlhausen, L. A., and Zahnle, D. L., 1988. Further evaluation of long-chain alkenones as indicators of paleoceanographic conditions. *Geochim. Cosmochim. Acta*, **52**: 2303–10.

Prahl, F. G., de Lange, G. J., Lyle, M., and Sparrow, M. A., 1989. Post-depositional stability of long-chain alkenones under contrasting redox conditions. *Nature*, **341**: 434–7.

Prezelin, B. B. and Boczar, B. A., 1986. Molecular bases of cell absorption and fluorescence in phytoplankton: potential applications to studies in optical oceanography. *Prog. Phycol. Res.*, **4**: 349–464.

Proctor, L. M. and Fuhrman, J. A., 1990. Viral mortality of marine bacteria and cyanobacteria. *Nature*, **343**: 60–2.

Prospero, J. M., Savoie, D. L., Saltzman, E. S., and Larsen, R., 1991. Impact of oceanic sources of biogenic sulphur on sulphate aerosol concentrations at Mawson, Antarctica. *Nature*, **350**: 221–3.

Rayns, D. G., 1962. Alternation of generation in a coccolithophorid, *Cricosphaera carterae* (Braarud and Fagerland) Braarud. *J. Mar. Biol. Assn. U.K.*, **42**: 481–4.

Sherr, E. and Sherr, B., 1988. Role of microbes in pelagic food webs: A revised concept. *Limnol. Oceanogr.*, **33**: 1225–7.

Sieburth, J. M., 1979. *Sea Microbes*. Oxford University Press, New York. 491 pp.

Sikes, C. S. and Wilbur, K. M., 1982. Functions of coccolith formation. *Limnol. Oceanogr.*, **27**: 18–26.

Smayda, T. J., 1966. A quantitative analysis of the phytoplankton of the Gulf of Panama. III. General ecological conditions, and the phytoplankton dynamics at 8 45'N, 79 23'W from November 1954 to May 1957. *Bull. Inter-Amer. Trop. Tuna Commn.*, **11**: 355–612.

Smayda, T. J., 1970. The suspension and sinking of phytoplankton in the sea. *Oceanogr. Mar. Biol. Ann. Rev.*, **8**: 353–414.

Smayda, T. J. and Bienfang, P. K., 1983. Suspension properties of various phyletic groups of phytoplankton and tintinnids in an oligotrophic, subtropical system. *Mar. Ecol.*, **4**: 289–300.

Sournia, A., 1982. Form and function in marine phytoplankton. *Biol. Rev.*, **57**: 347–94.

Sunda, W. G. and Huntsman, S. A., 1983. Effect of competitive interactions between manganese and copper on cellular manganese and growth in estuarine and oceanic species of the diatom *Thalassiosira. Limnol. Oceanogr.*, **28**: 924–34.

Suttle, C. A., Chan, A. M., and Cottrell, M. T., 1990. Infection of phytoplankton by viruses and reduction of primary productivity. *Nature*, **347**: 467–9.

Tappan, H., 1980. *The Paleobiology of Plant Protists*. W.H. Freeman and Co.

Thomsen, H. A., 1981. Identification by electron microscopy of nanoplanktonic coccolithophorids (Prymnesiophyceae) from West Greenland, including the description of *Papposphaera sarion*. sp. nov. *Br. Phycol. J.*, **16**: 77–94.

Vairavamurthy, A., Andreae, M. O., and Iverson, R. L., 1985. Biosynthesis of dimethylsulfide and dimethylpropiothetin by *Hymenomonas carterae* in relation to sulfur source and salinity variations. *Limnol. Oceanogr.*, **30**: 59–70.

Venrick, E. L., 1982. Phytoplankton in an oligotrophic ocean: Observations and questions. *Ecol. Monogr.*, **52**: 129–54.

Verbeek, J. W., 1989. Recent calcareous nannoplankton in the southernmost Atlantic. *Polarforschung*, **59**: 45–60.

Wal, P. van der, de Jong, L., Westbroek, P. and de Bruijn, W.C. 1983. Calcification in the coccolithophorid alga *Hymenomonas carterae. Ecol. Bull.* 35:251–8.

Wal, P. van der, de Vrind, J.P.M., de Vrind-de Jong, E.W., and Borman, A.H. 1987. Incompleteness of the coccosphere as a possible stimulus for coccolith formation in *Pleurochrysis carterae* (Prymnesiophyceae). *J. Phycol.*, 23:218–21.

Watabe, N. and Wilbur, K. M., 1966. Effects of temperature on growth, calcification, and coccolith form in *Coccolithus huxleyi* (Coccolithineae). *Limnol. Oceanogr.*, **11**: 567–575.

Wheeler, P. A., North, B. B., and Stephens, G. C., 1974. Amino acid uptake by marine phytoplankters. *Limnol. Oceanogr.*, **19**: 249–59.

Wilbur, K. M. and Watabe, N., 1963. Experimental studies on calcification in molluscs and the alga *Coccolithus huxleyi. Ann. N.Y. Acad. Sci.*, **109**: 82–112.

Winter, A., 1985. Distribution of living coccolithophores in the California Current System, southern California Borderland. *Mar. Micropaleontol.*, **9**: 385–93.

Winter, A., Reiss, Z., and Luz, B. 1979. Distribution of living coccolithophore assemblages in the Gulf of Elat ('Aqaba). Mar. Micropaleontol., **4**: 197–223.

Young, J. R. and Westbroek, P., 1991. Genotypic variation in the coccolithophorid species *Emiliania huxleyi*. Mar. *Micropaleontol.*, **18**: 5–23.

4 Composition and morphology of coccolithophore skeletons

WILLIAM G. SIESSER AND AMOS WINTER

Introduction

Until a few years ago, it was widely believed that calcareous nannoplankton first appeared in the Early Jurassic. Discoveries in the Alps (Di Nocera and Scandone, 1977; Jafar, 1983) and in the Indian Ocean (Bralower *et al.*, 1991) now show that the geological record of these fossils goes back to the Carnian Stage (earliest Late Triassic), approximately 230 Ma.

Several interesting questions immediately spring to mind. Why did these nannoplankton suddenly begin to produce skeletons at that point in time? For that matter, why was it advantageous for them to produce a skeleton at all? And, once skeletons began to form, why was calcite advantageous, rather than silica or some other material?

We have no answers to these questions at the moment, but some ideas bearing on the matter have recently been put forward by Brasier (1986), Westbroek *et al.* (1989), Lowenstam and Weiner (1989), Simkiss and Wilbur (1989) and Pentecost (1991). Whatever the forcing conditions for the development of skeletons may have been, and whatever functional purpose skeletons serve (see chapter 5) the skeleton has clearly been an evolutionary success. The proof of this is the profusion of calcareous nannoplankton that have flourished in oceanic waters from Late Triassic time until today. Not incidentally – vast amounts of calcium and carbon have been recycled from the oceans back to the lithosphere through the sedimentation of nannofossil skeletons. These skeletons have been incorporated in chalks primarily, but also in limestones and other calcareous rocks.

In this chapter we shall discuss the coccolithophore skeleton. In particular, we shall examine the mineralogy and elemental composition of the component coccoliths, the morphology of coccoliths, and the arrangement of coccoliths on coccospheres.

Composition of coccoliths

It has been well established that invertebrate marine animals and protists extract elements from ambient sea water in order to form their skeletons. The skeletal composition of most hard-part secreting organisms has been determined in the last few decades, largely as a result of the development and application of sophisticated analytical instruments. Mineralogy and major, minor, and trace elements in the skeletons of most organisms can now be routinely analysed.

Calcareous nannoplankton are an exception. Their extremely small size makes detailed mineralogic and/or elemental analyses difficult, and therefore few studies have been made. Early petrographic work (see chapter 1) had suggested that calcite was the mineral making up coccoliths. Later rudimentary chemical and X-ray diffraction analyses of bulk nannofossil ooze added further support to this idea; Isenberg *et al.* (1963), using better chemical and X-ray diffraction techniques, eventually confirmed that coccoliths are composed of calcite.

Wilbur and Watabe (1963) reported traces of aragonite, precipitated along with calcite, in coccoliths produced in laboratory cultures under normal marine conditions. In cultures using a nitrogen-poor solution, vaterite was also reported along with the aragonite and calcite. Vaterite and aragonite have not been found by other analyses of nannofossil sediment, or in cultured specimens of *Emiliania huxleyi* (Young *et al.*, 1991). It is likely that, even if originally present, these unstable minerals would invert very quickly to calcite. Older accounts of aragonite in discoasters (Tan, 1927) and in a single Cretaceous specimen (Hart *et al.*, 1965) can probably be discounted owing to the analytical techniques used.

The calcite produced by coccolithophores is the 'low-Mg' variety (high-Mg calcite contains >4% $MgCO_3$; low-Mg calcite contains <4% $MgCO_3$ – see Siesser, 1971). Siesser (1977), using an electron microprobe to analyze individual coccoliths, did not find detectable magnesium in the calcite. The absence of magnesium is particularly interesting, since many benthic marine organisms easily incorporate this element in their calcitic skeletons. It is probably not coincidental that the other common planktonic calcareous organisms (planktonic foraminifers) in the ocean also build skeletons of low-Mg calcite. This may be because planktonic organisms prefer the lighter calcite phase (low-Mg) to the heavier phase (high-Mg) for flotation purposes. Or it may be simply that planktonic organisms have less need for, or are less prone to extract, magnesium from ambient waters than are benthic organisms. Thompson and Bowen (1969) and Thompson (1972) presented elemental analyses of bulk nannofossil oozes. They found that the oozes contained magne-

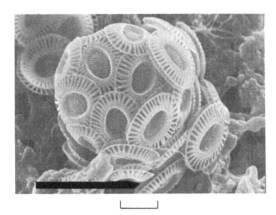

Fig. 1. Coccosphere of the heterococcolith *Emiliania huxleyi*. Crystal elements differ markedly in size and shape. Scale bar = 2 μm.

Fig. 2. Holococcolith of *Helladosphaera cornifera*. Note the similarity is size and shape of crystal elements. Scale bar = 1 μm.

sium in average values of 0.11% (Atlantic Ocean oozes), 0.17% (Pacific Ocean oozes) and 0.13% (Indian Ocean oozes). Strontium averaged 0.15% (Atlantic), 0.16% (Pacific), and 0.21% (Indian). Comparatively high average values of barium (160–510 ppm) and manganese (265–1500 ppm) were also reported. There is an obvious problem with drawing conclusions as to the composition of one sedimentary component (the nannofossils) based on bulk analysis of an ooze containing a number of different components (foraminifers, clay, etc.). The other components may be contributing a part (or all) of any given element.

Analysis of individual nannoplankton skeletons is clearly preferable to bulk-sediment analyses. Siesser (1977) analyzed coccoliths of several different species from each Tertiary epoch at DSDP Sites 360, 361 and 362 (off southwestern Africa). Analyses for the elements Al, Ca, Fe, K, Mg, Mn, Na, P, and Si were conducted. Only calcium and phosphorus were detectable within the microprobe's detection limits (500 ppm). Judging from these analyses, the calcite making up nannoplankton skeletons seems to incorporate little in the way of minor or trace elements. The phosphorus content was fairly uniform in Paleocene and Oligocene coccoliths, but a progressive increase in phosphorous content occurred from middle Miocene time onward. Siesser (1977) attributed this increase to the greater availability of phosphorus after upwelling began in the Benguela Current during the Miocene.

An acidic polysaccharide has been extracted from the coccoliths of *E. huxleyi* cells grown in culture (Fichtinger-Schepman *et al.*, 1981; van Emburg *et al.*, 1986). Westbroek *et al.* (1989) concluded that this polysaccharide plays an important regulatory role in the biomineralization process. The polysaccharide is believed to inhibit the early crystallization of $CaCO_3$ within the cell until changing cellular morphological and biochemical conditions are suitable for coccolithogenesis. Later, after a coccolith has formed, further crystal growth is arrested by the polysaccharide (see Pienaar, chapter 2, for additional details on this interesting process).

Morphology of coccoliths

Coccoliths are divided into two general groups: heterococcoliths and holococcoliths. Heterococcoliths are constructed of crystal elements that differ in size and shape (Fig. 1), whereas the crystal elements making up holococcoliths are essentially identical in size and shape (Fig. 2), and are considerably smaller (less than 0.1 μm) than the elements forming heterococcoliths. Holococcoliths appear to be much less common than heterococcoliths, both among living coccolithophores and in the fossil record. The construction of holococcoliths probably causes them to disintegrate more rapidly than the comparatively robust heterococcoliths.

Electron diffraction and high-resolution transmission electron microscopy studies have shown that each coccolith element of *Emiliania huxleyi* is a single crystal (Mann and Sparks, 1988). Black (1963) showed that the calcite in coccoliths may crystallize in rhombohedral, prismatic, tabular or scalenohedral forms, although most coccoliths utilize the rhombohedral form. A smaller number of species build their coccoliths of hexagonal prisms. Two extinct groups, the discoasters and stephanoliths, used the tabular and scalenohedral forms, respectively.

Holococcoliths are composed of hexagonal prisms, rhombohedral crystals, or both. In holococcoliths, the original crystal form, complete with crystal faces and angles, can be seen (Fig. 3). A typical arrangement of holococcolith crystals consists of parallel prisms packed tightly together (Fig. 2). This arrangement is sometimes modified by the regular absence of a prism from the pack, forming a sieve-like structure (Fig. 4). Holococcoliths constructed of rhombohedral crystals characteristically have rhombohedra neatly stacked, or arranged in an offset 'checkerboard' pattern with gaps between the squares (Fig. 3).

The majority of heterococcoliths are built of rhombohedral crystals, with only a few species choosing to construct their skeletons of hexagonal prisms. In a heterococcolith (in

Fig. 3. Holococcolith (motile phase of *Coccolithus pelagicus*) showing euhedral crystals. Note the offset 'checkerboard' pattern of crystal elements. Scale bar = 1 μm. Reproduced with permission from Nishida, S. (1979), *Atlas of Pacific Nannoplanktons*. Micropaleontol. Soc. of Osaka.

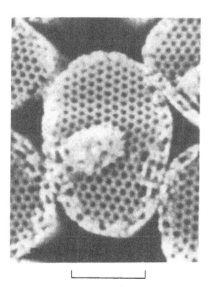

Fig. 4. Holococcolith of *Sphaerocalyptra gracillima*, showing sieve-like arrangement of crystal elements. Scale bar = 1 μm. Reproduced with permission from Nishida, S. (1979), *Atlas of Pacific Nannoplanktons*. Micropaleontol. Soc. of Osaka.

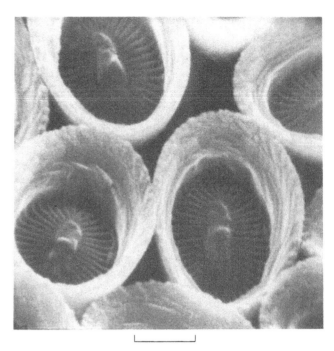

Fig. 5. Simple heterococcolith wall structure of *Coronosphaera mediterranea (= Syracosphaera mediterranea)*. Scale bar = 1 μm.

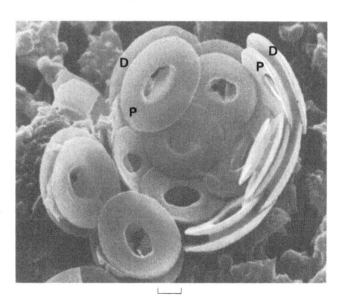

Fig. 6. Interior view of a broken coccosphere of *Gephyrocapsa* sp. Distal (D) and proximal (P) shields of placoliths are joined by a short central column. Note interlocking of adjacent edges of placolith shields. Scale bar = 2 μm.

marked contrast to a holococcolith), the living cell forces the crystals it secretes into shapes in which faces and angles are suppressed (Figs. 1; 5–14; 16–26).

The simplest arrangement of rhombohedral crystals on a heterococcolith is in the form of a single disc with a 'wall' of rhombohedral crystals at the circumference (e.g., the genus *Syracosphaera*, Fig. 5). Other rhombohedral crystals grow inward from the foot of the wall towards the center of the disc. Black (1963) noted that coccoliths of this type are

probably formed by accretion upon an organic scale and decided that crystal growth never extends beyond the margin of the scale; once an outer wall is built, further growth of crystals can only take place inwards.

A more complex arrangement taken by heterococcoliths consists of two circular or elliptical 'shields' joined by a central cylindrical column; there is no peripheral wall (Fig. 6). In

53

Fig. 7. Coccosphere of *Coccolithus pelagicus*. Central area is open and spanned by a short bridge. Crystal elements of the distal shields override one another. Scale bar = 6 μm.

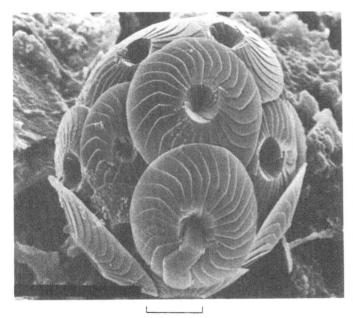

Fig. 8. Coccosphere of *Calcidiscus leptoporus*. Central area is almost closed, with only a small hole in the center. Scale bar = 3 μm.

this construction, the central column is either a solid pillar or a hollow tube. In the case of the hollow tube, the wall of the tube is also thought to be composed of rhombohedral crystals, either joined like staves in a barrel, or spiraling obliquely down the wall of the tube (Black, 1963). The joined 'shields' are designated 'proximal' and 'distal'. The proximal shield is on the inner (concave) side of the coccolith, and the distal shield is on the outer (convex) side of the coccolith (Figs. 6

Fig. 9. Coccosphere of *Gephyrocapsa oceanica*. A net-like structure fills the central area of each placolith, and a prominent bridge spans the central area. Scale bar = 2 μm.

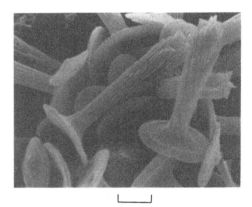

Fig. 10. Prominent central process of *Rhabdosphera clavigera*. Scale bar = 2 μm. Photo courtesy of J. Alcober, University of Valencia.

and 13). Most heterococcoliths also have a distinctive open or closed 'central area' (Figs. 7 and 8) from which shield elements radiate outward. The elements of a shield tend to taper near the central area in order to fit better around a circular or elliptical center, whereas they tend to expand towards the circumference, giving them a 'petal-like' appearance. The crowding together of elements often causes overriding and imbrication of adjacent elements (e.g., Fig. 7). 'Nets', 'pores', 'crosses', 'bridges' 'crossbars' and other structures commonly occur in the central area of a shield (Fig. 9). A "central process" (also called a "spine" or "knob") sometimes projects upward from a coccolith (e.g., Fig. 10).

It is clear from the foregoing discussion that the living cell exerts a very real control over the shape and structure of the coccoliths it secretes. Environmental influences play a lesser role in determining coccolith morphology. Intraspecific variation in the morphology of *E. huxleyi* has been investigated by Young and Westbroek (1991). They found that cultured specimens showed variation in the degree of coccolith completion, size, degree of calcification, and malformation.

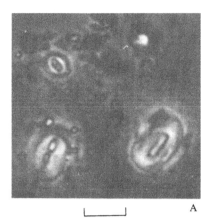
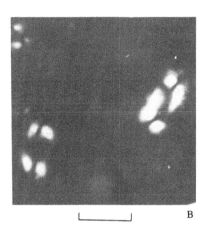

A   B

Fig. 11. Light micrographs of *Helicosphaera carteri*: A. plane- polarized light; B. cross-polarized light. Specimen on the left in each photo is a proximal view; specimen on the right is a distal view. Different extinction patterns are shown by the distal and proximal shields in cross-polarized light. Scale bars = 10 μm.

Young and Westbroek (1991) concluded that a range of environmental parameters influenced the variations.

The general process of coccolithogenesis has been examined in detail by several authors (e.g., see Tappan, 1980; chapter 2, this volume). Young (1989) has, however, recently suggested a model for heterococcolith development that increases our understanding of the structure and geometry ultimately obtained by coccoliths. In the first stages of Young's model, crystal nucleation forms a proto-coccolith ring around the edge of an organic base-plate scale. The number, spacing and orientation of crystal elements are determined during the nucleation stage. Uniform element growth then proceeds from the proto-coccolith ring. This model is not applicable to holococcoliths, whose elements probably formed by extra-cellular calcification (Manton and Leedale, 1963; Young, 1989). This model is supported by the recent discovery of an ontogenetic sequence in which proto-coccolith rings, early growth stages, and fully developed coccoliths of *Watznaueria* occur (Young and Bown, 1991).

Forcing calcite crystals into a variety of orientations results in differently arranged optic axes. Since calcite exhibits a high degree of double refraction, even the tiny crystals of calcite making up coccoliths will give strong extinction patterns when viewed between the crossed nicols of a polarizing microscope. This phenomenon was noted by Sorby as early as 1861. Kamptner (1954), realizing the potential usefulness of this feature, pointed out that coccoliths from different nannoplankton species will show diagnostic extinction patterns. Thus was born a tool which could be used to differentiate species which otherwise look alike in ordinary light.

As examples, the distal shield of a heterococcolith may have an optic-axis arrangement different from that of the proximal shield, or several cycles of elements may be present on the same shield with axes arranged differently among the different cycles. Collars, bridges, central processes, and other central-area structures may also have varying arrangements of axes. The result is a variety of interference figures seen between the crossed nicols of a polarizing microscope. Species identifications are routinely made on the basis of taxonomically diagnostic extinction patterns (Fig. 11).

A few taxa are not birefringent between crossed nicols. The most important among these is the extinct genus, *Discoaster*, which ranged from the Paleocene to the latest Pliocene. A discoaster normally comes to rest flat on a microscope slide, and, since its optic axis is then perpendicular to the plane of the slide and parallel to the light ray, it appears dark between crossed nicols. This has led some workers to reason that individual discoasters consist of single calcite crystals. Black's (1972) study showed, however, that each ray of a discoaster is an independent crystal, with its c-axis oriented perpendicular to the plane of the disc.

Coccolith shape classification

A number of terms have been introduced to describe the overall shapes which coccoliths may take. These terms can replace more lengthy descriptive phrases and avoid unnecessary descriptive repetition. A number of the shape terms found in the literature are, however, not widely used. The following is a list of the more common shape terms used for the coccoliths of living coccolithophores. Terms used for fossil forms may be found in Hay (1977), Tappan (1980), and Gartner (1981).

CALYPTROLITH – holococcolith; basket-shaped, open proximally (e.g., *Homozygosphaera triarcha*, Fig. 12A)

CANEOLITH – disc- or bowl-shaped; lath-filled central area (e.g., *Syracosphaera pulchra*, Fig. 12B).

CERATOLITH – horseshoe- or wishbone-shaped (e.g., *Ceratolithus cristatus*, Fig. 12C).

CRIBRILITH – disc-shaped; numerous central-area perforations (e.g., *Pontosphaera syracusana*, Fig. 12D).

CYRTOLITH – disc-shaped; convex outward, often with a projecting central process (e.g., *Discosphaera tubifera*, Fig. 12E).

DISCOLITH – disc-shaped; thickened, raised rim. May or may not have central-area perforations. Definitions for this term vary; synonymous in part with cribrilith.

55

A. CALYPTROLITH

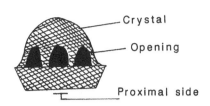

Homozygosphaera triarcha

B. CANEOLITH

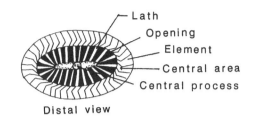

Distal view

Syracosphaera pulchra

C. CERATOLITH

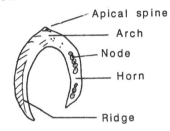

Ceratolithus cristatus

D. CRIBRILITH

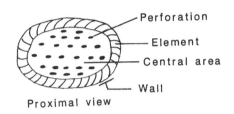

Proximal view

Pontosphaera syracusana

E. CYRTOLITH

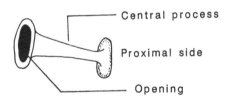

Discosphaera tubifer

F. HELICOLITH

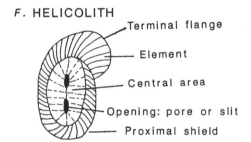

Helicosphaera carteri

HELICOLITH – spiraling, overlapping marginal flange (e.g., *Helicosphaera carteri*, Fig. 12F).

LOPADOLITH – basket-, cup-, or vase-shaped, with high rim, opening distally (e.g., *Scyphosphaera* sp., Fig. 12G).

PENTALITH – five four-sided crystals joined to form a pentagon (e.g., *Braarudosphaera bigelowii*, Fig. 12H).

PLACOLITH – two shields joined by a central column (e.g., *Coccolithus pelagicus*, Fig. 12I).

PRISMATOLITH – solid or perforate polygonal prism (e.g., *Thoracosphaera heimii* (a calcareous dinoflagellate), Fig. 12J).

RHABDOLITH – single shield, surmounted by a club-shaped central process (e.g., *Rhabdosphaera clavigera*, Fig. 12K).

SCAPHOLITH – rhombohedral shape, with parallel laths joining in the central area (e.g., *Anoplosolenia brasiliensis*, Fig. 12L).

Morphological terms for the various parts of these coccoliths are shown in Fig. 12A to 12L.

Morphology of coccospheres

Figures 13 to 24 show how various coccolith types (placolith, caneolith, etc.) are arranged around cells to form coccospheres. Placoliths lock together by inserting the leading

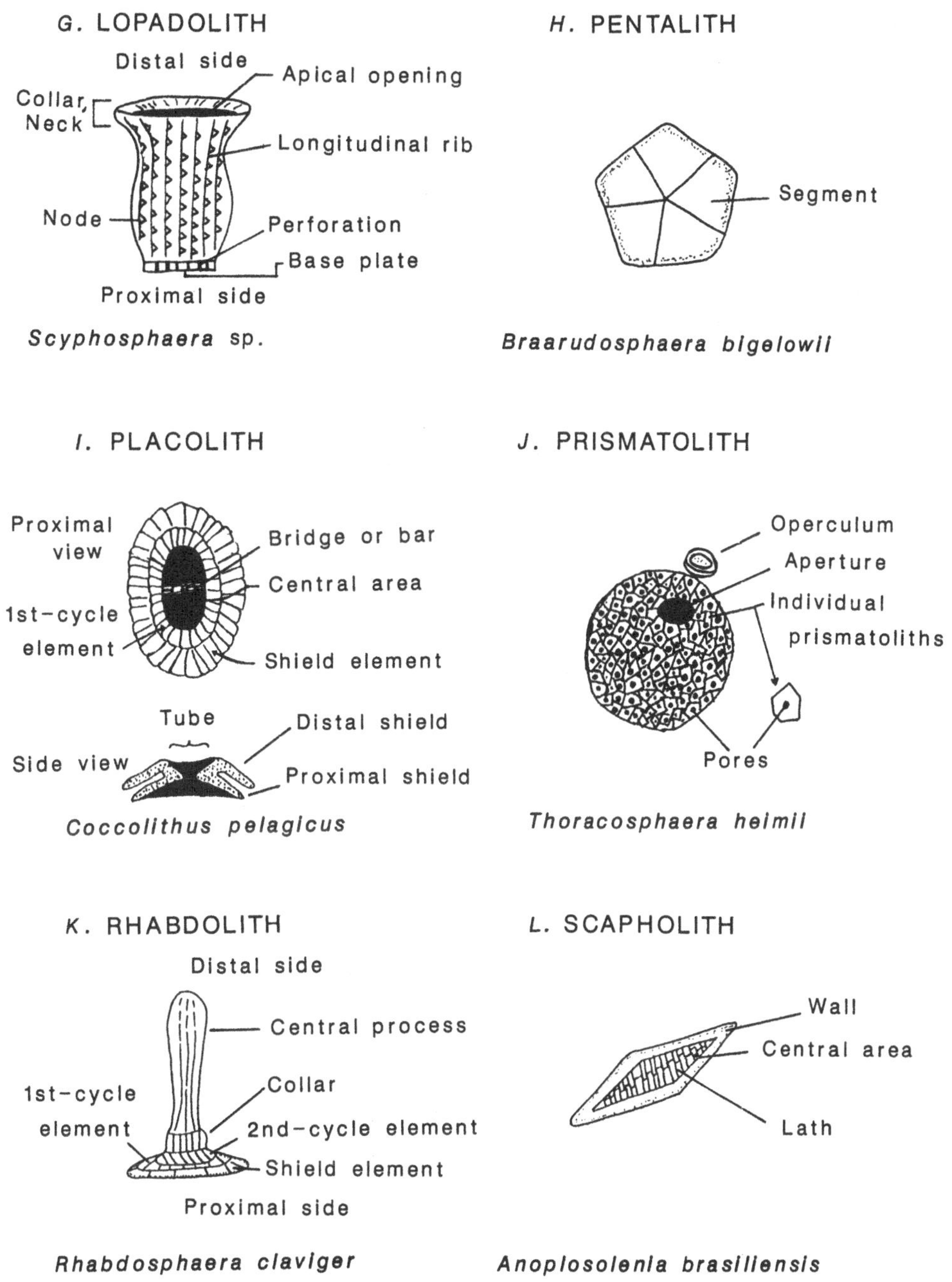

G. LOPADOLITH

Distal side — Apical opening

Collar, Neck

— Longitudinal rib

Node

— Perforation

Base plate

Proximal side

Scyphosphaera sp.

H. PENTALITH

— Segment

Braarudosphaera bigelowii

I. PLACOLITH

Proximal view — Bridge or bar

— Central area

1st-cycle element

— Shield element

Tube — Distal shield

Side view — Proximal shield

Coccolithus pelagicus

J. PRISMATOLITH

— Operculum

Aperture

Individual prismatoliths

Pores

Thoracosphaera heimii

K. RHABDOLITH

Distal side

— Central process

Collar

1st-cycle element

2nd-cycle element

— Shield element

Proximal side

Rhabdosphaera claviger

L. SCAPHOLITH

— Wall

Central area

Lath

Anoplosolenia brasiliensis

edge of one distal shield into the gap between the distal and proximal shields of adjacent placoliths (Figs. 6 and 13). More coccospheres constructed of placoliths are found in sediment than any other shape-form, suggesting that interlocking plates is the strongest architectural arrangement for coccospheres. Helicoliths have an arrangement somewhat similar to placoliths, with the distal flaring margin lapping over and locking into adjacent helicoliths (Fig. 14). Some caneoliths interjoin weakly with their neighbors along adjacent edges.

Calyptroliths, caneoliths, cribriliths, cyrtoliths, lopadoliths, rhabdoliths and scapholiths abut, and occasionally

Fig. 12. Sketches of A. calyptrolith; B. caneolith; C. ceratolith; D. cribrilith; E. cyrtolith; F. helicolith; G. lopadolith; H. pentalith; I. placolith; J. prismatolith; K. rhabdolith; and L. scapholith shape-forms and their morphologic parts.

overlap one another on their coccospheres, but do not seem to have any structural interconnection among adjacent coccoliths (Figs. 15–22). Presumably they are held in place on living coccolithophores simply by the organic membrane.

Pentaliths join together tightly to form an imperforate coccosphere (Fig. 23). For prismatoliths, individual ele-

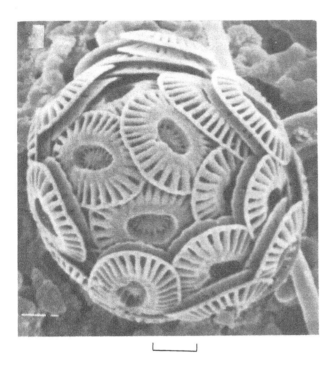

Fig. 13. Coccosphere of *Emiliania huxleyi*, formed by layers of interlocking placoliths. Both distal and proximal shields of placoliths can be seen at the sides of the coccosphere. Scale bar = 2 μm. Reproduced with permission from Nishida, S. (1979), *Atlas of Pacific Nannoplanktons*. Micropaleontol. Soc. of Osaka.

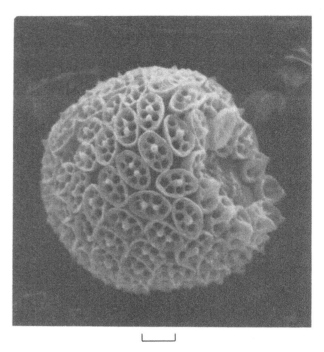

Fig. 15. Coccosphere of *Homozygosphaera vercellii*, with abutting calyptroliths. Scale bar = 2 μm. Photo courtesy of J. Alcober, University of Valencia.

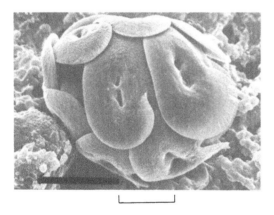

Fig. 14. Coccosphere of *Helicosphaera carteri* with interlocking lopadoliths. Scale bar = 5 μm.

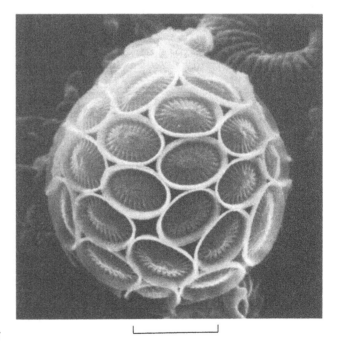

Fig. 16. Coccosphere of *Syracosphaera anthos*, constructed of closely packed caneoliths. Scale bar = 2 μm.

ments fit together in a jigsaw fashion to form the spherical or elongated genus *Thoracosphaera* (traditionally studied with the calcareous nannoplankton, although now known to be a calcareous dinoflagellate – see chapter 6. Some of the prisms are perforate (e.g., *T. heimii,* Fig. 24), but most are imperforate. Ceratoliths do not form a coccosphere. Norris (1971) found living cells of *Ceratolithus cristatus* in the Indian

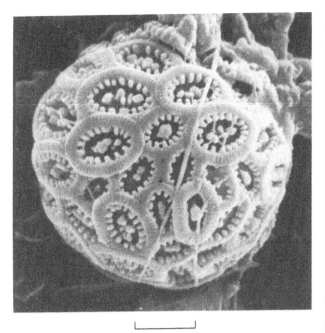

Fig. 17. Coccosphere of *Syracosphaera molischii*, with overlapping caneoliths. Scale bar = 2 μm.

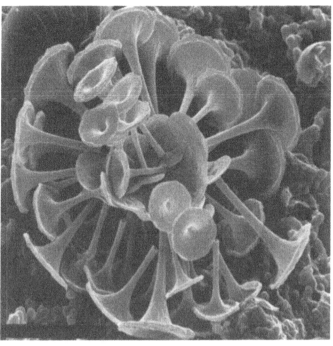

Fig. 19. Coccosphere of *Discosphaera tubifera*, constructed of cyrtoliths . Note that the proximal plates of adjacent cyrtoliths overlap. Scale bar = 3 μm.

Fig. 18. Coccosphere of *Pontosphaera syracusana*, contructed of cribriliths. Scale bar = 2 μm. Reproduced with permission from Nishida, S. (1979), *Atlas of Pacific Nannoplanktons.* Micropaleontol. Soc. of Osaka.

Fig. 20. Coccosphere of *Scyphosphaera apsteinii*. This species is dimorphic, showing both lopadoliths of *S. apsteinii* and cribriliths of *Pontosphaera* cf. *P. japonica* on the same coccosphere. Scale bar = 10 μm. Photo courtesy of J. Alcober, University of Valencia.

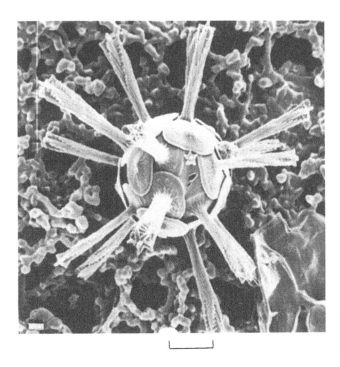

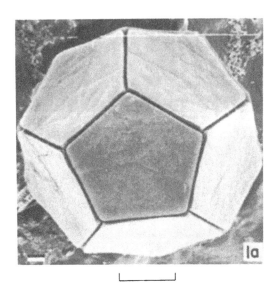

Fig. 23. Coccosphere of *Braarudosphaera bigelowii*, constructed of tightly joined pentaliths. Scale bar = 3 μm. Reproduced with permission from Nishida, S. (1979), *Atlas of Pacific Nannoplanktons*. Micropaleontol. Soc. of Osaka.

Fig. 21. Coccosphere of *Rhabdosphaera clavigera*. The coccosphere is constructed mostly of overlapping rhabdoliths, but also has some coccoliths without central processes. Scale bar = 3 μm. Reproduced with permission from Nishida, S. (1979), *Atlas of Pacific Nannoplanktons*, Micropaleontol. Soc. of Osaka.

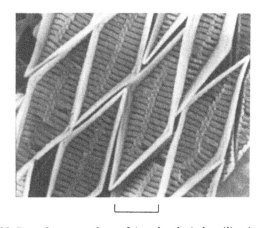

Fig. 22. Part of a coccosphere of *Anoplosolenia brasiliensis*, constructed of tightly packed scapholiths. Scale bar = 1 μm. Photo courtesy of J. Alcober, University of Valencia.

Fig. 24. Coccosphere of *Thoracosphaera heimii* composed of individual prismatoliths, each of which is pierced by a pore. Scale bar = 2 μm.

Ocean, in which a single horseshoe-shaped ceratolith surrounds the nucleus of the cell.

Understanding coccosphere construction is complicated by the 'dimorphism' shown by some species. One part of the coccosphere of these species carries one type of coccolith, whereas another part will carry an altogether different type of coccolith. *Acanthoica quattrospina*, for example, has spine-bearing coccoliths around the polar end of the coccosphere, but spineless coccoliths around the remainder of the coccosphere (Fig. 25). Furthermore, the two 'species' *Scyphosphaera apsteinii* and *Pontosphaera japonica* are

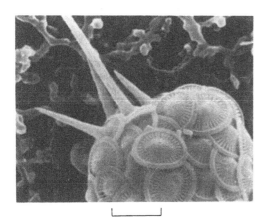

Fig. 25. Part of a coccosphere of *Acanthoica quattrospina* (= *A. acanthifera*). This dimorphic species has spine-bearing coccoliths at the polar end, and spineless coccoliths over the rest of the coccosphere. Scale bar = 3 μm. Photo courtesy of J. Alcober. University of Valencia.

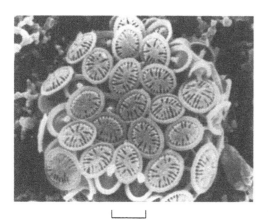

Fig. 26. Coccosphere of *Scyracosphaera histrica*. This is a dithecate species with an endothecal layer of caneoliths and an exothecal layer of cyrtoliths. Scale bar = 2 μm. Photo courtesy of J. Alcober, University of Valencia.

now known to occur on the same coccosphere (Fig. 20), as are *Helicosphaera carteri* and *H. wallichii* (Nishida, 1979).

In addition to the potential dimorphic problem, some living coccolithophores exhibit 'dithecatism'; that is, they are constructed of a double layer of coccoliths, each layer of which may contain coccoliths of markedly different shape. Figure 26 shows the dithecate genus *Syracosphaera*, which has an endothecal (internal) layer of caneoliths and an exothecal (external) layer of cyrtoliths.

References

Bralower, T., Bown, P.R., and Siesser, W.G., 1991. Significance of Upper Triassic calcareous nannofossils from the Southern Hemisphere (ODP Leg 122, Wombat Plateau, N.W. Australia). *Mar. Micropaleontol.*, **17**: 119–154.

Black, M., 1963. The fine structure of the mineral parts of coccolithophoridae. *Linnean Soc. London Proc.*, **174**: 41–6.

Black, M., 1972. Crystal development in Discoasteraceae and Braarudosphaeraceae (planktonic algae) *Paleontology*, **15**: 476–89.

Brasier, M., 1986. Why do lower plants and animals biomineralize? *Paleobiology*, **12**: 241–50.

Di Nocera, S. and Scandone, P., 1977. Triassic nannoplankton limestones of deep basin origin in the central Mediterranean region. *Palaeogeogr., Palaeoclimatol., Palaeoecol.*, **21**: 101–11.

Fichtinger-Schepman, A.M.J., Kanerling, J.P., Versluis, C. and Vliegenthart, J.F.G., 1981. Structural studies of the methylated acidic polysaccharide associated with coccoliths of *Emiliania huxleyi* (Lohmann) Kamptner. *Carbohydrate Res.*, **93**: 105–23.

Gartner, S., 1981. Calcareous nanofossils in marine sediments. In *The Sea: The Oceanic Lithosphere*, ed. C.Emiliani, vol. 7; 1145–77. John Wiley, New York.

Hart, G.F., Pienaar, R.N., and Caveney, R., 1965. An aragonite coccolith from South Africa. *S. Afr. J. Sci.*, **61**: 425–6.

Hay, W.W., 1977. Calcareous nannofossils. In *Oceanic Micropaleontology*, ed. A.T.S. Ramsay, pp. 1055–200. Academic Press, London.

Isenberg, H.D., Lavine, L.S., Moss, M.L., Kupferstein, D., and Lear, P.E., 1963. Calcification in a marine coccolithophorid. *Ann.N.Y. Acad. Sci.*, **109**: 49–64.

Jafar, S.A., 1983. Significance of Late Triassic calcareous nannoplankton from Austria and southern Germany. *Neues Jahr. Geologie Palaeontol. Abh.*, **166**: 218–59.

Kamptner, E., 1954. Untersuchungen uber den Feinbau der Coccolithen. *Arch. Protistenk.*, **100**: 1–90.

Lowenstam, H.A. and Weiner, S., 1989. *On Biomineralization*. Oxford University Press, New York. 324 p.

Mann, S. and Sparks, N.H.C., 1988. Single crystalline nature of coccolith elements of the marine alga *Emiliania huxleyi* as determined by electron diffraction and high-resolution transmission electron microscopy. *Proc. Roy. Soc. London. B*, **234**: 441–53.

Manton, I. and Leedale, G.F., 1963. Observations on the micro-anatomy of *Crystallolithus hyalinus* Gaarder & Markali. *Arch. Mikrobiol.*, **47**, 115–36.

Nishida, S., 1979. *Atlas of Pacific Nannoplanktons*. *Micropaleontol. Soc. Osaka*, Spec. Paper No. 3.

Norris R.E., 1971. Extant calcareous nannoplankton from the Indian Ocean. In *Proceeding II Planktonic Conf.*, *Roma 1970*, ed. A. Farinacci, vol. II, pp. 899–909. Edizioni Technoscienza, Rome.

Pentecost, A., 1991. Calcification processes in algae and cyanobacteria. In *Calcareous Algae and Stromatolites*. ed. R. Riding, pp. 3-30. Springer-Verlag, Berlin.

Siesser, W.G., 1971. Mineralogy and diagenesis of some South African coastal and marine sediments. *Mar. Geol.*, **10**: 15–38.

Siesser, W.G., 1977. Chemical composition of calcareous nannofossils. *S. Afr. J. Sci.*, **73**: 283–5.

Simkiss, K. and Wilbur, K.M., 1989. *Biomineralization. Cell Biology and Mineral Deposition*. Academic Press, San Diego.

Sorby, H.C., 1861. On the organic origin of the so-called 'crystalloids' of the Chalk. *Ann. Mag. Nat. Hist.*, Ser. 3, 8: 193–200.

Tan, S.H., 1927. Discoasteridae incertae sedis. *Proc. Konig. Akad. Wet. Amsterdam Sect. Sci.*, **30**: 411–9.

Tappan, H., 1980. *The Paleobiology of Plant Protists*. Chapter 9:

Haptophyta, coccolithophores, and other calcareous nannoplankton, pp.678–803. Freeman and Company, San Francisco.

Thompson, G., 1972. A geochemical study of some lithified carbonate sediments from the deep sea. *Geochim. Cosmochim. Acta*, **36**: 1237–53.

Thompson, G. and Bowen, V.T., 1969. Analyses of coccolith ooze from the deep tropical Atlantic. *J. Mar. Res.* **27**: 32–8.

Van Emburg, P.R., De Jong, E.W., and Daems, W. Th., 1986. Immunochemical localization of a polysaccharide from biomineral structures (coccoliths) of *Emiliania huxleyi*. *J. Ultrastr. Mol. Structure Res.*, **94**:246–59.

Westbroek, P., Young, J.R. and Linschooten, K., 1989. Coccolith production (biomineralization) in the marine alga *Emiliania huxleyi*. *J. Protozool.*, **36**: 368–73.

Wilbur, K.M. and Watabe, N., 1963. Experimental studies on calcification in molluscs and the alga *Coccolithus huxleyi*. *Ann. N.Y. Acad. Sci.*, **109**: 82–112.

Young, J.R., 1989. Observations on heterococcolith rim structure and its relationship to developmental processes. In *Nannofossils and Their Biostratigraphic Applications*. ed. J. Crux and S. E. van Heck, pp. 1-20. British Micropalaeontology Association.

Young, J.R. and Bown, P.R., 1991. An ontogenetic sequence of coccoliths from the Late Jurassic Kimmeridge clay of England. *Palaeontology*, **34**: 843–50.

Young, J.R. and Westbroek, P., 1991. Genotypic variation in the coccolithophorid species *Emiliania huxleyi*. *Mar. Micropaleontol.*, **18**: 5–23.

Young, J.R., Didymus, J.M., and Mann, S., 1991. On the reported presence of vaterite and aragonite in coccoliths of *Emiliania huxleyi*. *Botanica Marina*, **34**: 589–91.

5 Functions of coccoliths

JEREMY R. YOUNG

Introduction

Coccoliths are beautiful and elaborate structures; their production is an important feature of coccolithophore biochemistry; and coccoliths must greatly affect the physiological ecology of the organism. However, nothing is known for certain about the functions of coccoliths. This is part of a wider problem since phytoplankton ecology is poorly understood. Since little is understood about how individual phytoplankton cells interact with their environment, there can be no clear understanding of the role of the cell-covering structures, the coccoliths in this interaction.

The two most widely suggested types of function are protection and flotation-regulation. Protection against predation has often been assumed to be the function of phytoplankton cell-coverings but it is uncertain whether they actually are effective in reducing predation. Nonetheless, cell-coverings may protect the cells from osmotic, chemical, or physical shocks, or from ultraviolet light. Flotation is important since all phytoplankton need to stay within the photic zone, and nutrient absorption may make movement through the water advantageous. Since cell-coverings can greatly alter cell buoyancy and cell-shape, significant flotation-related functions are possible. More specialized possible functions include light concentration, which is possible for cell-coverings which significantly increase the area over which light is collected. Finally, since coccoliths, unlike the cell-coverings of most other phytoplankton, are formed of calcite, it is possible that the chemical process of coccolith formation may aid photosynthesis.

Many workers have speculated on the possible significance of coccoliths: Lohmann (1913), Braarud et al. (1952), Hay (1968), Gartner and Bukry (1969), Honjo (1976), Sikes and Wilbur (1982), Westbroek et al. (1983), Covington (1985), Manton (1986), Bown (1987), and Young (1987). In particular, Manton (1986) discussed the parallels between coccoliths and uncalcified scales, and argued that protection-related functions could account for a wide range of morphologies. Sikes and Wilbur (1982) conducted various culture experiments in an attempt to test some putative functions. Reviews of form and function in the marine phytoplankton are given by Taylor (1980) and Sournia (1982). Relevant aspects of diatom and dinoflagellate biology are reviewed by Sargeant et al. (1987), Taylor (1987), Round et al. (1990) and Margulis et al. (1989).

Despite this work, there is no understanding as to what the function of coccospheres is, quite probably because they do not have a single function. Coccospheres are very diverse structures, as the plates in this volume show (see chapter 7), and it is reasonable to anticipate that, like most biological structures, coccoliths have been adapted to perform a range of functions. Coccolith morphology alone, however, does not provide enough clues to guess which of the various possible functions any particular coccolith actually performs. It is usually possible with a little ingenuity to imagine two or three possible functions for almost any morphotype. Also, it is difficult to distinguish functional aspects of coccolith morphology from aspects related to coccolith-formation processes. I have, therefore, tried here to use the ecological distribution of coccolith morphotypes to provide an additional constraint on functional interpretations.

The first part of this paper consists of a discussion of various possible functions of coccoliths: protection related, biochemical, flotation related, and light regulatory. This is followed by a brief analysis of coccolithophore biogeography in order to identify ecologically constrained groups of species. Finally, the applicability of the various functions to the ecologically defined groups is discussed.

Possible functions

Coccoliths as cell-covering components

Phytoplanktonic organisms are composed of two parts, the protoplasm and the cell-covering. The protoplasm consists of the nucleus, chloroplasts, mitochondria, golgi body, and other organelles together with the interconnecting endoplasmic reticulum, and cytoskeletal microtubules in a fluid matrix also containing storage products and dissolved phases (see chapter 2). Together these interact as a complex biochemical system. The cell-covering separates this system from the environment, providing cohesion and physical protection and moderating interchange of material between the cell and the environment. In the simplest cases the cell-covering is a single 'cell-membrane' (synonyms: plasmalemma, plasma

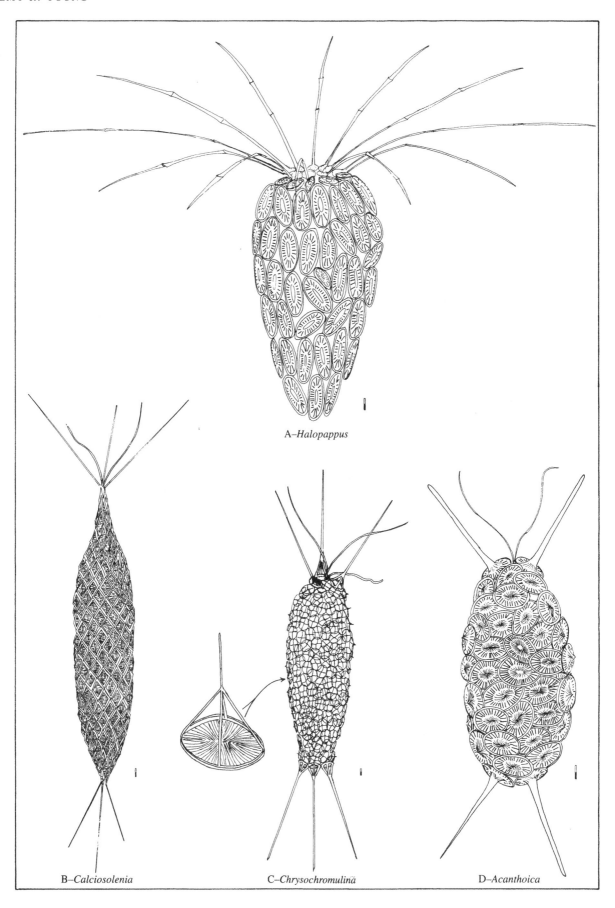

A–*Halopappus*

B–*Calciosolenia*

C–*Chrysochromulina*

D–*Acanthoica*

membrane). More commonly the cell-covering is extended by additional layers or components outside the cell-membrane. Since these are borne outside the cell they cannot perform an active role in the internal biochemistry. The very widespread occurrence of robust cell-coverings in phytoplankton suggests that they perform similar functions in all groups. Examples include diatom frustules and dinoflagellate thecae.

The cell-covering of most coccolithophores consists of two components, resistant organic scales and calcareous coccoliths. There are also a number of Prymnesiophytes which do not calcify (e.g., *Prymnesium*, *Chrysochromulina* and *Phaeocystis*). Most of these genera are not naked but have an extracellular cover of resistant organic scales (Fig. 1C). These scales are often elaborate and there can be strong similarities between individual scales and coccoliths. Two lines of evidence suggest that the similarities are significant.

First, coccoliths and organic scales are homologous, that is, formed by the same process. Both coccoliths and organic scales are formed in intracellular vesicles associated with the Golgi body. Moreover, in most studied species coccolith production is preceded by formation of an organic base-plate scale, around which crystal nucleation occurs (Outka and Williams, 1971; Young, 1989). In a few species, resistant base-plate scales are not formed, most notably *Emiliania huxleyi*, also *Gephyrocapsa oceanica* (Inouye and Pienaar, 1984). However, *E. huxleyi* has been shown to possess non-resistant scales that act as precursors to coccolith formation and that almost certainly are equivalent to the resistant base-plate scales of other species (Wilbur and Watabe, 1963; van der Wal *et al.*, 1983).

Second, coccoliths and organic scales appear in many cases to be analogous, that is, to carry out similar functions. The morphology of scales and coccoliths and of the tests that they form can be strikingly similar (Manton, 1986; Young, 1987, Fig. 1B–D). In such cases it is logical to predict that coccoliths and organic scales perform essentially similar functions, as cell-covering components.

The idea that coccoliths function as cell-covering components is supported by the observation that coccoliths are arranged as a continuous cover on the cell-surface – and often are modified in form so as to achieve this efficiently. There are two main coccolith arrangement patterns, directly abutting, and overlapping.

Fig. 1. Four motile prymnesiophytes.
A. *Halopappus adriaticus* (Syracosphaeraceae). B. *Calciosolenia* sp. (Calciosoleniaceae). C. *Chrysochromulina pringsheimii* (Hymenomonadaceae). D. *Acanthoica quattrospina* (Rhabdosphaeraceae). The diagrams are schematic reconstructions based on several micrographs.
These cells illustrate (1) The similarity of organic scales to coccoliths (C vs A, B, D). (2) Directly abutting vs overlapping arrangements (A, B vs C, D). (3) Various probable developments of spines for decelerated and/or orientated sinking (all four). Ecologically all the species would be included in the miscellaneous/intermediate group.

Directly abutting coccoliths are characteristic of various families, particularly the Syracosphaeraceae, Calyptrosphaeraceae, and Pontosphaeraceae (e.g., Fig. 1A,B). The coccoliths show various adaptations toward producing an accurately tessellating continuous cover. The coccolith walls are normally gently inclined outwards so that adjacent coccoliths on the curved coccosphere surface are in tight contact. The shape of individual coccoliths is often modified to reduce gaps between them; spectacularly so in the case of the genera *Anoplosolenia* and *Calciosolenia* which have diamond-shaped coccoliths (Fig. 1B).

Van der Wal *et al.* (1987) experimentally demonstrated that the continuity of coccolith cover is actively maintained in one such species. They found that after decalcification of a culture of *Pleurochrysis carterae* the calcification/coccolith production rate increased until the cells had produced complete new coccospheres.

Overlapping coccoliths occur on many of the most abundant coccolithophores, including the Coccolithaceae, Helicosphaeraceae, Noëlaerhabdaceae and Rhabdosphaeraceae. An overlapping arrangement ensures complete continuity of cover at some loss of economy as compared to the directly abutting arrangement. In some cases there is no obvious adaptation for strength, and the coccospheres tend to disintegrate on death. In the placolith-bearing group, however, the coccolith morphology consisting of two shields separated by a tube appears to be a direct adaptation of form for test rigidity. It is noticeable that these are also the most robust individual coccoliths.

The relationship of coccoliths to organic scales and the fact that coccospheres form a continuous cell-covering tend to suggest that coccoliths function as cell-covering components with a protective (*sensu lato*) function. Alternatively, calcification of organic scales may have allowed adoption of new functions unrelated to the original (?protective) roles of organic scales. The various possibilities are discussed in the following paragraphs, and summarized in Fig. 2.

Protection-related functions

PROTECTION FROM PREDATION

Grazing is one of the main controls on phytoplankton, strongly influencing species successions, the duration of blooms, and the magnitude of standing crops (Frost, 1980; Smayda, 1980). Since there are many different phytoplankton and zooplankton groups, selection should rapidly occur if even a mild degree of protection is afforded by coccoliths.

Coccospheres do not, however, appear very effective at stopping predation. Laboratory experiments have shown that coccolithophores form good foodstocks for a range of zooplankton (Boney, 1970; Honjo and Roman, 1978). Coccospheres are common in the guts and fecal pellets of

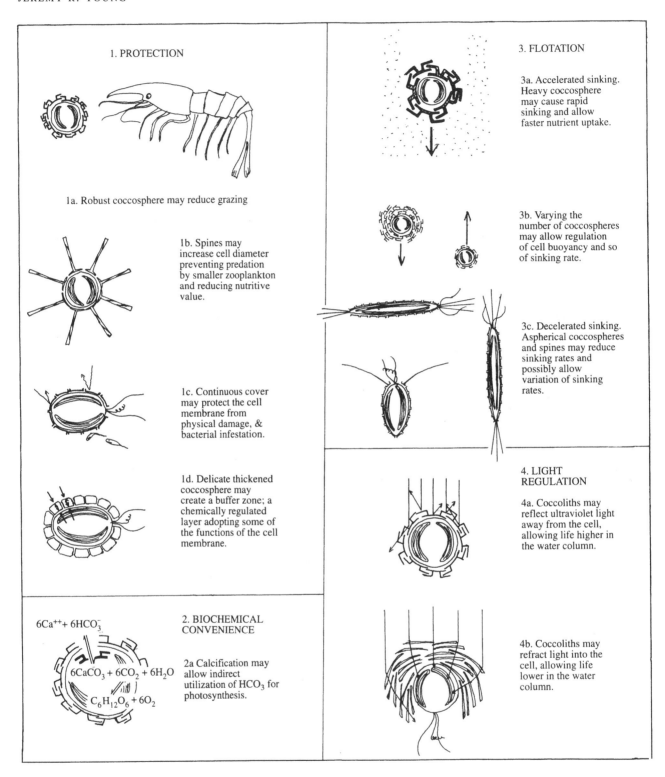

1. PROTECTION

1a. Robust coccosphere may reduce grazing

1b. Spines may increase cell diameter preventing predation by smaller zooplankton and reducing nutritive value.

1c. Continuous cover may protect the cell membrane from physical damage, & bacterial infestation.

1d. Delicate thickened coccosphere may create a buffer zone; a chemically regulated layer adopting some of the functions of the cell membrane.

$6Ca^{++}+ 6HCO_3^-$

2. BIOCHEMICAL CONVENIENCE

$6CaCO_3 + 6CO_2 + 6H_2O$

$C_6H_{12}O_6 + 6O_2$

2a Calcification may allow indirect utilization of HCO_3 for photosynthesis.

3. FLOTATION

3a. Accelerated sinking. Heavy coccosphere may cause rapid sinking and allow faster nutrient uptake.

3b. Varying the number of coccospheres may allow regulation of cell buoyancy and so of sinking rate.

3c. Decelerated sinking. Aspherical coccospheres and spines may reduce sinking rates and possibly allow variation of sinking rates.

4. LIGHT REGULATION

4a. Coccoliths may reflect ultraviolet light away from the cell, allowing life higher in the water column.

4b. Coccoliths may refract light into the cell, allowing life lower in the water column.

Fig. 2. Cartoon summary of putative functions.

salps, copepods, and other zooplankton (e.g., Norris, 1971; Roth *et al.*, 1975). Moreover Sikes and Wilbur (1982) found that the copepod *Calanus finmarchicus* grazed as rapidly on coccolith-bearing cells of *Emiliania huxleyi* as on naked cells. Furthermore, most zooplankton seem to be rather unselective grazers ingesting all particles within a given size range – and particularly so at the lowest size range as represented by coccolithophores.

Calcification evidently does not prevent grazing. It may, however, still be adaptive by making grazing more difficult and less efficient. This may be most important in unstable eutrophic environments (e.g., blooms) when slight variations in grazing pressure may be critical in determining which species grow fastest (Frost, 1980).

There are various ways in which a protective function may operate. First, an armoring effect might render coccospheres indigestible or cause them to be rejected as food particles. The elaborate adaptations of placolith form toward producing a rigid test suggest such a function, but there is little other direct evidence for it. Second, since filter feeders are size selective the grazing pressure on different cell sizes will vary. Increasing coccosphere size may thus lower grazing pressure. This is most economically achieved by spines. In *Rhabdosphaera clavigera*, the coccosphere diameter may be four times the cell diameter. Third, a low ratio of cell size to coccosphere size might reduce the feeding efficiency of the zooplankton, particularly if there is a large volume of calcite in the coccosphere. This effect would not benefit individual cells and so could only be adaptive in species which dominate populations, or at least size classes of populations.

CELL-MEMBRANE PROTECTION

Even if predation protection is not a function of phytoplankton cell-coverings, the coverings may still serve a protective function by isolating the cell-membrane from the immediate environment (Manton, 1986). The cell-membrane must be delicate since the flagella, ingested food particles, and even complete coccoliths, are able to pass through it. Thus, it may be vulnerable to damage, making physical protection of it important and a basic function of cell-coverings. A variant of this might be protection against bacterial infestation. Calcification may have occurred initially in order to enhance protection.

ENVIRONMENTAL BUFFERING AND NUTRIENT UPTAKE

As well as bounding the cell, the cell-membrane carries out a number of regulatory functions. It controls the passage of nutrients, waste products, carbon dioxide and other dissolved gases into and out of the cell. It is possible that some of these regulatory functions may in part be adopted by other components of the cell-covering, such as coccoliths. Sikes and Wilbur (1982) observed that calcified strains of *Emiliania huxleyi* were more tolerant of lowered salinity levels than naked strains and suggested that 'by enclosing a small volume [of water] within the coccosphere but outside the cell-membrane the cell [may] gain somewhat greater control over the immediate external environment'. Manton (1986) suggested that 'protection of the plasmalemma [i.e. cell-membrane] against environmental shocks of many kinds (mechanical, chemical, osmotic etc.) must be a primary function of periplasts [i.e., cell-cover components] in general'. She also suggested that trapped water was a major

mechanism by which this was achieved. Two extensions of this buffer-layer concept can be suggested. First, inclusion of suitable macromolecules could enhance any of the possible functions. Second, this might allow an additional function, nutrient absorption. Nutrient absorption is a central, often dominant feature of phytoplankton ecology and any adaptation to assist nutrient uptake or storage should be valuable. If the extra-cellular volume created by the coccosphere were filled by macromolecules capable of absorbing nutrient ions then it could act as a kind of nutrient sponge. This would both increase the effective surface area for nutrient absorption and act as a nutrient store.

These types of functions can provide an explanation for the form of numerous, otherwise anomalous, coccoliths. In particular many species produce two-layered coccospheres with a space between layers which could function as a water-trapping adaptation (cf. Manton, 1986). Examples include: (1) numerous holococcoliths with a domal form (e.g., *Calyptrolithophora*, *Calyptrosphaera*, *Daktylethra*, *Sphaerocalyptra*, Fig. 3A), (2) various Syracosphaeraceae which have two-layered coccospheres consisting of an inner layer of normal coccoliths with a low basin form, and an outer layer of discoidal or domal coccoliths, (e.g., *Syracosphaera anthos*, *S. pirus*, *S. pulchra*, Fig. 3B), and (3) *Umbellosphaera* and *Discosphaera* which have coccoliths with flaring processes that unite to form a continuous outer sphere (Fig. 3C). Bown (1987) described fossil coccolithophores with analogous morphologies.

Biochemical functions of coccolith formation

If, as argued above, coccoliths and organic scales are both homologous and analogous, then the differences between them may not be very significant. It is possible that calcification is little more than a biochemical convenience; that is, $CaCO_3$ may be a more convenient material than organic macromolecules. This may be because calcification is a convenient means of providing rigidity or simply because it is energetically economical to precipitate calcite.

An elegant biochemical explanation for the occurrence of calcification is that it is linked to photosynthesis (Paasche, 1962). The basic equations for calcification and photosynthesis are:

Calcification: $Ca^{2+} + 2HCO_3^- \rightarrow CaCO_3 + CO_2 + H_2O$
Photosynthesis: $6CO_2 + 6H_2O \rightarrow C_6H_{12}O_6 + 6O_2$

Since carbon dioxide (CO_2) is produced by calcification and used by photosynthesis it is possible for the two reactions to be linked. This should favor calcification and may make replacement of organic scales by calcareous coccoliths energetically advantageous. It is further possible that there is a direct linkage between photosynthesis and calcification and that calcification acts as a source of carbon dioxide for photosynthesis, or as a sink for hydroxyl ions (OH^-). In this

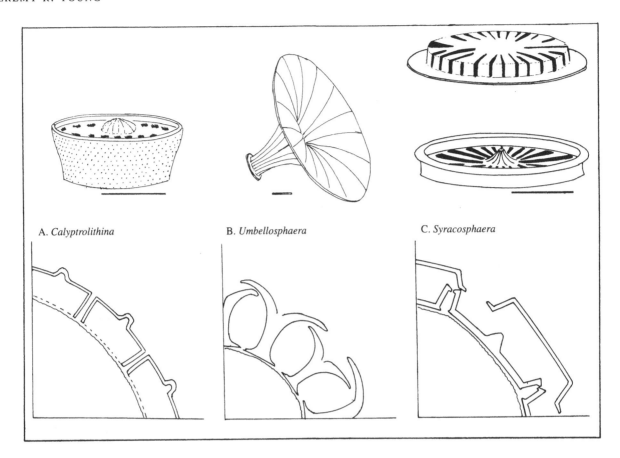

A. *Calyptrolithina*   B. *Umbellosphaera*   C. *Syracosphaera*

Fig. 3. Adaptations of coccolith morphology to produce two-layer coccospheres.
A. Holococcoliths - *Calyptrolithina*; B Umbelliform coccoliths - *Umbellosphaera*; C. exothecal coccoliths - *Syracosphaera*.

way, calcification would reduce the energy cost of photosynthesis (Sikes *et al.*, 1980; Westbroek *et al.*, 1983). Support for this suggestion comes from data on *Emiliania huxleyi* showing a close stoichiometric correspondence between calcification and photosynthesis rates (Paasche, 1962; Klaveness and Paasche, 1979; Sikes et al., 1980). However, *E. huxleyi* produces an unusually large amount of coccolith carbonate. Individual cells frequently have multiple layers of coccoliths and loose coccoliths are abundant in many laboratory cultures and oceanic samples. Numerous other species of coccolithophores have only a single layer of coccoliths and probably only produce extra coccoliths in response to a need to maintain this covering, as has been demonstrated for *Pleurochrysis carterae* (van der Wal *et al.*, 1987). So although calcification may have been adapted to aid photosynthesis in a few species such as *E. huxleyi* this is unlikely to have been the original function of biomineralization, and still less for the formation of a cell-covering of elaborately formed coccoliths.

Flotation-related functions of coccoliths

The bulk of oceanic phytoplankton occur in the surface mixed-layer above the thermocline (about 10 m to 150 m). Within this layer they (particularly non-motile forms) are essentially randomly distribution by turbulence, and show minimal depth zonation. Nonetheless, all phytoplankton cells need a flotation strategy in order to remain within the mixed-layer, and in shelf seas to stay above the sediment bottom. Also, controlling the orientation of cells may be useful for maximizing photosynthesis rates. Interacting with these flotation requirements is the possibility that motion through the water may assist nutrient absorption by forced-convection (Munk and Riley, 1952; Hutchinson, 1967; Sournia, 1980, Walsby and Reynolds 1980). This was strongly argued by Hutchinson, (1967, p.293): 'If an autotrophic organism were to remain suspended at rest in an undisturbed body of water, it would rapidly utilize the nutrients dissolved in its immediate vicinity. The rate of division of such a cell would depend on the rate at which nutrients could diffuse from the main mass of water far from the cell into the impoverished shell of water in its immediate vicinity. If, however, the cell started to sink through the water, it would continually encounter regions from which nutrients have not been removed'.

Three main flotation strategies are available to phytoplankton. (1) Passive floating/sinking – this is the basic strategy of non-motile cells which must use buoyancy regulation, turbulent mixing and differential reproduction to remain within the photic zone. Sinking in this case may be adaptive for rapid nutrient absorption. (2) Active swimming – flagellate cells can use their motility to actively maintain or change position, with the possibility of diurnal migration. Even random swimming would achieve the cell motion necessary for enhanced nutrient uptake. (3) Commensalism or symbiosis with other organisms. In this case the problem of flotation is adopted by the host organism, which also probably provides a nutrient source. This strategy is common among dinoflagellates and diatoms which can live symbiotically with radiolarians, planktonic foraminifers, and other microplankton. The only well documented possible case among coccolithophores is association of the species *Reticulofenestra sessilis* with centric diatoms (Gaarder and Hasle, 1962; Okada and McIntyre, 1977).

There is more literature on flotation than on most aspects of phytoplankton ecology, possibly in part because it is a quantifiable subject. Useful reviews include Hutchinson (1967), Smayda (1970), Walsby and Reynolds (1980), Sournia (1982), Reynolds (1986).

FORCED CONVECTION – NUTRIENT UPTAKE AND CELL MOVEMENT

Various workers have investigated mathematically the effect of forced convection from phytoplankton movement on nutrient uptake (Munk and Riley, 1952; Hutchinson, 1967; Pasciak and Gavis, 1974; Gavis, 1976). They have all concluded that it should be significant. Gavis (1976) calculated the likely magnitude of this effect, from a consideration of nutrient-diffusion rates and uptake kinetics. His results suggest that nutrient uptake rates in a moving cell will at most be about 40% higher than in a stationary cell, and usually only about 5–20% higher. An increase of this magnitude would only have a marginal effect on cell viability, but could give a significant competitive advantage in certain circumstances. The actual magnitude of the benefit from forced convection would depend on the nutrient uptake properties of the cell, and on the rates at which nutrient ions diffuse through the water (diffusivities). Nutrient diffusivities are poorly known (Gavis, 1976), and it is possible that the benefit from forced convection would be much stronger for some nutrients than others.

The velocities required to maximize forced convection can be roughly approximated from simple volume considerations. The concentrations of the principal nutrients nitrogen and phosphorus in sea water are typically 10^5 to 10^6 times lower than in phytoplankton cells (based on cell dry weight, excluding oxygen, being approximately 10% nitrogen, 1% phosphorus, and sea water concentrations of 1–10 μM nitrogen, 0.1–1 μM phosphorus). Cell division results in a halving of nutrient content per cell and so a doubling of the nutri-

ent content of the new cell is needed before division can occur again. Consequently, cells need to extract nutrients from a volume of water 10^5 to 10^6 times their volume prior to division. If the volume of water the cells extract nutrients from is approximately equivalent to the volume they pass through, then the distance they need to travel is:

$$s = (V/A \times N_c/N_w)$$

Where: s = distance the cell needs to move through the water to double its nutrient content; V = cell volume; A = cross sectional area; N_c = nutrient concentration in the cell; N_w = nutrient concentration in the water.

For a given division time these distances can be converted into velocities:

$$v = (V/A \times Nc/Nw)/t$$

Where: v = cell velocity; t = division time.

Solutions for this equation are given on the graph (Fig.4); this suggests that cells 4–20 μm in diameter (the size range of most coccolithophores) need to sink or swim between 0.1 m and 10 m if they are to maximize forced convection. If this occurred over one day then they would need to travel through the water at between 2–200 μm s^{-1}. This range of speeds can be compared to likely cell sinking and swimming rates, as discussed below.

SINKING RATES

It is generally assumed that phytoplankton cells are small enough for their sinking to occur by laminar flow (Hutchinson, 1967; Walsby and Reynolds, 1980). Under these conditions the viscous forces retarding sinking will be related to surface area and so will be proportional to the cell radius squared. The gravitational forces causing sinking will be independent of size but proportional to the density difference between the cell and the water. These relationships are formalized in the Stokes equation, which for spherical bodies is:

$$v_s = 2/9.g.r^2.\Delta\rho.n^{-1} \text{ Stokes equation}$$

Where: v_s is the terminal sinking velocity; g is gravitational attraction (9.8 m s^{-2}); r is the cell radius; $\Delta\rho$ is the excess-density of the cell (= cell density - sea water density); n is the viscosity of the water (*c.* 1.0 centipoise = 1×10^{-3} kg m^{-1} s^{-1}).

The density of naked cells is close to that of the surrounding water since dry weight is typically only about 10% of total weight. The organic macromolecules that make up most of the dry weight usually have densities around 1.3 to 1.7 g cm^{-3} (Walsby and Reynolds, 1980). Consequently naked cells are likely to have densities of 1.03 to 1.07 g cm^{-3}. Sea-water density is about 1.02 to 1.03 g cm^{-3}. So naked cells should have excess densities of 0.0 to 0.05 g cm^{-3}. Density can be reduced by flotation vacuoles and by low density fat bodies, although these have not been documented in coccolithophores. Density can be increased by biomineralization.

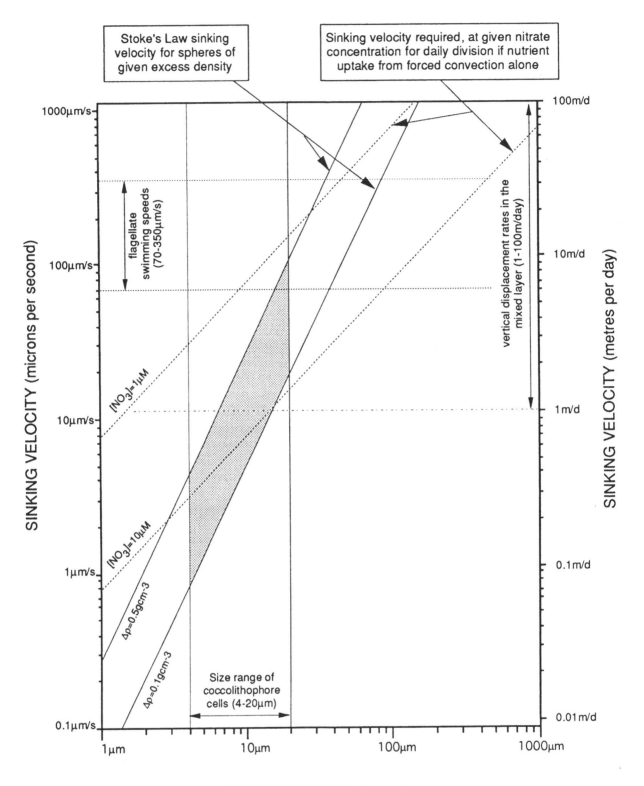

Fig. 4. Graph of sinking rates against cell size.
Vertical lines, range of coccolithophore cell diameters (4–20 μm).
Steeply inclined lines, predicted sinking rates for spheres with excess densities of 0.1 and 0.5 g cm^{-3} respectively, based on the Stokes equation. These represent the envelope of likely sinking rates for coccolithophores. *Gently inclined lines*, calculated sinking rates required if cells are to double their nutrient content by motion through the water column in one day (see text), with nutrient concentrations of 1 & 10 μM nitrate respectively.

Shaded area, Predicted cell size / sinking rate range of coccolithophores. a. Cell sinking rates too low to aid nutrient uptake. b. Cell sinking rates high enough to aid nutrient uptake. c. Cell sinking rates high enough to aid nutrient uptake but also likely to cause population to sink out of mixed layer. *Horizontal dashed line*, sinking speed above which turbulent mixing may not be sufficient to reliably maintain random distribution of populations. *Horizontal dotted lines*, reported range of swimming speeds for flagellate phytoplankton.

Calcite has a density of 2.7 g cm^{-3}, a heavily calcified coccolithophore might be 30% calcite by volume (my estimate from transmission electron micrographs), resulting in a density of 1.55 g cm^{-3}, and an excess density of 0.5 g cm^{-3}. Thus, heavy coccoliths can cause a ten-fold increase in excess-density and sinking rate. Eppley *et al.* (1967) observed that calcified *Emiliania huxleyi* cells sink 4.5 times faster than naked cells. Even lightly calcified coccolithophores are probably about 5% calcite by volume, causing an excess density of 0.08 g cm^{-3}.

The significance of these values can be investigated by plotting Stokes Law sinking rates vs. diameter for spheres of varying excess densities (Fig. 4). For spheres of 4–20 μm diameter, excess densities of 0.1 to 0.5 g cm^{-3} are likely to cause sinking at rates from 1 to 150 μm per second, equivalent to 1 to 10 m/day. These rates fall within the range of speeds needed to allow nutrient absorption by forced convection (Fig. 4). There is considerable uncertainty in these calculations, but it appears that sinking rates are of the appropriate order of magnitude. The graph also suggests that the minimum cell-size for this strategy is a couple of microns; this may provide a constraint on coccolithophore size.

TURBULENT MIXING

A sinking population of cells will gradually pass out of the photic zone. Intuitively this might be expected to make sinking terminally maladaptive, but reproduction and turbulent mixing can neutralize the effect of sinking. If we assume a random distribution of cells through the mixed-layer then:

Fallout rate = mixed-layer depth/sinking rate

For instance, if the mixed-layer is 100 m deep and the sinking rate is 2 m/day then the fallout rate will be 2% per day. Turbulent mixing cannot reduce the fallout rate, but by maintaining the random vertical cell distribution it can keep this rate constant. Reproduction can compensate for the fallout loss of cells. Under these conditions cell sinking will be supportable as long as the sinking rate is significantly lower than the vertical dispersion rates caused by turbulent mixing, and the growth rate is at least as high as the fallout rate. In general this means that passive sinking is more likely to be adaptive in eutrophic high turbulence environments and less likely to be adaptive in oligotrophic low turbulence environments.

Water-column mixing rates by turbulent mixing are strongly dependent on the mixed-layer depth and therefore on wind speeds. Denman and Gargett (1983) provide a compilation of data suggesting mixing times for upper mixed layers from between a week to less than an hour which they calculate to correspond to random vertical-dispersion rates from about 1 m/day to over 100 m/day. The higher rates would be adequate to maintain a random distribution of cells. The lower rates would not be, since they are close to the sinking rates of the cells. In general, it may

be reasonable to expect that cell sinking rates much in excess of 1 m/day will be sustainable only in high turbulence environments. As shown on the graph (Fig. 5.4) many coccolithophores are so small that they sink at rates of less than 1 m/day, even with considerable excess density. Coccolithophores with cell diameters over about 10 μm are, however, able to sink at rates of several meters per day. Such coccolithophores presumably need either high turbulence environments or motility to remain within the photic zone – unless like diatoms they can reduce the density of their cytoplasm. Denman and Gargett (1983) give mixing times in the thermocline of 7 to 35 days, equivalent to 0.14 to 0.7 m/day of vertical dispersion. These rates are too low to compensate for sinking losses. Consequently, thermocline dwelling cells are unlikely to be able to adopt passive sinking as a nutrient absorption strategy and non-motile thermocline dwellers cannot support high excess densities.

WATER TEMPERATURE EFFECTS

Temperature affects the density and the viscosity of water both of which influence sinking rates. From 25 °C to 10 °C (the typical temperature range in which coccolithophore live), sea water density increases from 1.023 to 1.026 g cm^{-3} (Fig. 5). This will have only a very slight effect on coccolithophore excess densities (calculated above to be 0.1 to 0.5 g cm^{-3}) and therefore on sinking speeds. Over the same temperature range the viscosity of sea-water increases from 0.95 to 1.38 centipoise (Fig. 5). According to the Stokes Law equation given previously, sinking speed is inversely proportional to viscosity so this variation would cause a 45% reduction in sinking speed. Both effects reduce sinking rates in cold water; consequently phytoplankton may be expected to adjust their buoyancy in cold water. Various dinoflagellates show test-shape variation with water temperature which has been interpreted in terms of this effect (Smayda, 1970). There are no obvious examples of this type of cell shape variation among coccolithophores, but there is considerable evidence for heavier calcification of placolith coccoliths in cold water. The living species with the largest coccoliths, *Coccolithus pelagicus*, is restricted to high latitudes. *Emiliania huxleyi* tends to be more heavily calcified in cold water (Burns, 1977; Jordan, 1988; Okada, 1989; Young and Westbroek 1991). High latitude Neogene fossil reticulofenestrid assemblages are normally dominated by coccoliths with closed central areas whereas low latitude assemblages are normally dominated by coccoliths with open central areas (Young, 1990). This correspondence could be due to non-functional causes related to the biomineralization process but it is suggestive of water-temperature control.

VARIABLE ACCELERATED SINKING

A subtle possibility is that phytoplankton cells may be able to adjust their density in response to ecological conditions.

Fig. 5. Variation of water properties with temperature. Upper graph viscosity (centipoises), lower graph density. Both data sets are for salinity of 34 ppt (Riley & Skirrow, 1975). According to the Stokes Equation cell sinking rates are inversely proportional to viscosity and proportional to excess density.

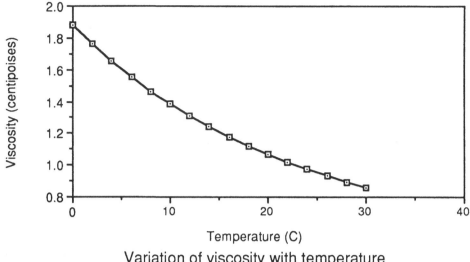

Variation of viscosity with temperature

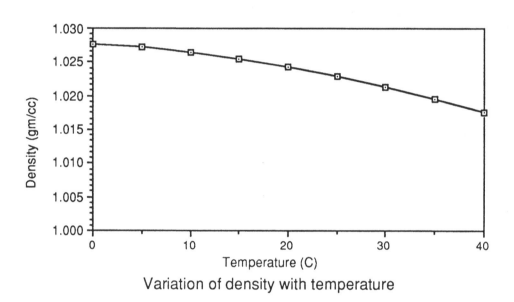

Variation of density with temperature

In particular, higher sinking rates for senescent cells than for actively growing cells have consistently been recorded (Eppley *et al.*, 1967; Smayda, 1970). This may be an adaptive response since new nutrient sources are most likely to be found at depth. For coccolithophores possible mechanisms include: increased weight of calcification of individual coccoliths, increasing the number of coccolith layers in the coccosphere and aggregation of cells. In culture, *Emiliania huxleyi* shows strong increase in the number of coccoliths per cell and in cell diameter after logarithmic growth ends (Young and Linschooten, unpub. data). Cell aggregation has been noted in *Emiliania huxleyi* (Cadee, 1985) and *Gephyrocapsa oceanica* (Burns, 1977; Winter *et al.*, 1979; Hallegraef, 1984), but there are no data to show what triggers this aggregation.

DECELERATED SINKING

Outside high turbulence environments accelerated sinking is liable to be maladaptive and the ballasting effect of coccolithophores may need compensating. So it is reasonable to expect adaptations of coccoliths toward reducing sinking rates.

Some non-motile coccolithophores do show such adaptations. Spines with flaring ends may increase cell diameter without greatly increasing weight and so may reduce excess density, and settling rate: *Rhabdosphaera clavigera* is a prominent example of this morphotype. Lohmann (1913) interpreted the equatorial beaker-shaped coccoliths of *Scyphosphaera* as an adaptation aimed at producing a slow sinking, discoidal-shaped coccosphere.

Somewhat paradoxically the most likely adaptations for decelerated sinking occur on motile coccolithophores. These are the polar spines that occur on several genera and the whorl coccoliths of *Ophiaster, Michaelsarsia* and *Calciopappus* (Fig.1). Experimental work on analogous spines on diatoms has shown that they do indeed decelerate sinking (Walsby and Reynolds, 1980). The coincidence of motility and these elaborate spines suggest that they may have relatively complex ecologies. Possibly they are able to vary their behavior in response to nutrient availability, alternating between active swimming and passive floating.

MOTILITY

The swimming speeds of flagellates can be observed by direct microscopy. Such observations have been summarized by Throndsen (1973) and Sournia (1982). They suggest a range of about 70 to 350 μm s^{-1}, with no obvious correlation with size. The coccolithophore *Pleurochrysis carterae* swims at similar speeds (pers. obs.). These rates are well in excess of those needed by all but the largest coccolithophores to maximize nutrient uptake by forced convection (Fig.4).

Swimming rates of 70 to 350 μm s^{-1} are equivalent to 6 to 30 m/day. This suggests that motility could be adaptive for diurnal movement in environments with a shallow thermocline, as in dinoflagellates (Levandowsky and Kaneta, 1987). Alternatively, random movement on this order may be adaptive in environments where nutrients are patchily distributed.

Coccoliths cannot, of course, directly contribute to motility but the coccoliths and coccospheres of motile species may show various morphological adaptations related to motility. It is noteworthy that most motile coccolithophores have relatively light coccoliths and so their cells may be nearly neutrally buoyant. This applies, for example, to most Calyptrosphaeraceae and Syracosphaeraceae, whose coccoliths must form a much lower percentage of total coccolithophore volume than those of typical placolith-bearing species. The reduction in excess-density should make swimming easier, whereas sinking is functionally redundant in motile species.

Motile coccolithophores need a flagellar opening in the coccosphere. In many species this is surrounded by coccoliths showing some degree of adaptation of form. In most cases these adaptations are in terms of detail of size and shape and probably are primarily to ensure continuity of the coccolith cover around the flagella bases.

Light regulation

Since phytoplankton are dependant on photosynthesis, light regulation is an attractive possible function for cell coverings. Two separate light regulatory functions have been proposed. Braarud *et al.* (1952), suggested that coccoliths could reflect ultraviolet light thus enabling them to live higher in the water column than other phytoplankton. Possible support

for this comes from remote-sensing work which has identified oceanic patches with high light reflectance caused by blooms of *Emiliania huxleyi* (Holligan *et al.*, 1983), but it is uncertain whether this effect is strong enough to benefit individual cells.

Gartner and Bukry (1969) noted that coccoliths would tend to refract light into the cell, since calcite has a higher refractive index than water. So they suggested the reverse role: that coccoliths might act as light gatherers, enabling coccolithophores to live deeper than other phytoplankton in the water column. This adaptation would be of most benefit to thermocline-dwelling species, which have to cope with permanently low light levels. It might also be of value for species in deep mixed-layer environments, which are temporarily exposed to low light levels.

Ecological distribution of coccosphere types

Identification of characteristic assemblages

There is a reasonable amount of information available on the oceanic scale distribution of nannoplankton derived from studies of living assemblages collected by filtering sea-water (e.g., McIntyre and Bé, 1967; Okada and Honjo, 1973; Nishida, 1979; Kleijne *et al.*, 1989). Supplementary data are available from studies of smaller areas (e.g., Schei, 1975; Winter *et al.*, 1979; Reid, 1980; Milliman, 1980; Mitchell-Innes and Winter, 1987; Verbeek, 1989), from studies of coccolith assemblages in core-top sediment samples (e.g., Geitzenauer *et al.*, 1977; Roth and Colbourne, 1982; Okada, 1983; Tanaka, 1991), from bottle-enrichment experiments (e.g., Hulburt and Corwin, 1969; Hulburt, 1985), and from sediment traps (e.g., Honjo, 1982; Samtleben and Bickert, 1990; Steinmetz, 1991). Using these data it is possible to separate three groups of species with distinctive morphologies and biogeographies. These groups each dominate coccolithophore assemblages in particular environments, as summarised in Fig. 6 and discussed below. It seems reasonable to assume that they have different ecological adaptations which may be reflected in coccolith functional morphology.

Since the groups are defined on ecological and gross morphological criteria, they cut across taxonomic divisions. To avoid confusion of these groups with taxonomic divisions, new names 'floriform' and 'umbelliform' are provided for two of them. A residual 'miscellaneous group' of species does not have an obvious distribution pattern, and only occasionally dominates assemblages.

In the discussion below, the groups are related first to the data of Okada and Honjo (1973), Honjo and Okada (1974), and Honjo (1977) on the distribution of coccolithophores in the North Pacific, since this is the best oceanic-scale survey

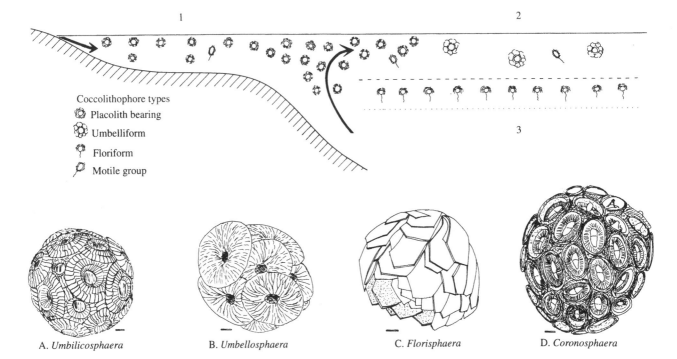

Coccolithophore types
- ⊚ Placolith bearing
- ⊛ Umbelliform
- ⚲ Floriform
- ⚲ Motile group

| A. *Umbilicosphaera* | B. *Umbellosphaera* | C. *Florisphaera* | D. *Coronosphaera* |

of coccolithophores available. Secondly, they are related to data from other sources.

PLACOLITH-BEARING COCCOLITHOPHORES

In equatorial waters (approximately 10 °N to 10 °S) Okada and Honjo (1973) recorded diverse assemblages, but all the common species (forming >90% of the total assemblage) are placolith-bearing species. These include the species *Emiliania huxleyi*, *Gephyrocapsa ericsonii*, *G. oceanica*, *Calcidiscus leptoporus*, *Umbilicosphaera sibogae*, *U. hulbertiana*, and *Oolithotus fragilis* (NB where appropriate more recent names are given for species, rather than the names in the original paper). The relative proportions of the species varies, so Okada and Honjo (1973) recognized three 'communities'. For present purposes, however, it is more important to note that all their communities are dominated by placolith-bearing species. Placolith-dominated assemblages also occurred at high latitudes (40–50° N), in association with abundant diatoms (Honjo, 1977) but *E. huxleyi* was the only important species.

Many other authors have recorded placolith-dominated assemblages from numerous locations with superficially erratic distribution. However, three recurring, and somewhat overlapping, types of distribution appear to be common. First, placoliths predominate in areas of upwelling. This includes both the equatorial divergence zones (e.g., Okada and Honjo, 1973, Nishida, 1979, Roth and Coulbourne, 1982), and areas of coastal upwelling (e.g., Roth and Coulbourne, 1982, Mitchell-Innes and Winter, 1987, Kleijne et al., 1989). Second, they are predominately bloom-forming coccolithophores (e.g., Birkenes and Braarud, 1952;

Fig. 6. Ecological distribution of coccolithophore types. *Above*: Cartoon to illustrate the main nannoplankton assemblage types as discussed in the text, representing a transect from continental shelf to mid-ocean at low latitudes. 1. Placolith-dominated assemblages in coastal and upwelling environments; 2. Umbelliform-dominated assemblages in oligotrophic mid-ocean environment; 3. Floriform-dominated assemblages in deeper stably stratified water. Species of the fourth, miscellaneous, group rarely dominate assemblages but are most common in intermediate environments. Arrows indicate nutrient flux. *Below*: Typical representatives of each of the ecological groups (mainly based on micrographs in Nishida, 1979).

Holligan *et al.*, 1983 – *Emiliania huxleyi*; Grindley and Taylor, 1970 – *Gephyrocapsa oceanica*; Milliman, 1980 – *Coccolithus pelagicus*; Honjo, 1982 – *Umbilicosphaera sibogae*). This occurrence pattern is particularly important in seasonally stratified higher latitude waters. Third, they normally dominate coastal and shallow-sea assemblages (Okada and Honjo, 1975; Okada, 1983).

These three environments are physically variable but ecologically similar in that they are all eutrophic (Kilham and Kilham, 1980), that is, they are environments in which light levels are adequate for algal growth and nutrient supply, and exceed the demands of the standing crop. Therefore, rapid population growth can occur.

UMBELLIFORM COCCOLITHOPHORES

Okada and Honjo (1973) found that in subtropical latitudes (approximately 10–30° N) upper water (0–100 m) communities were dominated by three species – *Umbellosphaera tenuis*, *U. irregularis*, and *Discosphaera tubifera*. These species have coccoliths with large processes which flare distally to produce a double-layered coccosphere. The coccospheres lack obvious flagelllar openings so the cells are probably non-motile. This coccosphere morphotype is termed here 'umbelliform'. Placolith-bearing species are present in the umbelliform-dominated assemblages but form only about 10–30% of the total population.

Umbelliform-dominated assemblages have also been reported by Nishida (1979) Okada and McIntyre (1979) and Kleijne *et al.* (1989), from similar latitudes in the Pacific, Atlantic, and Indian Oceans. In sediment assemblages these species are less dominant but have clear abundance peaks in these areas (data of Geitzenauer *et al.*, 1977, Conley, 1979, Roth and Coulbourne, 1982). The distribution of the umbelliform-dominated assemblages at least approximately coincides with the oligotrophic, nutrient-depleted, waters of the mid-ocean gyres, as noted by Kleijne *et al.* (1989). This suggests that they are adapted to oligotrophic conditions – in contrast to the eutrophic-adapted placoliths. The absence of umbelliforms from higher latitude, oligotrophic waters suggest that they cannot tolerate either the lower temperatures or the deeper mixing and consequent low light levels of these environments.

Indirect support for the eutrophic/high productivity vs. oligotrophic/low productivity division between umbelliform and placolith-bearing communities is provided by the reported abundances of the umbelliform-dominated assemblages which are lower (10^3 to 10^4 cells/l) than the placolith-dominated assemblages (10^4 to 10^7 cells/l), Honjo and Okada (1974), Nishida (1979), Kleijne *et al.* (1989). Also umbelliform coccoliths are relatively much less common in the sediments than in the surface waters (e.g., Roth and Coulbourne, 1982). This may in part be due to differential preservation rates, but probably also indicates low reproduction and low coccolith production rates.

FLORIFORM COCCOLITHOPHORES

Okada and Honjo (1973) showed that the deep photic-zone assemblages (about 150–200 m) in low- to mid-latitudes (0–40° N) are dominated by a third group of species – *Florisphaera profunda*, *Thorosphaera flabellata*, and *Algirosphaera quadricornu*. This assemblage has been widely recognized elsewhere and is well established in the literature as a deep-water assemblage (e.g., Honjo, 1977; Reid, 1980; Sournia, 1982; Okada, 1983). Moreover none of these species occurs other than extremely rarely in surface-water assemblages. Most sediment assemblage surveys have omitted these species, since their coccoliths are easily overlooked; they have, however, been shown to be extremely abundant in low- to mid-latitude sediments (e.g., Okada, 1983, Biekart, 1989, Molfino and McIntyre, 1990, Tanaka, 1991).

The coccoliths of the characteristic species are of rather bizarre and diverse form. In all three cases, however, they contribute to form a coccosphere in which a dense asymmetrical mass of coccoliths surrounds a much smaller cell. These species also all have flagellar openings and are probably motile. In the absence of an existing name, the term 'floriform' is proposed here for this coccosphere morphotype.

The placolith-bearing species *Oolithus fragilis* creates a coccosphere in some ways similar to the floriform-coccospheres and also is more abundant in the deeper part of the placolith-domminted assemblages. It may be useful to consider *O. fragilis* an intermediate species. Other species with a tendency toward deeper waters include *Syracosphaera anthos* (syn. *Deutschlandia anthos*) and *Hayaster perplexus*. Coccospheres of common placolith-bearing species such as *Emiliania huxleyi* are also often reported from deep waters (e.g., Okada and Honjo, 1973, Reid, 1980). It is not clear whether these species are viable at depth, or whether they have accidentally sunk to these levels. Coccospheres of the umbelliform species do not seem to occur at depth even below waters in which they are abundant (Okada and Honjo, 1973; Reid, 1980; pers. obs.), although isolated coccoliths do occur. This suggests that umbelliform coccospheres do not sink prior to disintegration.

The absence of floriforms from surface waters suggest that they live below the mixed layer, in or under the thermocline. This environment is characterized not only by low light and temperature levels, but also usually by high nutrient levels – since the nutricline and thermocline normally coincide. The viability of the environments is critically dependent on light levels so it is unlikely to be productive at high latitudes, below deep mixed-layers, or below areas of high surface productivity. These factors may make the environment rather unstable.

MISCELLANEOUS GROUPS

In addition to the species which fit into the three major groups there are numerous other less common species. These rarely dominate assemblages and so have less clear biogeographical distribution. They constitute probably >80% of the species but <20% of the individual coccolithophores. They may have diverse ecologies but for simplicity they are treated here as a single 'miscellaneous group'. Most species of the following families can be included in this group: Syracosphaeraceae, Helicosphaeraceae, Calyptrosphaeraceae (holococcolith-bearing species), Rhabdosphaeraceae and Pontosphaeraceae. Most of these are motile but there some important nonmotile genera, such as *Pontosphaera*, *Rhabdosphaera*, and *Scyphosphaera*. The coccoliths of this group are highly variable in form but have a general tendency toward complex and delicate architecture. They do not have an obviously distinctive biogeography either individually or as a group but tend to be more important in intermediate environments than in

Table 1. *Summary of phytoplankton assemblages developed during ecological succession*

STAGES (Margalef 1967)	I	II/III	IV
Turbulence	High ———————————————————————————→		low
Stability	Low ————————————————————————————→		high
Nutrients - Nitrate	≥10 μM ———————————————————————→		≤1 μM
Typical environments	eutrophic	mestrophic	oligotrophic
	early bloom	later bloom	post blooms
	estuarine	coastal	open ocean
	upwelling core	upwelling fringes	open ocean
Production (carbon/day)	>50-5g/m²	5-0.5g/m²	0.5-0.05g/m²
GENERAL CHARACTER OF ALGAE			
Abundance (cells/litre)	10^5-10^7	10^4-10^5	≤10^3
Diversity	low	moderate -> high	high -> low
Division rate	≥1/day ————————————————————————→		≤1/week
Chlorophyll content	high ——————————————————————————→		low
Motile forms	rare ———————————————————————————		> common
Symbiosis & heterotrophy	rare ———————————————————————————		> common
Complex test forms	rare ———————————————————————————		> common
Ease of culturing	easy ——————————————————————————→		very hard
DIATOMS			
Relative importance	=========================+++++++++++++++————————————————————		
Size	small - medium	medium - large	medium - very large
Typical genera	*Thallassiosira*	*Chaetoceras*	*Hemiaulus*
DINOFLAGELLATES			
Relative importance	—————————————++++++++++++++++++==============================+++++++++		
Typical genera		*Ceratium*	*Ornithocercus*
COCCOLITHOPHORES			
Relative importance	——————————++++++++++++++++++++++++++++++++++++====================		
Typical genera		*Emiliania*	*Discosphaera*
ECOLOGICAL STRATEGY	**r** / opportunist ————————————————————→		K / specialist
	(growth rate maximising)		(efficiency maximizing)
COCCOLITHOPHORE TYPES			
Placolith bearing	——————————————+++++++++++++++++===============+++++++++++————————		
Miscellaneous group	——————————————+++++++++++++++++++++++++++————————		
Umbelliform	——————————————++++++++++++=========		
Floriforms		n/a	

Based on schemes in standard references (Margalef, 1958, 1967, 1978; Kilham and Kilham, 1988) except as noted below.

Production – my figures, mesotrophic and oligotrophic figures based on annual production estimates, hypertrophic figure based on biomass present in blooms.

Relative importance of phytoplankton groups – my synthesis of less explicit discussions (typical genera from Margalef, 1978).

Relative importance of coccolithophore types – my interpretation.

Relative importance symbols:

========= Dominant component of assemblage.

+++++++++ Significant component of assemblage.

————— Minor component of assemblage.

the extreme eutrophic or oligotrophic conditions. For instance, Okada and Honjo (1973) and Reid (1980) recorded assemblages with high abundances of these groups on the northern margins of, respectively, the North Pacific and North Atlantic gyres.

It may be reasonable to regard the group as a whole as characteristic of intermediate/normal nutrification conditions. Individual species almost certainly have varied adaptations to niches defined by different parameters to the dominant eutrophic–oligotrophic division.

Ecological interpretation

The variation among the groups of coccolithophores can be related rather directly to the influential work on phytoplankton ecological succession and evolutionary strategies of Margalef (1958, 1967, 1978; Table 1). Using observations on the Ria of Vigo, (Spanish Mediterranean coast) Margalef (1958) showed that storm-induced turbulence caused eutrophication of the surface waters. This resulted in a bloom of a limited number of diatom species. As the nutrient content and turbulence declined and environmental stability increased, a characteristic succession of phytoplankton communities was developed with concomitant changes in abundance and diversity, ending in a low abundance dinoflagellate dominated community. This pattern could be repeated several times in one year. Subsequent work has shown that this type of succession, including the broad taxonomic details, is developed in numerous different ecological situations (Margalef, 1967, 1978; Guillard and Kilham, 1977; Kilham and Kilham, 1980; Harris, 1986; Kilham and Hecky, 1988). These include temperate-latitude seasonal successions (even though energy availability rather than turbulence alone triggers the initial bloom) and upwelling situations, where the succession may occur in migrating patches of upwelled water rather than in one place. As a result, the geographical distribution of assemblages can mirror the successional sequence.

There are consistent biological differences among the species at the ends of the succession spectrum (Table 1). The late-stage species, relative to the early-stage species, tend to have lower division rates, more involved life cycles, smaller chloroplasts, larger cells, more complex test architecture, are more often motile, and may be heterotrophic or symbiotic. This has been interpreted in terms of evolutionary ecology as an example of r–K differentiation (Guillard and Kilham, 1977; Margalef, 1978, Kilham and Kilham, 1980; Harris, 1986; Kilham and Hecky, 1988). The complex adaptations of the late-stage taxa are thus interpreted as evolutionary specializations typical of K-selected taxa adapted to stable but difficult environments. In contrast, the small, simple, fast-reproducing, early-stage taxa are interpreted as r-selected opportunists adapted to rapid exploitation of temporarily available resources.

On the basis of the biogeographical distribution described above it would seem reasonable to predict that placolith-bearing species are characteristic of coccolithophore assemblages in eutrophic environments because they are relatively early succession opportunistic/r-selected species. Conversely since the umbelliforms are characteristic of oligotrophic assemblages they appear to be ideal late-succession, strongly K-selected species. The miscellaneous group might be tentatively interpreted as having a spread of adaptations with a tendency toward weak K-selection. Coccolithophores have sometimes been characterized as a uniformly oligotrophic K-selected group. This is probably because their maximum relative importance is at the end oligotrophic stage of the succession and they are unimportant in the earliest eutrophic stage (Table 1). However, placoliths are frequently a major component of normal eutrophic assemblages, Margalef (1978) included *Coccolithus* (= *Emiliania*) as a characteristic species of mid-succession eutrophic conditions.

In addition to their broad biogeographic distribution coccolithophores show some of the other characteristic features of the r–K gradient (Table 1). As previously noted the abundance of coccolithophores is consistently higher in placolith-dominated assemblages than in other assemblages, as suggested by the model. Diversity of assemblages is lowest for strongly eutrophic assemblages, but also low in extreme oligotrophic conditions. It is highest in intermediate conditions, when there is a significant abundance of the miscellaneous group (e.g., Honjo, 1976). This again is in agreement with the general model.

A basic prediction of the r- vs. K-division is that r-selected species should be able to respond to eutrophication, whereas K-selected species should not. This can be field tested, Hulburt and Corwin (1969) added nutrients to bottle samples of natural phytoplankton assemblages from the Caribbean and compared the results to natural eutrophication caused by seasonal nutrient input from the Amazon River. In both cases 'Coccolithus huxleyi' (=*Emiliania huxleyi*, and also probably includes *Gephyrocapsa* spp.) responded strongly to nutrient enrichment, with a major increase in population. In contrast, *Umbellosphaera tenuis*, *Discosphaera tubifera* and *Syracosphaera pulchra* did not.

To be cultured successfully, species must be tolerant of raised nutrient levels and an abnormal environment – r-selected characters. Ease of culturing a species can, therefore, be used as an indicator of its degree of opportunism/r-selection. Most placolith-bearing species have been cultured – *Calcidiscus leptoporus*, *Coccolithus pelagicus*, *C. neohelis*, *E. huxleyi*, *Gephyrocapsa oceanica*, *Reticulofenestra sessilis*, *Umbilicosphaera sibogae*, and *U. hulbertiana*. None of umbelliforms and only a few of the miscellaneous group have been cultured – *Calyptrosphaera sphaeroidea*, *Syracosphaera pulchra*, and a few non-oceanic species (Klaveness, 1973; Honjo and Roman, 1978; Inouye and Pienaar, 1984; Dudley *et al.*, 1986).

Table 2. *Applicability of functions to coccolith types*

	Placolith bearing	Miscellaneous group	Umbelliform	Floriform
MAIN ENVIRONMENT	Eutrophic ——————————>		Oligotrophic	Deep water
Turbulence	High —————————————>		Low	Low
Light	Lower —————————————>		High	Low
Nutrients	High —————————————>		Low	High
POSSIBLE FUNCTIONS				
Cell-membrane protection	y + y → Y	y + y → Y	y + y → Y	y + y → Y
Predator protection	? + y → ?	y + n → N	y + ? → ?	? + ? → ?
Buffering / water trapping	? + n → N	y + y → Y	y + y → Y	y + ? → ?
Accelerated sinking	y + y → Y	? + n → N	n + y → N	n + y → N
Decelerated sinking	n + n → N	y + y → Y	y + n → N	y + n → N
Light concentration	? + y → ?	? + n → N	n + n → N	y + y → Y
Photosynthesis linkage	? + y → ?	n + n → N	n + n → N	y + y → Y

For each case of the consideration is: Would this function be ecologically useful to the group? Is coccolith morphology suitable to perform the function? Is the function therefore likely to be applicable? Symbols: Ecological utility + morphological suitability → possibility of function. y/Y – Yes; n/N – No; ? – uncertain.

Various other predicted biological differences between r- and K-selected species are listed in Table 1. Of these, test forms certainly become more complex in the miscellaneous and umbelliform groups. Motility increases in the miscellaneous group, but not in the umbelliforms. Cell-size is not obviously different in the three groups. We do not have any independent data on reproduction rates, chlorophyll content, or adoption of heterotrophy.

Overall, the concept of succession-related assemblages appears to be supported by the biogeographical distribution, and r- and K-adaptation is evidenced by the culture work. Other evidence is only indirectly supportive but the concepts are probably worth exploring as possible clues to coccolithophore ecology and functional morphology.

Discussion – application of functions to ecological groups

In the preceding section four groups of coccolithophores have been identified that have distinctive morphologies, distribution patterns, and inferred ecological strategies. This provides some new constraints on functional interpretations of coccolith morphology; therefore it is now possible to reconsider each of the putative functions of coccoliths as applied to these groups. Table 2 summarizes the likely applicability of each of more reasonable functions to the coccolith groups, and they are briefly discussed here.

Simple cell-membrane protection

The delicacy of the cell-membrane means that it is likely to require some degree of protection. This is supported by the near ubiquity of cell-coverings in phytoplankton. It is perhaps the only viable functional interpretation of simple organic-scale cell-coverings and seems likely to be a basic function of coccoliths. Cell-membrane protection would seem equally possible as a function in all of the ecological groups, although it cannot explain the diversity of coccolith (or scale) morphologies

Protection from predators

The possible importance of this function is not aided by the ecological analysis, since it is difficult to predict in which of the environments predator protection would be most important. Cropping pressure is very high in eutrophic environments but ecological theorists have generally regarded predator protection as a classical K-selected strategy for optimizing the survival of individuals. In phytoplankton this is typified by toxic dinoflagellates, which occur late in the ecological succession. Phytoplankton inhabiting deeper-water layers should also benefit from protection since numerous zooplankton species are known to occupy deeper water at least part of the time. So predator protection could be useful in any environment.

Likely adaptations of coccoliths for providing predator protection include structures which produce robust coccospheres and structures which produce coccospheres with diameters significantly greater than the cell-diameter. Placolith coccoliths produce robust coccospheres, whereas both umbelliform and floriform coccospheres have much greater diameters than the cells they enclose. So all three of these coccolith types could have predation protection functions. It is, however, probably possible to eliminate predation protection as a function from most of the miscellaneous group species, many of which have delicate coccospheres that do not significantly alter the coccosphere size.

Chemical buffering and nutrient absorption

Functions of a trapped layer of water, or macromolecular soup, within the coccosphere might be expected to be more important in the late succession, low-nutrient, environments. The miscellaneous group and umbelliform coccospheres that characterize these environments also have coccoliths that show likely adaptations toward this role, that is, formation of double- layered coccospheres. In the case of many members of the miscellaneous group (e.g., many Calyptrosphaeraceae and Syracosphaeraceae) this seems to be the only likely explanation for their coccolith form.

Accelerated sinking

This can only be adopted as a strategy to aid nutrient absorption in high-turbulence environments. It can, therefore, only be applicable to species which inhabit these environments, that is, placolith-bearing coccolithophores. These also form the heaviest coccospheres. Accelerated sinking must be an effect of placolith coccospheres and is likely to be a function of their production.

Decelerated sinking

In low turbulence environments the ballasting effect of coccoliths is liable to cause coccolithophores to sink out of the photic zone and so adaptations to decelerate sinking are likely to be needed. The most likely examples of these are the spines and whorl coccoliths of various of the miscellaneous-group coccolithophores which inhabit low- to moderate-turbulence environments. Umbelliform and floriform coccospheres inhabit low turbulence environments but do not show any such adaptations, indeed they have heavy coccospheres which should cause accelerated sinking. Floriform coccolithophores may use motility to avoid sinking. Umbelliforms do not appear to be motile; possibly they are able to decrease cell density instead.

Light concentration and biochemical linkage of calcification to photosynthesis

These are quite different functions but they both should be adaptive in low-light environments. Also both functions would require formation of large numbers of coccoliths. So it is not possible to distinguish between these functions on the broad ecological-morphological criteria established here. Floriform coccoliths are the most likely candidates for these functions since deep-living species are almost certainly light-limited rather than nutrient-limited. Also they form particularly large coccospheres, relative to cell size.

The functions are unlikely to be applicable to many other species because light is less often limiting than nutrients and because few other species produce excessive numbers of coccoliths. The placolith-bearing species *Emiliania huxleyi* is a possible candidate since it does produce excess coccoliths and has a tendency toward colder water, deeply mixed environments where light may be limiting.

Summary

Of the different possible functions that have been proposed for coccolith formation a considerable number seem conceptually reasonable: cell-membrane protection, protection from predation, cellular buffering, acceleration of sinking, deceleration sinking, light concentration, and calcification as an aid to photosynthesis.

Coccolith morphology is very diverse and so are the ecological niches occupied by coccolithophores, this makes it difficult to discuss the functional morphology of coccoliths in general. Instead, I have attempted to distinguish groups of coccolithophores with distinctive coccosphere types and discrete ecological strategies. These are: (1) placolith-bearing r-selected species adapted to eutrophic conditions, (2) 'umbelliform' K-selected species adapted to low-latitude oligotrophic waters, (3) a miscellaneous group of mainly motile species most characteristic of intermediate conditions, and (4) deep-dwelling 'floriform' species. Likely functions can be identified for each group on the basis of coccolith morphology and ecology (Table 2).

This analysis suggests that it is unlikely that there is a universal function or functions of coccoliths. Instead it is probable that coccoliths act as cell-coverings, affect the light entering the cell, and increase cell-density, whereas their biosynthesis will have wide ranging influences on cell biochemistry. Within different habitats it is virtually certain that particular adaptations of coccolith form, and patterns of ecological behavior, related to minimizing or exploiting these effects will have evolved.

Acknowledgements

This paper was written while I was on a Clyde Petroleum PLC funded post-doctoral fellowship at the Natural History Museum. I am grateful to many colleagues and friends for discussion of ideas on the subject of functional morphology. In particular I am grateful to Ric Jordan, Paul Pearson, Bill Siesser, and Amos Winter for their comments on the manuscript.

References

Biekart, J.W., 1989. The distribution of calcareous nannoplankton in Late Quaternary sediments collected by the Snellius II Expedition in some south east Indonesian basins. *Proc. K. Neder. Akad. Wetensch. Ser.B.*, **92**: 77–141.

Birkenes, E. , and Braarud, T., 1952. Phytoplankton in the Oslo Fjord during a 'Coccolithus huxleyi summer'. *Avh. Norske Vidensk. Akad. Oslo*, **2**: 1–23.

Boney, A.D., 1970. Scale bearing phytoflagellates: an interim review. *Oceanogr. Mar. Biol . A. Rev.*, **8**: 251–305.

Bown, P.R., 1987. Taxonomy, evolution and biostratigraphy of Late Triassic–Early Jurassic calcareous nannofossils. *Spec. Papers in Palaeontol.*, **38**: 1–118.

Braarud, T., Gaarder, K.R., Markali, J. and Nordli, E. 1952. Coccolithophorids studied in the electron microscope. *Nytt. Mag. Bott.*, **1**: 129–34.

Burns, D.A., 1977. Phenotypes and dissolution morphotypes of the genus *Gephyrocapsa* Kamptner and *Emiliania huxleyi* (Lohmann). *N.Z.J. Geol. Geophys.*, **20**: 143–55.

Cadee, G.C., 1985. Macroaggregates of *Emiliania huxleyi* in sediment traps. *Mar. Ecol. Progr. Ser.*, **24**: 193–6.

Conley, S.M., 1979. Recent coccolithophores from the Great Barrier Reef-Coral Sea region. *Micropaleontology*, **25**(1): 20–43.

Covington, M., 1985. New morphologic information on Cretaceous nannofossils from the Niobrara Formation (Upper Cretaceous) of Kansas. *Geology*, **13**: 683–6.

Denman, K.L. and Gargett, A.E., 1983. Vertical mixing and advection of phytoplankton in the upper ocean. *Limnol. Oceanogr.*, **28**: 801–15.

Dudley, W.C., Blackwelder, P., Brand, L., and Duplessy, J.C., 1986. Stable isotope composition of coccoliths. *Mar. Micropaleontol.*, **10**: 1–8.

Eppley, R.W., Holmes, R.W., and Strickland, J.D.H., 1967. Sinking rates of marine phytoplankton measured with a fluorometer. *J. Expt. Mar. Biol. Ecol.*, **1**: 191–208.

Frost, B.W., 1980. Grazing. In: *The Physiological Ecology of Phytoplankton*, ed. I. Morris, pp. 465–92. University of California Press, Berkeley.

Gaarder, K.R. and Hasle, G.R. 1962. On the supposed symbiosis between diatoms and coccolithophorids in Brenneckella. *Nytt Mag. Bott.*, **9**: 145–9.

Gartner, S. and Bukry, D., 1969. Tertiary holococcoliths. *J. Palaeontol.*, **43**: 1213–21.

Gavis, J., 1976. Munk and Riley revisited: nutrient limited diffusion transport and rates of phytoplankton growth. *J. Mar. Res.*, **34**: 161–79.

Geitzenauer, K.R., Roche, M.B., and McInytre, A., 1977. Coccolith biogeography from North Atlantic and Pacific surface sediments. In: *Oceanic Micropalaeontology*, ed. A.T.S. Ramsey, pp. 973–1008. Academic Press, London.

Grindley, J.R. and Taylor, F.J.R., 1970. Factors affecting plankton blooms in False Bay. *Trans. Roy. Soc. S. Africa*, **39**: 201–10.

Guillard, R.R.L. and Kilham, P., 1977. The ecology of marine planktonic diatoms. In: *The Biology of Diatoms*. ed. D. Werner, pp. 372–469. Blackwell Scientific Publications, Oxford.

Hallegraef, G.M., 1984. Coccolithophorids (calcareous nannoplankton) from Australian waters. *Botanica Mar.*, **27**: 229–47.

Harris, G.P. 1986. *Phytoplankton Ecology: Structure, Functions, and Fluctuations*. Chapman and Hall, London.

Hay, W.W., 1968. Coccoliths and other calcareous nannofossils in marine sediments in Cyrenaica. In *Geology and Archaeology of Northern Cyrenaica, Libya*, ed. F.T. Barr, pp. 149–157. Petroleum Exploration Society of Libya.

Holligan, P.M., Viollier, M., Harbour, D.S., Camus, P. and Champagne-Phillippe, M., 1983. Satellite and ship studies of coccolithophore production along a continental shelf edge. *Nature*, **304**: 339–42.

Honjo, S. 1976. Coccoliths: production, transportation, and sedimentation. *Mar. Micropaleontol.*, **1**: 65–79.

Honjo, S., 1977. Biogeography and provincialism of living coccolithophorids. In: *Oceanic Micropalaeontology*, ed. A.T.S. Ramsey, pp. 951-972. Academic Press, London.

Honjo, S. 1982. Seasonality and interaction of biogenic and lithogenic particulate flux at the Panama basin. *Science*, **219**: 883–4.

Honjo, S. and Roman, M.R., 1978. Marine copepod fecal pellets: production, preservation and sedimentation. *J. Mar. Res*, **36**: 45–57.

Honojo, S. and Okada, H. 1974. Community structure of coccolithophores in the photic layer of the mid-Pacific. *Micropalaeontol.*, **20**: 209–30.

Hulbert, E.M., 1985. Adaptation and niche breadth of phytoplankton species along a nutrient gradient. *J. Plankton Res.*, **7**: 581–94.

Hulbert, E.M. and Corwin, N. 1969. Influence of the Amazon River outflow on the ecology of the western tropical Atlantic III – the plankton flora between the Amazon River and the Windward Islands. *J. Mar. Res.* **27**: 55–72.

Hutchinson, G.E., 1967. *A Treatise on Limnology*, vol.2. *Introduction to Lake Biology and the Limnoplankton*. J. Wiley & Sons, New York.

Inouye, I. and Pienaar, R.N., 1984. New observations on the coccolithophorid *Umbilicosphaera sibogae* var. *foliosa* (Prymnesiophyceae) with reference to cell covering, cell structure, and flagellar apparatus. *Br. Phycol. J.*, **19**: 357–69.

Jordan, R.W., 1988. Coccolithophore communities in the North East Atlantic. Unpub. Ph.D. thesis, University of Surrey.

Kilham, P. and Hecky, R.E. 1988. Comparative ecology of marine and freshwater phytoplankton. *Limnol. Oceanogr.*, **33**: 776–95.

Kilham, P. and Kilham, S.S. 1980. The evolutionary ecology of phytoplankton. In: *The Physiological Ecology of Phytoplankton*, ed. I. Morris, pp. 571–97. University California Press, Berkeley.

Klaveness, D. 1973. The microanatomy of *Calyptrosphaera sphaeroidea*, with some supplementary observations on the motile stage of *Coccolithus pelagicus*. *Norw. J. Bot.*, **20**: 151–62

Klaveness, D. and Paasche, E., 1979. Physiology of coccolithophorids. In *Biochemistry and Physiology of Protozoa*, 2nd edn, vol. 1, ed. M. Levandovsky, S.H. Hunter, pp. 191-213. Academic Press, London.

Kleijne, A., Kroon, D., and Zeveboom, W., 1989. Phytoplankton and foraminiferal frequencies in northern Indian Ocean and Red Sea surface waters. *Neth. J. Sea Res.*, **24**: 531–9.

Levandowsky, M. and Kaneta, P. 1987. Behaviour in dinoflagellates. In *The Biology of Dinoflagellates*, Bot. Monogr. 21, ed. F.J.R. Taylor, pp. 360–97. Blackwell Scientific Publications, Oxford.

Lohmann, H., 1913. Beitrage zur Charakterisierung des Tier- und Pflanzen-lebens in den von Deutschland wahrend Fahrt nach Buenos Ayres durch fahrenen Gebieten des Atlantischen Ozeans, III Teil. *Int. Rev. Ges. Hydrobiol. Hydrographie*, V: 343–504.

McIntrye, A. and Bé, A.W.H., 1967. Modern coccolithophores of the Atlantic Ocean – I. Placoliths and cyrtoliths. *Deep-Sea Res.*, **14**: 561–97.

Manton, I. 1986. Functional parallels between calcified and uncalcified periplasts. In *Biomineralization in Lower Plants and Animals*, ed. B.S.C. Leadbeater, and R. Riding, Systematics Assn. Spec. Vol., 30. pp. 157–72.

Margalef, R., 1958. Temporal succession and spatial hetrogeneity in phytoplankton. In *Perspectives in Marine Biology* ed. A.A. Buzzati-Traverso, pp. 323–51. Int. Union Biol. Sci. Publs, B/27.

Margalef, R., 1967. The food web in the pelagic environment. *Helgolander Wiss. Meeresunters.*, **15**: 548–59.

Margalef, R., 1978. Life forms of phytoplankton as survival alternatives in an unstable environment. *Oceanol. Acta*, **1**: 493–509.

Margulis, L., Melkonian, M., and Corliss, J.O., 1989. *Handbook of Protoctista*. Jones and Bartlett, Woods Hole.

Milliman, J.D., 1980. Coccolithophorid production and sedimentation, Rockall Bank. *Deep-Sea Res.*, **A27**: 959–63.

Mitchell-Innes, B.A. and Winter, A. 1987. Coccolithophores: a major phytoplankton component in mature upwelled waters off the Cape Peninsula, South Africa in March 1983. *Mar. Biol.*, **95**: 25–30.

Molfino, B. and McIntrye, A. 1990. Precessional forcing of nutricline dynamics in the equatorial Atlantic. *Scienc*, **249**: 766–9.

Munk, W.H. and Rilery, G., 1952. Absorption of nutrients by aquatic plants. *J. Mar. Res.*, **11**: 215–40.

Nishida, S. 1979. *Atlas of Pacific Nannoplanktons*. Micropalaeontol. Soc. Osaka, Spec. Paper No. 3.

Norris, R.E., 1971. Extant calcareous nannoplankton from the Indian Ocean. In *Proc. II Planktonic Conf.*, Roma, vol. 2, ed. A Farinacci, pp. 899-910. Edizioni Tecnoscienza, Rome.

Okada, H., 1983. Modern nannofossil assemblages in sediments of coastal and marginal seas along the western Pacific Ocean. *Utrecht Micropaleontol. Bull.*, **30**: 171–87.

Okada, H., 1989. Morphometric and floral variations of nannoplankton in relation to their living environment. *INA Newsl.*, **11**(2): 87–8.

Okada, H. and Honjo, S., 1973. The distribution of oceanic coccolithophorids in the Pacific. *Deep-Sea Res.*, **20**: 355–74.

Okada, H. and Honjo, S., 1975. The distribution of coccolithophorids in marginal seas along the western Pacific Ocean and in the Red Sea. *Mar. Biol.*, **31**: 271–85.

Okada, H. and McIntyre, A., 1977. Modern coccolithophoridae from the Pacific and North Atlantic Oceans. *Micropaleontol.*, **23**: 1–55.

Okada, H. and McIntyre, A., 1979. Seasonal distribution of modern coccolithophores in the western North Atlantic Ocean. *Mar. Biol.*, **54**: 319–28.

Outka, D.E. and Williams, D.C., 1971. Sequential coccolith morphogenesis in *Hymenomonas carterae*. *J. Protozool.*, **18**: 285–97.

Paasche, E., 1962. Coccolith formation. *Nature*, **193**: 1094–5.

Pasciak, W.J. and Gavis, J., 1974. Transport limitation of nutrient uptake in phytoplankton. *Limnol. Oceanogr.*, **19**: 881–8.

Reid, F.M.H., 1980. Coccolithophorids from the North Pacific central gyre with notes on their vertical and seasonal distribution. *Micropalaeontol.*, **26**: 151–76.

Reynolds, C.S. 1986. Diatoms and the geochemical cycling of silicon. In *Biomineralization in Lower Plants and Animals*, ed. B.S.C. Leadbeater, and R. Riding, pp.189-203. Systematics Assn. Spec. vol. 30.

Round, F.E., Crawford, R.M. and Mann, D.G., 1990. *The Diatoms, Biology and Morphology of the Genera*. Cambridge Univ. Press, Cambridge.

Roth, P.H. and Coulborne, W.T., 1982. Floral and solution patterns of coccoliths in surface sediments of the North Pacific. *Mar. Micropaleontol.*, **7**: 1–52.

Roth, P.H., Mullin, M.M., and Berger, W.H. 1975. Coccolith sedimentation by fecal pellets: laboratory experiments and field observations. *Bull. Geol. Soc. Amer.*, **86**: 1079–84.

Samtleben, C. and Bickert, T., 1990. Coccoliths in sediment traps from the Norwegian Sea. *Mar. Micropaleontol.*, **16**: 39–64.

Sargeant, W.A.S., Lacalli, T. and Gaines, G., 1987. The cysts and skeletal elements of dinoflagellates: speculations on the ecological causes for their morphology and development. *Micropaleontol.*, **33**: 1–36.

Schei, B., 1975. Coccolithophorid distribution and ecology in coastal waters of north Norway. *Norw. J. Bot.*, **22**: 217–25.

Sikes, C.S., Roer, R.D. and Wilbur, K.M., 1980. Photosynthesis and coccolith formation: Inorganic carbon sources and net inorganic reaction of deposition. *Limnol. Oceanogr.* **25**: 248–61.

Sikes, C.S. and Wilbur, K.M., 1982. Functions of coccolith formation. *Limnol. Oceanogr.*, **27**: 18–26.

Smayda, T.J., 1970. The suspension and sinking of phytoplankton in the sea. *Oceanogr. Mar Bio l. Assn.*, **8**: 285–97.

Smayda, T.J., 1980. Phytoplankton species succession. In *The Physiological Ecology of Phytoplankton*. ed. I. Morris, pp. 493–570. University of California Press, Berkeley.

Sournia, A., 1980. Form and function in marine phytoplankton. *Biol. Rev. Camb. Phil. Soc.*, **57**: 347–94.

Sournia, A., 1982. Is there a shade flora? *J. Plankton Res.*, **4**: 391–9.

Steinmetz, J.C., 1991. *Calcareous Nannoplankton Biocoenosis: Sediment Trap Studies in the Equatorial Atlantic, Central Pacific and Panama Basin*. Ocean Biocoenosis Series, 1, Woods Hole Oceanogr. Inst.

Tanaka, Y., 1991. Calcareous nannoplankton thanatocoenoses in surface sediments from seas around Japan. *Sci. Repts. Tohoku Univ.*, 2nd Ser., **61**: 127–98.

Taylor, F.J.R., 1980. Basic biological features of phytoplankton cells. In: Morris, I. (Ed.) *The Physiological Ecology of Phytoplankton*. Univ. California Press, Berkeley, pp. 3–56.

Taylor, F.J.R. (Ed.): *The Biology of Dinoflagellates*. Bot. Monogr. 21, pp. 360–97. Blackwell Scientific Publications, Oxford.

Throndsen, J., 1973. Motility in some marine nanoplankton flagellates. *Norw. J. Zool.*, **21**: 193–200.

Verbeek, J.W., 1989. Recent calcareous nannoplankton in the southernmost Atlantic. *Polarforschung*, **59**: 45–60.

Wal, P. van der, Jong, E. W de, Westbroek, P., Bruijn, W.C. de, and Mulder-Stapel, A.A. 1983. Ultrastructural polysaccharide localization in calcifying and naked cells of the coccolithophorid *Emiliania huxleyi*. *Protoplasma*, **118**, 157–68.

Wal, P. van der, Vrind-de Jong, E.W. de, Borman, A.H., and Vrind, J.P.M. de, 1987. Incompleteness of the coccosphere as a possible stimulus of coccolith formation in *Hymenomonas carterae*. *J. Phycol.*, **23**: 218–21.

Walsby, A.F. and Reynolds, C.S., 1980. Sinking and floating. In: Morris, I. (Ed.) *The Physiological Ecology of Phytoplankton*. Univ. California Press, Berkeley, pp.1-625.

Westbroek, P., Jong, E.W. de, Wal, P. van der, Borman, A.H., Vrind, J.P.M. de, Emburg, P.E. van, and Bosch, L., 1983. Coccolith formation, wasteful or functional? *Ecol. Bull.*, **35**: 291–9.

Wilbur, K.M. and Watabe, N., 1963. Experimental studies on calcification in molluscs and the alga *Coccolithus huxleyi. Ann. N. Y. Acad. Sci.*, **103**: 82–112.

Winter, A., Reiss, Z., and Luz, B., 1979. Distribution of living coccolithophore assemblages in the Gulf of Elat ('Aqaba). *Mar. Micropaleontol.*, **4**: 197–223.

Young, J.R., 1987. Possible functional interpretations of coccolith morphology. In *Proc. Int. Nannoplankton Assn. Meeting, Vienna 1985,* ed. H. Stradner, and K. Perch-Nielsen, *Abh. Geol. Bundesanst.*, **39**: 305–313.

Young, J.R., 1989. Observations on heterococcolith rim structure and its relation to developmental processes. In *Nannofossils and Their Applications,* pp. 1–20. ed. Crux, J. and Heck, S.E. van (Eds.), Br. Micropaleontol. Soc. Ser.

Young, J.R., 1990. Size variation in Neogene reticulofenestrid coccoliths from Indian Ocean DSDP cores. *J. Micropalaeontol.*, **9**: 71–86.

Young, J.R. and Westbroek, P. 1991. Genotypic variation in the coccolithophorid species *Emiliania huxleyi. Mar. Micropaleontol.*, **18**: 5–23.

6 A classification system for living coccolithophores

RIC W. JORDAN AND ANNELIES KLEIJNE

Introduction

Since Wallich described the first living coccolithophores in 1877, coccolith morphology (see Chapter 4) has always been the most important factor in the identification of this group. During the late nineteenth century it was still uncertain whether these coccoliths were of inorganic (Ehrenberg, 1836; Harting, 1872; Barrois, 1876) or biological (Sorby, 1861; Wallich, 1861) origin. Nor was it certain whether they were individual cells (Gümbel, 1870; Carter, 1871; Schwarz, 1894), parts of larger animals (Wallich, 1861; Huxley, 1868) or plants (Thomson, 1874; Murray and Blackman, 1898). Eventually, coccoliths were positively identified as the calcitic armature of minute algae, although Lohmann (1902) included them in the Mastigophoren, a superclass of protozoans which contained both animal and plant flagellates. The terms coccolithophorid and coccolithophore are interchangeable and both derive from the invalid genus *Coccolithophora* Lohmann (synonym of *Coccolithus*), but may just as well be applied to *Coccolithus* Schwarz (syn. *Coccosphaera* Wallich, in part), the valid name for the first discovered extant genus.

In early classification schemes (Lohmann, 1902) the coccolithophores were placed alongside the chrysomonads, in a group possessing many flagellar types. Those chrysomonads bearing two equal flagella were separated by Pascher (1910) into the order Isochrysidales (based on the Hymeno-monadaceae of Senn, 1900), whilst those possessing an additional 'modified third flagellum' were later placed in the Prymnesiales (Papenfuss, 1955). Christensen (1962) noticed that species belonging to these two orders bore smooth flagella (cf. the other chrysomonads) and furthermore that the possession of a third flagellum was a unique character. On this basis he created a new class, the Haptophyceae, to contain the Isochrysidales and Prymnesiales. In 1976 the Haptophyceae was reviewed by Hibberd following changes in the International Code of Botanical Nomenclature (ICBN), which proposed that all taxa above the rank of family must be based on generic names. The invalid class name was therefore changed to the Prymnesiophyceae, after the genus *Prymnesium* (Hibberd, 1976).

As haptophytes the coccolithophores were originally assigned to the Isochrysidales (Papenfuss, 1955) and then later to the Prymnesiales (Round, 1973). In his review of the Prymnesiophyceae Hibberd (1980) divided the class into four orders, following the creation of the Pavlovales for *Pavlova* and similar genera (Green, 1976) and the re-separation of the coccolithophores into the Coccosphaerales (Haeckel, 1894; Parke and Green, 1976). Several coccolithophores have now been studied in more detail and some of the motile forms possess the 'third flagellum', called a haptonema (Parke *et al.*, 1955), whilst in others it is either reduced or absent. The fact that some species display similar features to the Prymnesiales strengthens the relationship between the two groups. Furthermore, it is now clear that members of the Isochrysidales display a range of haptonematal development, although never as a long external structure, and thus the initial reason for their separation from the Prymnesiales is somewhat debateable (Green and Pienaar, 1977).

In the system presented below the coccolithophores are placed in the order Coccolithophorales (Schiller, 1925 – after the class name of Lemmermann, 1908), however, Tappan (1980) suggests that this name is incorrect as it is based on the invalid taxon *Coccolithophora* (synonym of *Coccolithus*) and claims it to be synonymous with the class Coccolithophyceae (Rothmaler, 1951). Like the term coccolithophore, the Coccolithophorales equally applies to *Coccolithus*, the valid name for *Coccolithophora*.

Previous classification schemes

Classification schemes for coccolithophores began around the start of the twentieth century, following phytoplankton studies in the North Atlantic (Murray and Blackman, 1898) and Mediterranean Sea (Lohmann, 1902). With the discovery of further taxa from a variety of locations, the schemes became progressively more comprehensive (Kamptner, 1928, 1941; Schiller, 1930; Deflandre, 1952), but were largely based on cell and coccolith shape, and associated apparatus (e.g. flagella). It was not until the advent of the electron microscope that detailed coccolith morphology could be investigated and more reliable characters found. At present, separation at the family level is by the proximal disc construction of the main coccolith type, at the generic level characters like central area structures, flagellar position, and additional coccolith types are more important. In general, the species differ on small-scale variations on the generic theme. Recent taxonomic works have generally been restricted to specific geographic areas (Parke and Green, 1976; Okada and McIntyre, 1977; Winter *et al.*, 1979) or to particular families or genera (Manton and colleagues, 1975–85; Gaarder and Heimdal, 1977; Heimdal and Gaarder, 1980, 1981; Norris, 1983, 1984, 1985; Kleijne, 1991). Furthermore, recent reviews (see Haq, 1978; Tappan, 1980; Perch-Nielsen, 1985; Chrétiennot-Dinet, 1990) have mainly concentrated on separating to family and genus level, or where species lists are included they are either incomplete or contain dubious and fossil taxa. In the system presented by us below, the list contains all those coccolithophores (marine, littoral, and freshwater) that are presently considered extant.

Thoracosphaerids

Until very recently the calcareous nannoplankton genus *Thoracosphaera* was generally considered to be a *bona fide* coccolithophore taxon. However, observations on fossil (Fütterer, 1976, 1977) and living (Inouye and Pienaar, 1983)

material have shown conclusively that this genus belongs with the dinoflagellates. For this reason it has been removed from the classification system presented below.

This classification scheme

In compiling this classification system it has been necessary to omit the non-calcified members of the Prymnesiophyceae, which are not considered as coccolithophores although they may be closely related. In fact several authors have noticed the similarities in scale structure between the genus *Chrysochromulina* and some fossil coccoliths (Manton and Leedale, 1963; Black, 1968), and between *Navisolenia* and the calcified rhomboliths of *Calciosolenia* and *Anoplosolenia* (Lecal, 1965a; Leadbeater and Morton, 1973). To avoid the inclusion of dubious taxa, the list only incorporates those species which have been verified by electron microscopy in the last 40 years. Recent transfers and observations, and current opinions are included in the subsection entitled 'Taxonomic notes'. SEM micrographs of many of the species listed in the following classification scheme can be found in chapter 7.

Classification system

Kingdom: Protista Haeckel, 1866
Phylum: Prymnesiophyta Hibberd, 1976

Class: Prymnesiophyceae Hibberd, 1976
Cells have smooth flagella (i.e. no hairs) and may possess a fully developed or reduced haptonema. Cells covered with organic and/or calcite scales.

Order: Coccolithophorales Schiller, 1926
Cells produce coccoliths at some stage in their life cycle.

(A) Heterococcolithophores   Taxonomic notes

Family: **BRAARUDOSPHAERACEAE** Deflandre, 1947
Cell covered by twelve pentaliths to form a regular dodecahedron. The pentalith is
generally constructed of five crystals, each radiating from a central point.

Genus: *BRAARUDOSPHAERA* Deflandre, 1947
B. bigelowii (Gran & Braarud, 1935) Deflandre, 1947   (1)

Family: **CALCIOSOLENIACEAE** Kamptner, 1937   (2)
Spindle-shaped cell covered by scapholiths (rhomboliths). Scapholiths are diamond-
shaped and composed of upright wall elements. Each wall element is associated with a
lamella which partially spans the short axis of the central area.

Genus: *ANOPLOSOLENIA* Deflandre, 1952
A. brasiliensis (Lohmann, 1919) Deflandre, 1952

Genus: *CALCIOSOLENIA* Gran, 1912
C. murrayi Gran, 1912

Family: **CERATOLITHACEAE** Norris, 1965
Cell partially covered by a single horseshoe-shaped coccolith and numerous hoop-shaped coccoliths.

Genus: *CERATOLITHUS* Kamptner, 1950
C. cristatus Kamptner, 1950 var. *cristatus* (3)
C. cristatus var. *telesmus* (Norris, 1965) Jordan & Young, 1990

Family: **COCCOLITHACEAE** Poche, 1913
Cell covered by overlapping placoliths. Each placolith consists of two shields of sub-horizontal elements and a connecting tube.

Genus: *CALCIDISCUS* Kamptner, 1950
C. leptoporus (Murray & Blackman, 1898) Loeblich & Tappan, 1978 f. *leptoporus* (4)
C. leptoporus f. *rigidus* (Gaarder, in Heimdal & Gaarder, 1980) Kleijne, 1991

Genus: *COCCOLITHUS* Schwarz, 1894
C. neohelis McIntyre & Bé, 1967a (5)
C. pelagicus f. *braarudii* (Gaarder, 1962) Kleijne, 1991 (4)
C. pelagicus f. *hyalinus* (Gaarder & Markali, 1956) Kleijne, 1991
C. pelagicus (Wallich, 1877) Schiller, 1930 f. *pelagicus*

Genus: *HAYASTER* Bukry, 1973
H. perplexus (Bramlette & Riedel, 1954) Bukry, 1973 (6)

Genus: *NEOSPHAERA* Lecal-Schlauder, 1950 (7)
N. coccolithomorpha Lecal-Schlauder, 1950

Genus: *OOLITHOTUS* Reinhardt, in Cohen and Reinhardt, 1968
O. fragilis var. *cavum* (Okada & McIntyre, 1977) Jordan & Young, 1990
O. fragilis (Lohmann, 1912) Martini & Müller, 1972 var. *fragilis*

Genus: *UMBILICOSPHAERA* Lohmann, 1902
U. angustiforamen Okada & McIntyre, 1977
U. calvata Steinmetz, 1991
U. hulburtiana Gaarder, 1970
U. maceria Okada & McIntyre, 1977
U. scituloma Steinmetz, 1991
U. sibogae var. *foliosa* (Kamptner, 1963) Okada & McIntyre, 1977 (8)
U. sibogae (Weber-Van Bosse, 1901) Gaarder, 1970 var. *sibogae*

Family: **HELICOSPHAERACEAE** Black, 1971, emend. Jafar & Martini, 1975
Cell covered by helicoliths in a spiral arrangement. Coccosphere may have an apical opening. Helicolith with narrow spiral flange and differentiated central portion. Central openings may be present on the distal side.

Genus: *HELICOSPHAERA* Kamptner, 1954
H. carteri (Wallich, 1877) Kamptner, 1954 var. *carteri* (9)
H. carteri var. *hyalina* (Gaarder, 1970) Jordan & Young, 1990
H. carteri var. *wallichii* (Lohmann, 1902) Theodoridis, 1984
H. pavimentum Okada & McIntyre, 1977

Family: **HYMENOMONADACEAE** Senn, 1900
Cell covered by tremaliths. Each tremalith has a ring of distinct wall elements which form
a distal tube. A short basal rim is also present. The central area is covered by an organic
base plate. In appearance the tremalith resembles the basal portion of a pappolith.

Genus: *HYMENOMONAS* Stein, 1878
H. coronata Mills, 1975
H. globosa (Magne, 1954) Gayral & Fresnel, 1976
H. lacuna Pienaar, 1976
H. roseola Stein, 1878

Genus: *OCHROSPHAERA* Schussnig, 1930
O. neapolitana Schussnig, 1930

Family: **NOELAERHABDACEAE** Jerkovic, 1970 (10)
Cell covered by overlapping placoliths. Placoliths (each composed of two shields)
characterized by a reticulum or grill covering the proximal part of the central area
opening.

Genus: *EMILIANIA* Hay & Mohler, in Hay *et al.*, 1967
E. huxleyi (Lohmann, 1902) Hay & Mohler, in Hay *et al.*, 1967 var. *huxleyi*
E. huxleyi var. *corona* (Okada & McIntyre, 1977) Jordan & Young, 1990

Genus: *GEHYROCAPSA* Kamptner, 1943 (11)
G. caribbeanica Boudreaux & Hay, in Hay *et al.*, 1967
G. crassipons Okada & McIntyre, 1977
G. ericsonii McIntyre & Bé, 1967b
G. muellerae Bréhéret, 1978 (12)
G. oceanica Kamptner, 1943
G. ornata Heimdal, 1973

Genus: *RETICULOFENESTRA* Hay, Mohler & Wade, 1966
R. parvula (Okada & McIntyre, 1977) Biekart, 1989 var. *parvula*
R. parvula var. *tecticentrum* (Okada & McIntyre, 1977) Jordan & Young, 1990
R. punctata (Okada & McIntyre, 1977) Jordan & Young, 1990
R. sessilis (Lohmann, 1912) Jordan & Young, 1990 (13)

Family: **PAPPOSPHAERACEAE** Jordan & Young, 1990 (14)
Cell covered by pappoliths. Each pappolith is constructed of a rim of two lath types, one small
and confined to proximal part of wall, the other is vertically expanded. Central process often present.

Genus: *PAPPOMONAS* Manton & Oates, 1975
P. flabellifera var. *borealis* Manton, Sutherland & McCully, 1976
P. flabellifera Manton & Oates, 1975 var. *flabellifera*
P. virgulosa Manton & Sutherland, 1975
P. weddellensis Thomsen, in Thomsen *et al.*, 1988

Genus: *PAPPOSPHAERA* Tangen, 1972
P. lepida Tangen, 1972
P. obpyramidalis Thomsen, in Thomsen *et al.*, 1988
P. sagittifera Manton, Sutherland & McCully, 1976
P. sarion Thomsen, 1981
P. simplicissima Thomsen, in Thomsen *et al.*, 1988
P. thomsenii Norris, 1983

Family: **PLEUROCHRYSIDACEAE** Fresnel & Billard, 1991
Cell covered by cricoliths. Each cricolith is composed of a ring of distinct elements. Each element has an upper and lower rim and thus a cricolith is really a special form of placolith (Manton & Leedale, 1969). The central area is covered by an organic base plate.

Genus: *CRICOSPHAERA* Braarud, 1960
C. elongata Droop, 1955
C. gayraliae Beuffe, 1978
C. quadrilaminata Okada & McIntyre, 1977

Genus: *PLEUROCHRYSIS* Pringsheim, 1955
P. carterae (Braarud & Fagerland, 1946) Christensen, 1978 var. *carterae*
P. carterae var. *dentata* Johansen & Doucette, in Johansen *et al.*, 1988
P. placolithoides Fresnel & Billard, 1991
P. pseudoroscoffensis Gayral & Fresnel, 1983
P. roscoffensis (Dangeard, 1934) Fresnel & Billard, 1991
P. scherffelii Pringsheim, 1955

Family: **PONTOSPHAERACEAE** Lemmermann, 1908
Cell completely or mostly covered by discoliths. Discoliths are flattened plates perforated by roundish pores. Rim may be present as a low thick or high thin wall. Lopadoliths, if present, are situated equatorially and are vase-shaped. Each is a hollow structure with ornamentations on the wall and perforations on the proximal surface.

Genus: *PONTOSPHAERA* Lohmann, 1902 (15)
P. discopora Schiller, 1925
P. japonica (Takayama, 1967) Nishida, 1971
P. syracusana Lohmann, 1902
P. turgida Müller, in Müller *et al.*, 1974

Genus: *SCYPHOSPHAERA* Lohmann, 1902
S. apsteinii Lohmann, 1902 f. *apsteinii*
S. apsteinii f. *dilatata* Gaarder, 1970

Family: **RHABDOSPHAERACEAE** Ostenfeld, 1899 (16)
Cell covered by rhabdoliths. A rhabdolith consists of a bicyclic sub-horizontal rim and a central area with one to three concentric cycles of elements. The central area may bear a distal protrusion or process.

Genus: *ACANTHOICA* Lohmann, 1903, emend. Schiller, 1913 and Kleijne, 1992 (17)
A. acanthifera Lohmann, 1912 ex Lohmann, 1913a
A. biscayensis Kleijne, 1992
A. jancheni Schiller, 1925
A. maxima Heimdal, in Heimdal and Gaarder, 1981
A. quattrospina Lohmann, 1903 (17)

Genus: *ALGIROSPHAERA* Schlauder, 1945, emend. Norris, 1984
A. oryza Schlauder, 1945
A. quadricornu (Schiller, 1914) Norris, 1984
A. robusta (Lohmann, 1902) Norris, 1984

Genus: *ANACANTHOICA* Deflandre, 1952
A. acanthos (Schiller, 1925) Deflandre, 1952
A. cidaris (Schlauder, 1945) Kleijne, 1992

Genus: *CYRTOSPHAERA* Kleijne, 1992
C. aculeata (Kamptner, 1941) Kleijne, 1992
C. cucullata (Lecal-Schlauder, 1951) Kleijne, 1992
C. lecaliae Kleijne, 1992

Genus: *DISCOSPHAERA* Haeckel, 1894
D. tubifera (Murray & Blackman, 1898) Ostenfeld, 1900

Genus: *PALUSPHAERA* Lecal, 1965b, emend. Norris, 1984
P. vandeli Lecal, 1965b, emend. Norris, 1984

Genus: *RHABDOSPHAERA* Haeckel, 1894
R. clavigera Murray & Blackman, 1898 var. *clavigera*
R. clavigera var. *stylifera* (Lohmann, 1902) Kleijne & Jordan, 1990
R. xiphos (Deflandre & Fert, 1954) Norris, 1984

Family: **SYRACOSPHAERACEAE** Lemmermann, 1908 (**18**)
Cell completely or mostly covered by caneoliths. Irrespective of shape the two coccolith types have a distal arrangement of wall elements, each joined at their bases by lamellae (lamellae are absent in one genus). In trough-shaped coccoliths and rhombolith-like caneoliths these lamellae are arranged across the short axis of the coccolith, whereas in most caneoliths they are radial. Only in caneoliths do the lamellae converge in the central area to form an overlapping random arrangement of elements or an organized central structure. This central structure may or may not produce a distally protruding central spine. Cyrtoliths and deviating coccoliths may form a complete or partial layer respectively on top of the caneoliths. Pole spines or ring and link coccoliths may also be associated with caneoliths (see Chapter 4 for more detail).

Genus: *ALISPHAERA* Heimdal, 1973
A. capulata Heimdal, in Heimdal and Gaarder, 1981
A. ordinata (Kamptner, 1941) Heimdal, 1973
A. spatula Steinmetz, 1991
A. unicornis Okada & McIntyre, 1977 (**19**)

Genus: *ALVEOSPHAERA* Jordan & Young, 1990 (**20**)
A. bimurata (Okada & McIntyre, 1977) Jordan & Young, 1990

Genus: *CALCIOPAPPUS* Gaarder & Ramsfjell, 1954, emend. Manton & Oates, 1983
C. caudatus Gaarder & Ramsfjell, 1954
C. rigidus Heimdal, in Heimdal and Gaarder, 1981

Genus: *CORONOSPHAERA* Gaarder, in Gaarder and Heimdal, 1977
C. binodata (Kamptner, 1927) Gaarder, in Gaarder and Heimdal, 1977
C. maxima (Halldal & Markali, 1955) Gaarder, in Gaarder and Heimdal, 1977
C. mediterranea (Lohmann, 1902) Gaarder, in Gaarder and Heimdal, 1977

Genus: *MICHAELSARSIA* Gran, 1912, emend. Manton *et al.*, 1984
M. adriaticus (Schiller, 1914) Manton *et al.*, 1984
M. elegans Gran, 1912, emend. Manton *et al.*, 1984

Genus: *OPHIASTER* Gran, 1912, emend. Manton & Oates, 1983
O. formosus Gran, 1912, *sensu* Gaarder, 1967, emend. Manton & Oates, 1983 var. *formosus*
O. formosus var. *inversus* Manton and Oates, 1983
O. hydroideus (Lohmann, 1903) Lohmann, 1913b, emend. Manton & Oates, 1983

O. minimus Manton & Oates, 1983
O. reductus Manton & Oates, 1983

Genus: *SYRACOSPHAERA* Lohmann, 1902   **(21)**
S. ampliora Okada & McIntyre, 1977
S. anthos (Lohmann, 1912) Jordan & Young, 1990   **(21)**
S. borealis Okada & McIntyre, 1977
S. corolla Lecal, 1966   **(27)**
S. corrugis Okada & McIntyre, 1977   **(21)**
S. epigrosa Okada & McIntyre, 1977
S. exigua Okada & McIntyre, 1977
S. halldalii f. *dilatata* (Heimdal, in Heimdal and Gaarder, 1981) Jordan & Young, 1990
S. halldalii Gaarder, in Gaarder and Hasle, 1971 f. *halldalii*   **(21)**
S. histrica Kamptner, 1941
S. lamina Lecal-Schlauder, 1951
S. molischii Schiller, 1925   **(21)**
S. nana (Kamptner, 1941) Okada & McIntyre, 1977
S. nodosa Kamptner, 1941
S. orbiculus Okada & McIntyre, 1977
S. ossa (Lecal, 1966) Loeblich & Tappan, 1968
S. pirus Halldal & Markali, 1955
S. prolongata Gran, 1912, ex Lohmann, 1913b
S. pulchra Lohmann, 1902
S. rotula Okada & McIntyre, 1977

GENERA INCERTAE SEDIS   **(22)**

Genus: *CALCIARCUS* Manton, Sutherland & Oates, 1977   **(23)**
C. alaskensis Manton, Sutherland & Oates, 1977

Genus: *FLORISPHAERA* Okada & Honjo, 1973   **(24)**
F. profunda var. *elongata* (Okada & Honjo, 1973) Okada & McIntyre, 1977
F. profunda Okada & Honjo, 1973 var. *profunda* Okada & McIntyre, 1977

Genus: *JOMONLITHUS* Inouye & Chihara, 1983
J. littoralis Inouye & Chihara, 1983

Genus: *POLYCRATER* Manton & Oates, 1980   **(25)**
P. galapagensis Manton & Oates, 1980

Genus: *THOROSPHAERA* Ostenfeld, 1910   **(26)**
T. flabellata Halldal & Markali, 1955

Genus: *TURRILITHUS* Jordan, Knappertsbusch, Simpson & Chamberlain, 1991
T. latericioides Jordan, Knappertsbusch, Simpson & Chamberlain, 1991

Genus: *UMBELLOSPHAERA* Paasche, in Markali and Paasche, 1955   **(27)**
U. irregularis Paasche, in Markali and Paasche, 1955
U. tenuis (Kamptner, 1937) Paasche, in Markali and Paasche, 1955

Genus: *WIGWAMMA* Manton, Sutherland & Oates, 1977   **(28)**
W. annulifera Manton, Sutherland & Oates, 1977
W. antarctica Thomsen, in Thomsen *et al.*, 1988
W. arctica Manton, Sutherland & Oates, 1977

W. scenozonion Thomsen, 1980a
W. triradiata Thomsen, in Thomsen *et al.*, 1988

B) Holococcolithophores: **(29)**

Family: **CALYPTROSPHAERACEAE** Boudreaux & Hay, 1969
Cell covered by holococcoliths, each composed of microcrystals arranged in an organised manner.

i) Monomorphic species

Genus: *CALICASPHAERA* Kleijne, 1991
C. blokii Kleijne, 1991
C. concava Kleijne, 1991
C. diconstricta Kleijne, 1991

Genus: *CALYPTROSPHAERA* Lohmann, 1902   **(30)**
C. cialdii Borsetti & Cati, 1976
C. dentata Kleijne, 1991
C. galea Lecal-Schlauder, 1951
C. heimdalae Norris, 1985
C. oblonga Lohmann, 1902
C. sphaeroidea Schiller, 1913

Genus: *DAKTYLETHRA* Gartner, in Gartner and Bukry, 1969
D. pirus (Kamptner, 1937) Norris, 1985

Genus: *FLOSCULOSPHAERA* Jordan & Kleijne, in Kleijne *et al.*, 1991
F. calceolariopsis Jordan & Kleijne, in Kleijne *et al.*, 1991
F. sacculus Kleijne & Jordan, in Kleijne *et al.*, 1991

Genus: *GLISCOLITHUS* Norris, 1985
G. amitakarenae Norris, 1985

Genus: *HOMOZYGOSPHAERA* Deflandre, 1952
H. arethusae (Kamptner, 1941) Kleijne, 1991
H. spinosa (Kamptner, 1941) Deflandre, 1952
H. triarcha Halldal & Markali, 1955
H. vavilovii Borsetti & Cati, 1976
H. vercellii Borsetti & Cati, 1979

Genus: *PERIPHYLLOPHORA* Kamptner, 1937   **(31)**
P. mirabilis (Schiller, 1925) Kamptner, 1937

Genus: *SYRACOLITHUS* Deflandre, 1952
S. bicorium Kleijne, 1991
S. catilliferus (Kamptner, 1937) Deflandre, 1952
S. confusus Kleijne, 1991
S. dalmaticus (Kamptner, 1927) Loeblich & Tappan, 1963
S. ponticuliferus (Kamptner, 1941) Kleijne & Jordan, 1990
S. quadriperforatus (Kamptner, 1937) Gaarder, 1962
S. schilleri (Kamptner, 1927) Loeblich & Tappan, 1963   **(32)**

ii) Dimorphic species

Genus: *ANTHOSPHAERA* Kamptner, 1937, emend. Kleijne, 1991   **(33)**
A. fragaria Kamptner, 1937, emend. Kleijne, 1991
A. lafourcadii (Lecal, 1967) Kleijne, 1991
A. periperforata Kleijne, 1991

Genus: *CALYPTROLITHINA* Heimdal, 1982
C. divergens (Halldal & Markali, 1955) Heimdal, 1982
C. divergens f. *tuberosa* (Heimdal, in Heimdal and Gaarder, 1980) Heimdal, 1982
C. multipora (Gaarder, in Heimdal and Gaarder, 1980) Norris, 1985
C. wettsteinii (Kamptner, 1937) Kleijne, 1991

Genus: *CALYPTROLITHOPHORA* Heimdal, in Heimdal and Gaarder, 1980
C. gracillima (Kamptner, 1941) Heimdal, in Heimdal and Gaarder, 1980
C. hasleana (Gaarder, 1962) Heimdal, in Heimdal and Gaarder, 1980
C. papillifera (Halldal, 1953) Heimdal, in Heimdal and Gaarder, 1980

Genus: *CORISPHAERA* Kamptner, 1937   **(31)**
C. gracilis Kamptner, 1937
C. strigilis Gaarder, 1962
C. tyrrheniensis Kleijne, 1991

Genus: *HELLADOSPHAERA* Kamptner, 1937   **(31)**
H. cornifera (Schiller, 1913) Kamptner, 1937
H. pienaarii Norris, 1985

Genus: *PORICALYPTRA* Kleijne, 1991
P. aurisinae (Kamptner, 1941) Kleijne, 1991
P. gaarderii (Borsetti & Cati, 1976) Kleijne, 1991
P. isselii (Borsetti & Cati, 1976) Kleijne, 1991
P. magnaghii (Borsetti & Cati, 1976) Kleijne, 1991

Genus: *PORITECTOLITHUS* Kleijne, 1991
P. maximus Kleijne, 1991
P. poritectus (Heimdal, in Heimdal and Gaarder, 1980) Kleijne, 1991
P. tyronus Kleijne, 1991

Genus: *SPHAEROCALYPTRA* Deflandre, 1952
S. adenensis Kleijne, 1991
S. quadridentata (Schiller, 1913) Deflandre, 1952

Genus: *ZYGOSPHAERA* Kamptner, 1937, emend. Heimdal, 1982   **(31)**
Z. amoena Kamptner, 1937   **(34)**
Z. bannockii (Borsetti & Cati, 1976) Heimdal, 1982
Z. hellenica Kamptner, 1937
Z. marsilii (Borsetti & Cati, 1976) Heimdal, 1982

iii) Partially calcified species   **(35)**

Genus: *BALANIGER* Thomsen & Oates, 1978
B. balticus Thomsen & Oates, 1978

Genus: *QUATERNARIELLA* Thomsen, 1980b
Q. obscura Thomsen, 1980b

Genus: **TRIGONASPIS** Thomsen, 1980c
T. diskoensis Thomsen, 1980c
T. melvillea Thomsen, in Thomsen *et al.*, 1988
T. minutissima Thomsen, 1980c

Genus: **TURRISPHAERA** Manton, Sutherland & Oates, 1976
T. arctica Manton, Sutherland & Oates, 1976
T. borealis Manton, Sutherland & Oates, 1976
T. polybotrys Thomsen, 1980d

Taxonomic Notes

(1) *Braarudosphaera bigelowii*
B. bigelowii is the sole living representative of the genus, although several species have been reported from the fossil record. The test of *B. bigelowii* is characteristically dodecahedral in shape with each pentagonal coccolith (pentalith) fitting perfectly with each of its five neighbors. The lack of test perforations may suggest that the coccosphere is really a resting stage or cyst (Tappan, 1980).

(2) Calciosoleniaceae
In 1937 Kamptner erected the family Calciosoleniaceae to accommodate a single genus, *Calciosolenia*, which bears rhombolith-type coccoliths. Deflandre (1952) later added several other morphologically similar genera, although only *Anoplosolenia* is recognized in recent taxonomic works. Since then most authors have regarded the calciosolenids as members of the Syracosphaeraceae, whereas Manton and Oates (1985) and Tappan (1980) have retained them in their own family. Tappan (1980), however, placed them in the Order Eiffelithales instead of the Syracosphaerales. In the classification system presented here, the two calciosolenid genera (*Calciosolenia* and *Anoplosolenia*) are separated from the Syracosphaeraceae, although the rhombolith nature of their coccoliths is regarded as a variation from the caneolith-type construction. It must be noted that structures resembling rhomboliths and pole spines, the two key characters defining the calciosolenids, are found in some members of the Syracosphaeraceae as well. The rhombolith-like coccoliths are present in the flagellar field area of cells of *Calciopappus* and *Michaelsarsia*, whilst pole spines are also found in *Calciopappus*. Despite this, the Calciosoleniaceae has a longer stratigraphic record and appear different in the light microscope under crossed nicols (B. Prins, pers. comm.). For these reasons the family is retained here as a valid taxon.

(3) *Ceratolithus cristatus*
In 1965 Norris figured a coccosphere of this species surrounded by a large number of ring-shaped structures of unknown composition. Similar specimens have recently been observed over the Mid-Atlantic Ridge (R. Jordan, unpubl. obs).

(4) *Calcidiscus leptoporus* and *Coccolithus pelagicus*
Parke and Adams (1960) demonstrated from life cycle studies that motile cells of *Crystallolithus hyalinus* emerged from the non-motile cells of *Coccolithus pelagicus* in pure culture. Recently it has been shown that, in addition to *Crystallolithus hyalinus*, cells of *Crystallolithus braarudii* can also be released from cells of *Coccolithus pelagicus* clones (Rowson *et al.*, 1986). According to Kleijne (1991) *Crystallolithus rigidus* represents a stage in the life cycle of *Calcidiscus leptoporus*. The taxonomic position of the holococcolithophores may therefore be viewed with increasing doubt. The genera *Coccolithus* (Schwarz, 1894) and *Calcidiscus* (Kamptner, 1950) have priority over *Crystallolithus* (Gaarder and Markali, 1956) and therefore the genus *Crystallolithus* has been eliminated (Kleijne, 1991).

(5) *Coccolithus neohelis*
This species was first observed in living material from the western Atlantic Ocean (McIntyre and Bé, 1967a) and later isolated and cultured from the Mediterranean (Fresnel, 1986). The coccoliths are placoliths but appear to have a different morphology to those seen in other species of *Coccolithus*. The cruciform bridge in the central area of the coccoliths prompted Reinhardt (1972) to transfer it to the genus *Cruciplacolithus* Hay and Mohler. However, this is not satisfactory. The overall appearance of the coccoliths suggests to us an affinity with the genus *Umbilicosphaera*, although the presence of the bridge and the grill-like crystal growths in the central area of *C. neohelis* are not features presently exhibited by species of *Umbilicosphaera*.

(6) *Hayaster perplexus*
This species was originally described from the fossil record as a discoaster (Bramlette and Riedel, 1954), but more recently has been transferred to a coccolithophore genus (Bukry, 1973) and has been recorded several times from water samples from the subtropical deep photic zone (Okada and McIntyre, 1977; Reid, 1980; Hallegraeff, 1984; Jordan,

unpubl. obs). The coccoliths of this species are constructed of two shields, each bearing radial elements delimited by straight suture lines. These coccoliths have a similar morphology to placoliths and in particular to those seen in *Oolithotus* species, in which the distal shield is significantly larger than the proximal shield. For these reasons *H. perplexus* has been transferred from the *Genera Incertae Sedis* (Okada and McIntyre, 1977) to the family Coccolithaceae.

(7) Genus *Neosphaera*

The coccoliths are composed of radial, overlapping elements originating from a central area. The collar surrounding the central area may represent a rudimentary distal shield or may be a shield ornamentation as in *G. oceanica*. Okada and McIntyre (1977) placed the genus in the Rhabdosphaeraceae, however, it was not included in a later review of the family by Norris (1984). Earlier Deflandre (1952) had placed it in the Coccolithidae (= Coccolithaceae) along with *Coccolithus*, *Gephyrocapsa* and *Umbilicosphaera*. Tappan (1980) supported its inclusion in the Coccolithaceae, although she placed it in the subfamily Tergestielloideae with *Calcidiscus*. In a recent review of prymnesiophyte genera *Neosphaera* was placed in the *Genera Incertae Sedis* and was separated from *Cyclolithella* (generally considered a synonym of the former genus) on the basis of coccolith morphology (Chrétiennot-Dinet, 1990). The coccoliths of *Neosphaera* were referred to as being similar to tremaliths or discoliths, whilst those of *Cyclolithella* were of possible cricolith origin. From the micrographs the main difference between the two genera was in the size of the open central area, with *Neosphaera* possessing large openings and *Cyclolithella* smaller ones. It is our opinion that *Cyclolithella* is a synonym of *Neosphaera* and that the coccoliths, with central openings of variable size, are similar to placoliths and thus should be placed in the Coccolithaceae.

(8) *Umbilicosphaera sibogae* var. *foliosa*

Observations on this species in culture have shown that two coccolith types (other than abnormal forms) may be produced at the same time by one cell, one identifiable as *U. sibogae* var. *foliosa* and the other resembling *U. sibogae* var. *sibogae* (Inouye and Pienaar, 1984).

(9) *Helicosphaera carteri*

In modern *Helicosphaera* species there is a considerable degree of variation in both the size and orientation of the helicolith central openings. However, variability also occurs within the same species and sometimes on the same coccosphere and therefore some authors have proposed that *H. wallichii* (Theodoridis, 1984) and *H. hyalina* (Jordan and Young, 1990) should be regarded as varieties of *H. carteri*.

(10) Family *Noelaerhabdaceae*

In 1970 Jerkovic described this family to include his new genus, *Noelaerhabdus*, since then however, the genera *Emiliania*, *Gephyrocapsa* and *Reticulofenestra* have been added. Previous taxonomic lists have put these genera in a variety of families and more recent papers still include them in the Coccolithaceae (Okada and McIntyre, 1977), Gephyrocapsaceae (Black, 1971; Tappan, 1980) or Prinsaceae (Haq, 1978; Perch-Nielsen, 1985).

(11) Genus *Gephyrocapsa*

Fossil species of this genus are widely used as biostratigraphic markers in marine sediments (Samtleben, 1980). Moreover their living counterparts often dominate phytoplankton assemblages in tropical and temperate areas (Okada and Honjo, 1975; Friedinger and Winter, 1987; Kleijne *et al.*, 1989). However, the number of extant species within the genus is a point for debate. For example, Winter *et al.* (1978) reported finding living cells of *G. protohuxleyi* in water samples, when previously it had only been seen in Pleistocene sediments. Moreover, the appearance of its coccoliths resembles those of poorly preserved *G. ericsonii*. *Gephyrocapsa caribbeanica* is found in living communities (Okada and McIntyre, 1977) but has often been confused in the literature with *G. muellerae*, which was only described in 1978. On a note of caution, the occurrence in some samples of malformed, incomplete and partially dissolved specimens may lead to taxonomic difficulties (see Burns, 1977; Kleijne, 1990).

(12) *Gephyrocapsa muellerae*

Coccoliths of this species were initially described from Pleistocene sediments by Bréhéret in 1978 and since then the species has been used as an environmental guide indicating the presence of cold water (Weaver and Pujol, 1988). Recently its presence in filtered water samples from temperate waters was confirmed (Jordan and Kleijne, unpubl. obs) and the cells maintained in mixed culture conditions where they actively divided for some time (Jordan, unpubl. obs). It is possible that coccoliths of this species have been previously confused with *Gephyrocapsa caribbeanica* occurring in the water column or surface sediments (see Note 11).

(13) *Reticulofenestra sessilis*

Several workers (Lohmann, 1912; Lecal-Schlauder, 1949; Okada and McIntyre, 1977; Jordan, unpubl. obs) have reported the possible symbiotic relationship in tropical regions between *R. sessilis* (syn. *Pontosphaera sessilis*, *Crenalithus sessilis* and *Dictyococcites sessilis*) and diatoms of the genus *Thalassiosira*, in which the coccospheres are intimately associated with the frustule of the diatom. It is worth noting that cells of *R. sessilis* are rarely found outside of this arrangement. Combinations involving diatoms and other coccolithophore species are more likely to be the products of agglutination (Okada and McIntyre, 1977).

(14) Family Papposphaeraceae

On reviewing the genera *Papposphaera* and *Pappomonas*, Norris (1983) recommended that they both be transferred to the family Deflandriaceae as their coccolith structure resembled the Late Cretaceous fossil, *Deflandius intercisus*. Their inclusion in the Rhabdosphaeraceae (Parke and Green, 1976), Halopappaceae (Tappan, 1980) and the Genera *Incertae Sedis* (Okada and McIntyre, 1977) had previously proved unsatisfactory. However, the family Deflandriaceae (= Prediscosphaeraceae) is invalid and the rim and central process structure of its coccoliths differs quite radically from those of the Papposphaeraceae (Jordan and Young, 1990). Recently, associations between two different genera ('combination cells') have been found involving species of *Papposphaera* and *Turrisphaera*, and between *Pappomonas flabellifera* var. *borealis* and *Trigonaspis* cf. *diskoensis* (Thomsen *et al.*, 1991). The authors believed that these combination cells represented a stage in the life history of a single species and therefore recommended taxonomic changes. However, this was only based on indirect evidence and cells were not seen carrying both types of coccoliths as seen in the case of *Coccolithus/Crystallolithus* (see Note 4). For this reason we have retained all the taxa involved in these combinations until further observations.

(15) Genus Pontosphaera

The genus *Pontosphaera* was described by Lohmann (1902) to include five new species, with coccospheres consisting of disc-shaped coccoliths with a thickened rim and a flagellar pore. Since then some authors (Halldal and Markali, 1955; Reid, 1980) have assigned species to this genus, without taking into consideration the coccolith morphology. In this system only one of Lohmann's species is recognized as a true pontosphaerid, *P. syracusana*. The others have either been transferred to other genera or have never been satisfactorily identified. True pontosphaerids bear simple discoliths, each with a low, thick rim or a higher, but thinner, wall.

(16) Family Rhabdosphaeraceae

In a recent review of the Rhabdosphaeraceae, Norris (1984) removed those species which did not possess classical cyrtolith-type coccoliths and redefined the family to include members whose cells bore one or two types of cyrtolith (= rhabdolith); having (i) a rim of one or two cycles of radial elements around the periphery of the cyrtolith, and (ii) a proximal pore if a central process was present (except in *Algirosphaera* spp.). Norris (1984) referred to the peripheral element arrangement as a podorhabdid rim, however, this term should only be used when describing the coccoliths of the extinct family, Podorhabdaceae. Furthermore there is no

evidence to support the suggestion that the rim structures are synonomous or that the families should be merged. It is, therefore, recommended that the term rhabdosphaerid rim be used when describing the coccolith morphology of the Rhabdosphaeraceae (Jordan, 1991; Kleijne, 1992). As a consequence of the above definition the genera *Neosphaera* (see Note 7) and *Umbellosphaera* (see Note 27) are omitted from the family (cf. the system presented in Okada and McIntyre, 1977).

(17) Acanthoica quattrospina

The type species *A. coronata* is a synonym of *A. quattrospina*. The latter name is preferred because it has been widely used, and is known from SEM micrographs, whereas *A. coronata* has not. *A. quattrospina* is now the type species (A. Kleijne, 1992).

(18) Syracosphaeraceae

The family contains all those taxa (except the Calciosolenids, see Note 2) which bear caneolith or caneolith-like coccoliths. However, many differences occur in the morphology of their other coccoliths; apical spines (*Calciopappus*), ring and link coccoliths (*Michaelsarsia*), link coccoliths (*Ophiaster*) and cyrtoliths (*Syracosphaera* – in part). Apart from the latter genus the other genera are monothecate, but some may appear monomorphic (*Alisphaera*, *Alveosphaera* and some species of *Syracosphaera*), dimorphic (*Coronosphaera* and some species of *Syracosphaera*) or polymorphic (*Calciopappus*, *Michaelsarsia*, *Ophiaster* and some species of *Syracosphaera*). Some authors have separated these genera into several families, the Halopappaceae (containing *Calciopappus*, *Michaelsarsia* and *Ophiaster*), Deutschlandiaceae (including only *Syracosphaera anthos*) and Syracosphaeraceae (the rest). However, *Halopappus adriaticus* was transferred to *Michaelsarsia* (Manton *et al.*, 1984) and the type species *H. vahselii* (Lohmann, 1912) has not been recorded since its original discovery. Furthermore, *Deutschlandia anthos*, the type species of the genus *Deutschlandia* has been transferred to *Syracosphaera* (Jordan and Young, 1990). If these transfers are generally accepted then it looks rather doubtful that these two families will remain in use (see Note 21 for further discussion).

(19) Alisphaera unicornis

The specimen featured by Reid (1980, pl.4, figs 8–11) as *A. unicornis* possesses coccoliths with greater wall and spine dimensions than those originally described for *A. unicornis* by Okada and McIntyre (1977). In addition, Reid's specimen lacks the rectangular plates and thus the zig-zag fissure characteristic of the central area of *A. unicornis*. A further specimen, recently found in the North Atlantic, has confirmed that the coccolith structure differs quite markedly from that seen in *Alisphaera*. These specimens will be transferred to a new genus in the near future (R. Jordan and J. Chamberlain, unpubl. obs).

(20) *Alveosphaera bimurata*
Okada and McIntyre (1977, pl.7, fig.1) described a new species with scapholith-type coccoliths which they tentatively assigned to *Calciosolenia*. They noticed that the coccoliths differ from those of *C. murrayi* and that the coccosphere apparently lacks pole spines. Manton and Oates (1985) pointed out that the coccoliths of *C.? bimurata* are not parallelograms and that the species should be removed from *Calciosolenia* and transferred to a new genus within the family Syracosphaeraceae. Jordan and Young (1990) named this genus *Alveosphaera*.

(21) *Caneosphaera*, *Deutschlandia* and *Syracosphaera*
In 1977 the generic description of *Syracosphaera* was emended and some species transferred to two new genera, *Caneosphaera* and *Coronosphaera* (Gaarder and Heimdal, 1977). The genus *Caneosphaera* was erected to include those species (*C. molischii* and *C. halldalii*) which bear monomorphic coccospheres with complete caneoliths (Gaarder and Heimdal, 1977). Recently it has been pointed out that the coccosphere of *C. molischii* may bear deviating coccoliths around the flagellar field (= pseudo-dithecatism). In addition the coccoliths of both species were described as lacking an intermediate continuous or beaded mid-wall rim; however, the stomatal coccoliths of *C. halldalii* var. *dilatata* bear beaded mid-wall rims like those of *Syracosphaera exigua* and *S. histrica* (Heimdal and Gaarder, 1981). Thus the reliability of the generic characteristics of *Caneosphaera* seemed in doubt and the species of *Caneosphaera* have been transferred back to *Syracosphaera* (Jordan and Young, 1990). The caneoliths of both *Syracosphaera molischii* and *S. halldalii* have been observed with or without finger-like centripetal protrusions emerging from the top of the wall (Gaarder and Heimdal, 1977). At the same time, however, Okada and McIntyre (1977) described a new species, *S. protrudens*, with wall protrusions similar to the previously mentioned two species. *Syracosphaera protrudens* and another protrusion-bearing species, *S. elatensis* (Winter *et al.*, 1979), are junior synonyms of *S. halldalii* and *S. molischii*, respectively. Whether these forms bearing protrusions represent distinct varieties or separate species is still a matter for discussion, but in this system the morphotypes are retained in *S. molischii* and *S. halldalii*. Okada and McIntyre (1977) described another new species, *S. corrugis*, as a dithecate coccosphere. Its resemblance to *S. molischii* was, however, pointed out by Heimdal and Gaarder (1981). The former authors stated that both species exhibited dithecatism, but differed in the degree of caneolith rim corrugation, the shape of the caneolith central structure and the morphology of the exothecal cyrtoliths. Both species are actually monothecate but possess deviating coccoliths which partially cover the endothecal caneoliths.

Despite the difference in cyrtolith morphology *S. corrugis* was regarded as conspecific with *S. molischii* (Heimdal and Gaarder, 1981). In this classification system *S. corrugis* is retained as a valid taxon. The genus *Deutschlandia* was described to include the species, *D. anthos* (Lohmann, 1912). Later a new family, the Deutschlandiaceae, was erected to accommodate *Deutschlandia* and *Calciosolenia*, as both genera were supposedly unrelated to the members of other existing families at the time (Kamptner, 1928). This arrangement was initially accepted (Schiller, 1930), although the latter genus was transferred to its own family, the Calciosoleniaceae, a few years later (Kamptner, 1937). Kamptner (1937) thought that *Deutschlandia* bore more relationship to the syracosphaerids, but still kept it in the Deutschlandiaceae, whereas Deflandre (1952) relocated the genus in the Anthosphaerinae, a subfamily of the Syracosphaeridae. In more recent times the genus has been placed into either the Deutschlandiaceae (Reid, 1980; Hallegraeff, 1984) or the Syracosphaeraceae (Okada and McIntyre, 1977, as *S. variabilis*). Tappan (1980) included it in the Halopappaceae, stating Deutschlandiaceae was a synonym. The emended generic description of *Deutschlandia* (Heimdal and Gaarder, 1981) does not significantly differ from that of *Syracosphaera* (emend. Gaarder and Heimdal, 1977). *Deutschlandia* has incomplete caneoliths, and exothecal cyrtoliths each with a distally raised hollow cone in the center, whereas *Syracosphaera* species were described as having complete caneoliths, and cyrtoliths each with a central depression in the center. Since the emended description of *Syracosphaera* (Gaarder and Heimdal, 1977) several new species of the genus have been reported (see Okada and McIntyre, 1977), and the genus now holds a wider variety of cyrtolith types, and complete and incomplete caneoliths. *D. anthos* has been transferred to *Syracosphaera* by Jordan and Young (1990). The creation of several new genera from this 'mixed bag' will only cause more difficulties as the separating factors would be based on a number of presence or absence combinations. Therefore we propose to retain the species in one genus.

(22) *Genera Incertae Sedis*
Genera Incertae Sedis represents a collection of species that either cannot be fitted into the existing classification system or are still awaiting formal redescriptions.

(23) *Calciarcus*
There is little information at present as to the possible phylogenetic relationship between this genus and other calcified taxa. However, Thomsen *et al.* (1991) have recently discovered a 'combination cell' involving *C.* cf. *alaskensis* and *Wigwamma* cf. *annulifera* (see also Note 14).

(24) *Florisphaera*
The coccoliths of this genus, whilst simple, are rather

unusual. Each coccolith is a thick polygonal plate which appears devoid of surface ornamentation. Occasionally more elongate forms are seen which at first resemble a wall of the four-sided float coccolith of *Thorosphaera flabellata*. Like *F. profunda* var. *elongata*, the elongate plates of *T. flabellata* are pentagons with a single prominent point. Whether these two genera are related is not known at present.

(25) *Polycrater galapagensis*
This species was first described from aragonitic coccospheres in surface waters around the Galapagos Islands (Manton and Oates, 1980). This was, however, not the first report of aragonitic coccoliths (see Hart *et al.*, 1965; but also see Chapter 3). Furthermore, Wilbur and Watabe (1963) suggest that environmental conditions may determine the form of calcium carbonate which the coccolithophore cell eventually secretes.

(26) Genus *Thorosphaera*
Ostenfeld (1910) erected the genus *Thorosphaera* to accommodate a single species, *T. elegans*. This species was described from only one dimorphic coccosphere consisting of ordinary elliptical coccoliths and tubular to funnel-shaped coccoliths. A second, smaller but better preserved specimen was later found by Bernard (1939). From this specimen the 'funnel-shaped coccoliths' were described as six-sided baseless tubes and the 'elliptical coccoliths' as discoliths seen as oval, shallow dishes. Bernard (1939) thought the species was related to *Syracosphaera*. Deflandre (1942) transferred the species to *Scyphosphaera* on the basis of its dimorphism and equatorially-placed lopadolith-like coccoliths. In 1955 Halldal and Markali described a second species, *T. flabellata*, which was also dimorphic and appeared to fit the original generic description of Ostenfeld (1910). However, the inclusion of a second species into *Thorosphaera* might be viewed as incorrect as the type species, *T. elegans*, was at this time a member of *Scyphosphaera*. The type species has not been recorded in water samples since then and Reid (1980) suggested that it is, in fact, a synonym of *T. flabellata*, despite the former's priority over the latter. This dilemma will only be solved when further specimens of *T. elegans* are found and its true structure elucidated by scanning electron microscopy. Recently a specimen has been found in the North Atlantic which fits almost perfectly Ostenfeld's description and drawings of *T. elegans*. However, the dimensions are smaller by a factor of ten, and the four-sided tubes arise from dish-like bases (R. Jordan and J. Chamberlain, unpubl. obs).

(27) *Umbellosphaera* and *Syracosphaera corolla*
The genus *Umbellosphaera* as defined by Paasche (in Markali and Paasche, 1955) does not share any of the characteristics of the Syracosphaeraceae. In fact after its removal from the Rhabdosphaeraceae (Norris, 1984) its nearest relatives are most likely the placolith-bearing genera. The presence of micrococcoliths (seen in both species) can also be seen in *Hayaster perplexus* (Okada and McIntyre, 1977). The inclusion of *Syracosphaera corolla* into *Umbellosphaera* (Heimdal and Gaarder, 1981) provided the system with a few problems, namely that a caneolith-bearing species was removed from the family Syracosphaeraceae. In our opinion the presence of '*Umbellosphaera*'-like caneoliths in *S. corolla* does not warrant its removal from *Syracosphaera*.

(28) *Wigwamma*
The coccoliths seen in this genus are reminiscent of the cricolith or pappolith morphology, although their wall structure is slightly different (see comments made by Manton *et al.*, 1977, and Inouye and Chihara, 1983). See Note (23) for possible affinities with *Calciarcus*.

(29) Holococcolithophores
Holococcolithophore taxonomy has received a great deal of attention in recent times with notable contributions from Borsetti and Cati (1972, 1976, 1979), Okada and McIntyre (1977), Heimdal and Gaarder (1980), Norris (1985) and Kleijne (1991). The discovery that the life cycles of *Coccolithus pelagicus* and *Calcidiscus leptoporus* involved *Crystallolithus* spp. (see Note 4) has cast doubt on the status of holococcolithophore taxonomy, moreover, priority of generic names may cause some heterococcolithophore taxa to become synonomous with holococcolithophore ones.

(30) Genus *Calyptrosphaera*
This genus was described by Lohmann (1902) to include two new species, *C. oblonga* and *C. globosa*. However, as with many early authors a type species was not designated. Some workers recognize the latter species as the type (Deflandre, 1952; Heimdal and Gaarder, 1980), whilst Norris (1985) and Kleijne (1991) credit the former. This confusion may be due in part to the apparent rarity of *C. globosa*, which has only been observed in the scanning electron microscope on two occasions (Lecal and Bernheim, 1960, pl.20, fig.33 – unsatisfactory SEM.; Borsetti and Cati, 1976, pl.12, figs 6–7 as *C. aff. globosa*). The specimen of Borsetti and Cati (1976) has since been referred to as a synonym of *C. sphaeroidea* (Kleijne, 1991). Thus *C. globosa* is omitted from this list. Norris (1985) transferred *C. galea* to *Calyptrolithophora*; however, stomatal coccoliths have not been observed in this species and therefore we propose to retain it in *Calyptrosphaera*.

(31) Genera *Periphyllophora*, *Corisphaera*, *Helladosphaera* and *Zygosphaera*
The above taxa were mentioned by Kamptner (1936) as new genera before they were properly described by him in 1937

(Loeblich and Tappan, 1963) and as a consequence subsequent authors have used 1936 (Okada and McIntyre, 1977; Heimdal, 1982; Norris, 1985) as the date of initiation. We prefer to use 1937 (see also Deflandre, 1952) as it represents the description of the genera and their species.

(32) *Syracolithus schilleri*
Originally described as *Syracosphaera schilleri* (Kamptner, 1927), this species was subsequently placed into *Syracolithus*, a subgenus of *Syracosphaera* (Kamptner, 1941). *Syracolithus* was given full generic status on the grounds that it is monomorphic and possesses perforated discoliths (Deflandre, 1952) and *Syracosphaera (Syracolithus) schilleri* was named *Syracolithus schilleri* (Loeblich and Tappan, 1963). Okada and McIntyre (1977) later transferred it to *Homozygosphaera*, but it was returned to *Syracolithus* by Norris (1985), although incorrectly as a new combination.

(33) Genus *Anthosphaera*
Initially described by Kamptner (1937) to include one species, *A. fragaria*. The specimen was drawn (his pl.15, fig. 20) as a subcircular coccosphere exhibiting dimorphism with ordinary and stomatal coccoliths. The heterococcolithophores *Syracosphaera robusta* (Kamptner, 1941), *S. quadricornu* (Halldal and Markali, 1955) and *Algirosphaera oryza* (Gaarder and Hasle, 1971) were transferred to *Anthosphaera*. However, *A. fragaria* actually possesses holococcoliths, and Kleijne (1991) has shown that

Anthosphaera still has priority. In addition *A. fragaria* bears fragarioliths as stomatal coccoliths (Kleijne, 1991). The heterococcolith-bearing species were transferred from *Anthosphaera* to *Algirosphaera* by Norris (1984), in his review of the Rhabdosphaeraceae.

(34) *Zygosphaera amoena*
The genus *Zygosphaera* is characterized by having lamino-form ordinary and stomatal coccoliths (Heimdal, 1982). However, *Z. amoena* has zygoform stomatal and lamino-form ordinary coccoliths. Since the holococcolithophore genera are defined by their combination of coccolith kinds, this unique combination implies that a new genus has to be erected to contain this species (Kleijne, 1991).

(35) Partially calcified species
Inclusion of *Balaniger* and *Turrisphaera* into the Calyptrosphaeraceae was first proposed by Tappan (1980) on the basis of the crystalline nature of their coccoliths. Two further genera, *Trigonaspis* and *Quaternariella*, were later transferred by Norris (1985) in his review of the holococco-lithophores. It has been noted that these genera have a greater affinity with the prymnesiophyte genus *Chrysochromulina* than they do with other living and fossil coccolithophores (Manton *et al.*, 1976). However, recent observations suggest that *Trigonaspis* and *Turrisphaera* may be part of the life cycles of *Pappomonas* and *Papposphaera* respectively (see Note 14).

Appendix

The above classification system, although useful as a checklist, is largely dependent on species recognition. However, one of the biggest problems concerning species identification within any group is the lack of a comprehensive atlas of LM and SEM micrographs. Instead, the taxonomist is probably faced with a rather large collection of original descriptive papers which need to be carried to and fro between the microscope room and the office. But to compile an atlas like this would be extremely costly. The collection of SEM micrographs contained in chapter 7 is not complete but represents the majority of species covered in this chapter. The appendix of this chapter provides an EM reference list for all the living species covered in the book and is intended as a supplement to the atlas section. Although the problem is not altogether solved, the amount of identification papers is somewhat reduced.

SPECIES	REFERENCE
Braarudosphaera bigelowii	Borsetti & Cati, 1972; Nishida, 1979
Anoplosolenia brasiliensis	Nishida, 1979; Manton & Oates, 1985
Calciosolenia murrayi	Nishida, 1979; Manton & Oates, 1985
Ceratolithus cristatus var. *cristatus*	Borsetti & Cati, 1976
C. cristatus var. *telesmus*	Borsetti & Cati, 1976
Calcidiscus leptoporus f. *leptoporus*	Nishida, 1979; Kleijne, 1991
C. leptoporus f. *rigidus*	Heimdal & Gaarder, 1980
Coccolithus neohelis	McIntyre & Bé, 1967a; Fresnel, 1986
C. pelagicus f. *braarudii*	Gaarder, 1962
C. pelagicus f. *hyalinus*	Gaarder & Markali, 1956
C. pelagicus f. *pelagicus*	Nishida, 1979

Hayaster perplexus	Nishida, 1979; Reid, 1980
Neosphaera coccolithomorpha	Nishida, 1979
Oolithotus fragilis var. *cavum*	Okada & McIntyre, 1977
O. fragilis var. *fragilis*	Okada & McIntyre, 1977
Umbilicosphaera angustiforamen	Okada & McIntyre, 1977
U. calvata	Steinmetz, 1991
U. hulburtiana	Gaarder, 1970; Borsetti & Cati, 1972
U. maceria	Okada & McIntyre, 1977
U. scituloma	Steinmetz, 1991
U. sibogae var. *foliosa*	Okada & McIntyre, 1977; Inouye & Pienaar, 1984
U. sibogae var. *sibogae*	Borsetti & Cati, 1976; Okada & McIntyre, 1977
Helicosphaera carteri var. *carteri*	Borsetti & Cati, 1972
H. carteri var. *hyalina*	Gaarder, 1970; Borsetti & Cati, 1972
H. carteri var. *wallichii*	Okada & McIntyre, 1977; Nishida, 1979
H. pavimentum	Okada & McIntyre, 1977; Borsetti & Cati, 1979
Hymenomonas coronata	Mills, 1975
H. globosa	Gayral & Fresnel, 1976; 1979
H. lacuna	Pienaar, 1976; Gayral & Fresnel, 1979
H. roseola	Braarud, 1954; Stoermer & Sicko-Goad, 1977
Ochrosphaera neapolitana	Gayral & Fresnel-Morange, 1971; Fresnel in Chrétiennot-Dinet, 1990
Emiliania huxleyi var. *corona*	Okada & McIntyre, 1977; Nishida, 1979
E. huxleyi var. *huxleyi*	Okada & McIntyre, 1977; Nishida, 1979
Gephyrocapsa crassipons	Okada & McIntyre, 1977
G. ericsonii	Nishida, 1979
G. muellerae	Samtleben, 1980
G. oceanica	Okada & McIntyre, 1977; Nishida, 1979
G. ornata	Heimdal, 1973; Nishida, 1979
Reticulofenestra parvula var. *parvula*	Okada & McIntyre, 1977
R. parvula var. *tecticentrum*	Okada & McIntyre, 1977
R. punctata	Okada & McIntyre, 1977
R. sessilis	Reid, 1980; Hallegraeff, 1984
Pappomonas flabellifera var. *borealis*	Manton, Sutherland & McCully, 1976
P. flabellifera var. *flabellifera*	Manton & Oates, 1975
P. virgulosa	Manton & Sutherland, 1975
P. weddellensis	Thomsen *et al.*, 1988
Papposphaera lepida	Tangen, 1972
P. obpyramidalis	Thomsen *et al.*, 1988
P. sagittifera	Thomsen *et al.*, 1988
P. sarion	Thomsen, 1981
P. simplicissima	Thomsen *et al.*, 1988
P. thomsenii	Norris, 1983
Cricosphaera elongata	No SEM ever published
C. gayraliae	Beuffe, 1978
C. quadrilaminata	Okada & McIntyre, 1977
Pleurochrysis carterae var. *carterae*	Braarud et al., 1952; Outka & Williams, 1971
P. carterae var. *dentata*	Johansen *et al.*, 1988
P. pseudoroscoffensis	Gayral & Fresnel, 1983
P. roscoffensis	Gayral & Fresnel, 1976; Inouye & Chihara, 1979
P. scherffelii	Leadbeater, 1971
Pontosphaera discopora	Borsetti & Cati, 1979; Nishida, 1979
P. japonica	Reid, 1980; Hallegraeff, 1984
P. syracusana	Borsetti & Cati, 1976; Nishida, 1979
P. turgida	Borsetti & Cati, 1979
Scyphosphaera apsteinii f. *apsteinii*	Borsetti & Cati, 1972; Nishida, 1979
S. apsteinii f. *dilatata*	Gaarder, 1970
Acanthoica acanthifera	Kleijne, 1992
A. biscayensis	Kleijne, 1992
A. jancheni	Kleijne, 1992
A. maxima	Heimdal & Gaarder, 1981
A. quattrospina	Kleijne, 1992
Algirosphaera oryza	Reid, 1980
A. quadricornu	Borsetti & Cati, 1972
A. robusta	Halldal & Markali, 1954
Anacanthoica acanthos	Kleijne, 1992

A. cidaris	Kleijne, 1992
Cyrtosphaera aculeata	Kleijne, 1992
C. cucullata	Kleijne, 1992
C. lecaliae	Kleijne, 1992
Discosphaera tubifera	Borsetti & Cati, 1972
Palusphaera vandeli	Norris, 1984
Rhabdosphaera clavigera var. *clavigera*	Borsetti & Cati, 1972
R. clavigera var. *stylifera*	Borsetti & Cati, 1972
R. xiphos	Kleijne, 1992
Alisphaera capulata	Heimdal & Gaarder, 1981
A. ordinata	Heimdal, 1973
A. spatula	Steinmetz, 1991
A. unicornis	Okada & McIntyre, 1977; Borsetti & Cati, 1979
Alveosphaera bimurata	Borsetti & Cati, 1976; Okada & McIntyre, 1977
Calciopappus caudatus	Manton & Oates, 1983
C. rigidus	Heimdal & Gaarder, 1981
Coronosphaera binodata	Gaarder & Heimdal, 1977
C. maxima	Gaarder & Heimdal, 1977
C. mediterranea	Gaarder & Heimdal, 1977
Michaelsarsia adriaticus	Manton *et al.*, 1984
M. elegans	Manton *et al.*, 1984
Ophiaster formosus var. *formosus*	Manton & Oates, 1983
O. formosus var. *inversus*	Manton & Oates, 1983
O. hydroideus	Manton & Oates, 1983
O. minimus	Manton & Oates, 1983
O. reductus	Manton & Oates, 1983
Syracosphaera ampliora	Okada & McIntyre, 1977
S. anthos	Heimdal & Gaarder, 1981
S. borealis	Okada & McIntyre, 1977
S. corolla	Okada & McIntyre, 1977; Borsetti & Cati, 1979
S. corrugis	Okada & McIntyre, 1977; Winter *et al.*, 1979
S. epigrosa	Okada & McIntyre, 1977; Heimdal & Gaarder, 1981
S. exigua	Okada & McIntyre, 1977; Heimdal & Gaarder, 1981
S. halldalii f. *dilatata*	Heimdal & Gaarder, 1981
S. halldalii f. *halldalii*	Gaarder & Heimdal, 1977
S. histrica	Gaarder & Heimdal, 1977; Okada & McIntyre, 1977
S. lamina	Borsetti & Cati, 1976; Okada & McIntyre, 1977
S. molischii	Gaarder & Heimdal, 1977; Okada & McIntyre, 1977
S. nana	Okada & McIntyre, 1977
S. nodosa	Okada & McIntyre, 1977
S. orbiculus	Okada & McIntyre, 1977
S. ossa	Okada & McIntyre, 1977
S. pirus	Gaarder & Heimdal, 1977
S. prolongata	Okada & McIntyre, 1977
S. pulchra	Okada & McIntyre, 1977
S. rotula	Okada & McIntyre, 1977
Calciarcus alaskensis	Manton *et al.*, 1977
Florisphaera profunda var. *elongata*	Okada & Honjo, 1973
F. profunda var. *profunda*	Okada & Honjo, 1973
Jomonlithus littoralis	Inouye & Chihara, 1983
Polycrater galapagensis	Manton & Oates, 1980
Thorosphaera flabellata	Borsetti & Cati, 1976; Reid, 1980
Turrilithus latericioides	Jordan *et al.*, 1991
Umbellosphaera irregularis	Nishida, 1979; Reid, 1980
U. tenuis	Borsetti & Cati, 1972; Nishida, 1979
Wigwamma annulifera	Manton *et al.*, 1977
W. antarctica	Thomsen *et al.*, 1988
W. arctica	Manton *et al.*, 1977; Thomsen *et al.*, 1988
W. scenozonion	Thomsen, 1980a
W. triradiata	Thomsen *et al.*, 1988
Calicasphaera blokii	Kleijne, 1991

99

C. concava	Kleijne, 1991
C. diconstricta	Kleijne, 1991
Calyptrosphaera cialdii	Borsetti & Cati, 1976
C. dentata	Kleijne, 1991
C. galea	Norris, 1985
C. heimdalae	Norris, 1985
C. oblonga	Kleijne, 1991
C. sphaeroidea	Gaarder, 1962; Klaveness, 1973
Daktylethra pirus	Norris, 1985; Kleijne, 1991
Flosculosphaera calceolariopsis	Kleijne *et al.*, 1991
F. sacculus	Kleijne *et al.*, 1991
Gliscolithus amitakarenae	Norris, 1985
Homozygosphaera arethusae	Borsetti & Cati, 1972; Kleijne, 1991
H. spinosa	Norris, 1985; Kleijne, 1991
H. triarcha	Borsetti & Cati, 1972; Kleijne, 1991
H. vavilovii	Borsetti & Cati, 1976
H. vercellii	Borsetti & Cati, 1979
Periphyllophora mirabilis	Borsetti & Cati, 1972; Kleijne, 1991
Syracolithus bicorium	Kleijne, 1991
S. catilliferus	Borsetti & Cati, 1972; Kleijne, 1991
S. confusus	Kleijne, 1991
S. dalmaticus	Kleijne, 1991
S. ponticuliferus	Okada & McIntyre, 1977
S. quadriperforatus	Norris, 1985; Kleijne, 1991
S. schilleri	Borsetti & Cati, 1979
Anthosphaera fragaria	Gaarder, 1962; Kleijne, 1991
A. lafourcadii	Kleijne, 1991
A. periperforata	Kleijne, 1991
Calyptrolithina divergens f. *divergens*	Winter *et al.*, 1979; Kleijne, 1991
C. divergens f. *tuberosa*	Heimdal & Gaarder, 1980
C. multipora	Heimdal & Gaarder, 1980; Kleijne, 1991
C. wettsteinii	Kleijne, 1991
Calyptrolithophora gracillima	Borsetti & Cati, 1972
C. hasleana	Heimdal & Gaarder, 1980; Kleijne, 1991
C. papillifera	Heimdal & Gaarder, 1980; Kleijne, 1991
Corisphaera gracilis	Kleijne, 1991
C. strigilis	Heimdal & Gaarder, 1980; Kleijne, 1991
C. tyrrheniensis	Kleijne, 1991
Helladosphaera cornifera	Kleijne, 1991
H. pienaarii	Norris, 1985; Kleijne, 1991
Poricalyptra aurisinae	Borsetti & Cati, 1972; Reid, 1980
P. gaarderii	Borsetti & Cati, 1976
P. isselii	Borsetti & Cati, 1976; Kleijne, 1991
P. magnaghii	Borsetti & Cati, 1976; Kleijne, 1991
Poritectolithus maximus	Kleijne, 1991
P. poritectum	Kleijne, 1991
P. tyronus	Kleijne, 1991
Sphaerocalyptra adenensis	Kleijne, 1991
S. quadridentata	Borsetti & Cati, 1972; Winter *et al.*, 1979
Zygosphaera amoena	Kleijne, 1991
Z. bannockii	Borsetti & Cati, 1976; Kleijne, 1991
Z. hellenica	Kleijne, 1991
Z. marsilii	Borsetti & Cati, 1976; Kleijne, 1991
Balaniger balticus	Thomsen & Oates, 1978
Quaternariella obscura	Thomsen, 1980b
Trigonaspis diskoensis	Thomsen, 1980c
T. melvillea	Thomsen *et al.*, 1988
T. minutissima	Thomsen, 1980c
Turrisphaera arctica	Manton, Sutherland & Oates, 1976
T. borealis	Manton, Sutherland & Oates, 1976
T. polybotrys	Thomsen, 1980d

References

Barrois, C., 1876. Mémoire sur l'embryologie de quelques éponges de la Manche. *Annls. Sci. nat. (Zool.)*, Ser. 6, part 3, article **11**: 1–84.

Bernard, F., 1939. Coccolithophorides nouveaux ou peu connus observés à Monaco en 1938. *Arch. zool. exp. et gén., notes et revue*, **81**: 33–44.

Beuffe, H., 1978. Une Coccolithophoracée marine nouvelle: *Cricosphaera gayraliae* nov. sp. *Protistologica*, **14**: 451–8.

Biekart, J.W., 1989. The distribution of calcareous nannoplankton in late Quaternary sediments collected by the Snellius II expedition in some southeast Indonesian basins. *Proc. Kon. Ned. Akad. Wetensch.*, Ser. B, **92**(2): 77–141.

Black, M., 1968. Taxonomic problems in the study of coccoliths. *Palaeontology*, **11**: 793–813.

Black, M., 1971. The systematics of coccoliths in relation to the palaeontological record. In: *The Micropalaeontology of the Oceans*, ed. B.M. Funnell and W.R. Riedel, pp. 611–24. Cambridge University Press.

Borsetti, A.M. and Cati, F., 1972. Il nannoplankton calcareo vivente nel Tirreno centromeridionale. *Giorn. Geol.*, Ser. 2a, **38**: 395–452.

Borsetti, A.M. and Cati, F., 1976. Il nannoplankton calcareo vivente nel Tirreno centromeridionale, parte 2. *Giorn. Geol.*, Ser. 2a, **40**: 209–40.

Borsetti, A.M. and Cati, F., 1979. Il nannoplankton calcareo vivente nel Tirreno centromeridionale, parte 3. *Giorn. Geol.*, Ser. 2a, **43**: 157–74.

Boudreaux, J.E. and Hay, W.W., 1969. Calcareous nannoplankton and biostratigraphy of the late Pliocene-Pleistocene-Recent sediments in the submarex cores. *Revista Espanola de Micropaleontologia*, **1**: 249–92.

Braarud, T., 1954. Coccolith morphology and taxonomic position of *Hymenomonas roseola* Stein and *Syracosphaera carterae* Braarud and Fagerland. *Nytt Mag. Bot.*, **3**: 1–4.

Braarud, T., 1960. On the coccolithophorid genus *Cricosphaera* n. gen. *Nytt Mag. Bot.*, **8**: 211–12

Braarud, T. and Fagerland, E., 1946. A Coccolithophoride in laboratory culture. *Syracosphaera carterae* n. sp. *Avhand. Norsk. Vid.-Akad. Mat.-Nat. Kl.*, no.2: 1–10.

Braarud, T., Gaarder, K.R., Markali, J. and Nordli, E., 1952. Coccolithophorids studied in the electron microscope. Observations on *Coccolithus huxleyi* and *Syracosphaera carterae*. *Nytt Mag. Bot.*, **1**: 129–34.

Bramlette, M.N. and Riedel, W.R., 1954. Stratigraphic value of discoasters and some other microfossils related to Recent coccolithophores. *Jour. Pal.*, **28**: 385–403.

Bréhéret, J., 1978. Formes nouvelles quaternaires et actuelles de la famille des Gephyrocapsaceae (Coccolithophorides). *C. r. hebd. Seanc. Acad. Sci., Paris*, 287 (Sér. D): 447–9.

Bukry, D., 1973. Phytoplankton stratigraphy, Deep Sea Drilling Project Leg 20, western Pacific Ocean. In: *Init. Repts. Deep Sea Drilling Project*, ed. B.C. Heezen, I.D. MacGregor, *et al.*, Washington: U.S. Govt. Printing Office. **20**: 307–17.

Burns, D.A., 1977. Phenotypes and dissolution morphotypes of the genus *Gephyrocapsa* Kamptner and *Emiliania huxleyi* (Lohmann). *N.Z. J. Geol. Geophys.*, **20**: 143–55.

Carter, H.J. (1871) On '*Melobesia unicellularis*', better known as the coccolith. *Ann. Mag. nat. Hist.*, Ser. 4, **7**: 184–9.

Chrétiennot-Dinet, M.-J., 1990. Chlorarachniophycées, Chlorophycées, Chrysophycées, Cryptophycées, Euglenophycées, Eustigmatophycées, Prasinophycées, Prymnesiophycées, Rhodophycées, Tribophycées. *Atlas du Phytoplancton Marin*, vol. 3, edns du CNRS, Paris.

Christensen, T., 1962. Alger. In *Botanik II. Systematisk Botanik*, ed. T.W. Bocher, M. Lange and T. Sorensen, Munksgaard, Copenhagen.

Christensen, T., 1978. Annotations to a textbook of Phycology. *Bot. Tidsskrift*, **73**: 65–70.

Cohen, C.L.D. and Reinhardt, P., 1968. Coccolithophorids from the Pleistocene Caribbean deep-sea core CP-28. *Neues Jahrb. Geol. Pal., Abh.*, **131**: 289–304.

Dangeard, P., 1934. Sur l'épiphytisme d'une Coccolithinée rencontrée à Roscoff. *P. V. Soc. linn. Bordeaux*, 45–7.

Deflandre, G., 1942. Coccolithophoridées fossiles d'Oranie. Genres *Scyphosphaera* Lohmann et *Thorosphaera* Ostenfeld. *Soc. Hist. Nat. Toulouse, Bull.*, **77**: 125–37.

Deflandre, G., 1947. *Braarudosphaera* nov. gen. type d'une famille nouvelle de Coccolithophoridés actuels à éléments composites. *C. r. hebd. Seanc. Acad. Sci., Paris*, **225**: 439–41.

Deflandre, G., 1952. Classe des Coccolithophoridés (Coccolithophoridae Lohmann, 1902). In: *Traité de Zoologie*, ed. Grassé, P.-P., Paris: Masson et Cie., **1**: 439–70.

Deflandre, G. and Fert, C., 1954. Observations sur les coccolithophoridés actuels et fossiles en microscopie ordinaire et électronique. *Ann. Pal.*, **40**: 115–76.

Droop, M.R., 1955. Some new supra-littoral protista. *J. mar. biol. Ass. U.K.*, **34**: 233–45.

Ehrenberg, C.G., 1836. Bemerkungen über feste mikroskopische, anorganische Formen in den erdigen und derben Mineralien. *Ber. Verhandlung. Konigl. Preuss. Akad. Wissenschaft., Berlin*, pp. 84–5.

Fresnel, J., 1986. Nouvelles observations sur une Coccolithacée rare: *Cruciplacolithus neohelis* (McIntyre et Bé) Reinhardt (Prymnesiophyceae). *Protistologica*, **22**: 193–204.

Fresnel, J. and Billard, C., 1991. *Pleurochrysis placolithoides* sp. nov. (Prymnesiophyceae), a new marine coccolithophorid with remarks on the status of cricolith-bearing species. *Br. phycol. J.*, **26**: 67–80.

Friedinger, P.J.J. and Winter, A., 1987. Distribution of modern coccolithophore assemblages in the southwest Indian Ocean off southern Africa. *J. micropalaeontol.*, **6**: 49–56.

Fütterer, D., 1976. Kalkige Dinoflagellaten ('Calciodinelloideae') und die systematische Stellung der Thoracosphaeroideae. *N. Jb. Geol. Paläontol. Abh.*, **151**: 119–41.

Fütterer, D., 1977. Distribution of calcareous dinoflagellates in cenozoic sediments of site 366. Eastern north Atlantic. In *Initial Reports of the Deep Sea Drilling Project*, ed. Y. Lancelot, E. Siebold *et al.*, **41**: 709–37.

Gaarder, K.R., 1962. Electron microscope studies on holococcolithophorids. *Nytt Mag. Bot.*, **10**: 35–51.

Gaarder, K.R., 1967. Observations on the genus *Ophiaster* Gran (Coccolithineae). *Sarsia*, **29**: 183–92.

Gaarder, K.R., 1970. Three new taxa of Coccolithineae. *Nytt Mag. Bot.*, **17**: 113–26.

Gaarder, K.R. and Hasle, G.R., 1971. Coccolithophorids of the Gulf of Mexico. *Bull. Mar. Sci.*, **21**: 519–44.

Gaarder, K.R. and Heimdal, B.R., 1977. A revision of the genus *Syracosphaera* Lohmann (Coccolithineae). *'Meteor' Forsch-Ergebnisse*, Ser.D, **24**: 54–71.

Gaarder, K.R. and Markali, J., 1956. On the coccolithophorid *Crystallolithus hyalinus* n. gen., n. sp. *Nytt Mag. Bot.*, **5**: 1–5.

Gaarder, K.R. and Ramsfjell, E., 1954. A new coccolithophorid from northern waters. *Calciopappus caudatus* n. gen., n. sp. *Nytt Mag. Bot.*, **2**: 155–6.

Gartner, S. Jr. and Bukry, D., 1969. Tertiary Holococcoliths. *J. Paleont.*, **43**: 1213–21.

Gayral, P. and Fresnel, J., 1976. Nouvelles observations sur deux Coccolithophoracées marines: *Cricosphaera roscoffensis* (P. Dangeard) comb. nov. et *Hymenomonas globosa* (F. Magne) comb. nov. *Phycologia*, **15**: 339–55.

Gayral, P. and Fresnel, J., 1979. Révision du genre *Hymenomonas* Stein. A propos de l'étude comparative de deux Coccolithacées: *Hymenomonas globosa* (Magne) Gayral et Fresnel et *Hymenomonas lacuna* Pienaar. *Rev. Algol.*, N.S., XIV, **2**: 117-125.

Gayral, P. and Fresnel, J., 1983. Description, sexualité et cycle de développement d'une nouvelle Coccolithophoracée (Prymnesiophyceae): *Pleurochrysis pseudoroscoffensis* sp. nov. *Protistologica*, **19**: 245–61.

Gayral, P. and Fresnel-Morange, J., 1971. Résultats préliminaires sur la structure et la biologie de la Coccolithacée *Ochrosphaera neapolitana* Schussnig. *C.r. hebd. Seanc. Acad. Sci., Paris*, **273**: 1683–6.

Gran, H.H., 1912. Pelagic plant life. In *The Depths of the Ocean*, eds. J. Murray and J. Hjort, pp. 307–386. MacMillan: London.

Gran, H.H. and Braarud, T., 1935. A qualitative study of the phytoplankton in the Bay of Fundy and the Gulf of Maine (including observations on hydrography, chemistry and turbidity). *J. Biol. Bd. Can.*, **1**: 279–467.

Green, J.C., 1976. Notes on the flagellar apparatus and taxonomy of *Pavlova mesolychnon* van der Veer, and on the status of *Pavlova* Butcher and related genera within the Haptophyceae. *J. mar. biol. Ass. U.K.*, **56**: 595–602.

Green, J.C. and Pienaar, R.N., 1977. The taxonomy of the Order Isochrysidales (Prymnesiophyceae) with special reference to the genera *Isochrysis* Parke, *Dicrateria* Parke and *Imantonia* Reynolds. *J. mar. biol. Ass. U.K.*, **57**: 7–17.

Gümbel, C.W., 1870. Über Nulliporenkalk und Coccolithen. *Verhandl. Geol. R. A. Wien*, p.201.

Haeckel, E., 1866. *Generelle Morphologie der Organismen*. Reimer, Berlin.

Haeckel, E., 1894. *Systematische Phylogenie der Protisten und Pflanzen*. Reimer: Berlin.

Halldal, P., 1953. Phytoplankton investigations from Weathership M in the Norwegian Sea, 1948–49. *Hvalråd Skr.*, **38**: 1–91.

Halldal, P. and Markali, J., 1955. Electron microscope studies on coccolithophorids from the Norwegian Sea, the Gulf Stream and the Mediterranean. *Norske Vidensk.-Akad. Oslo, Avh., Mat.-Nat., Kl.*, **1**: 1–30.

Hallegraeff, G.M., 1984. Coccolithophorids (calcareous nanoplankton) from Australian waters. *Bot. mar.*, **27**: 229–47.

Haq, B.U., 1978. Calcareous nannoplankton. In *Introduction to Marine Micropaleontology*, ed. B.U. Haq and A. Boersma, pp. 79–107. Elsevier, North Holland.

Hart, G.F., Pienaar, R.N. and Caveney, R., 1965. An aragonite coccolith from South Africa. *S. Afr. J. Sci.*, **61**: 425–6.

Harting, P., 1872. Recherches de Morphologie synthétique. *Acad. des sciences Neerlandaises*, **14**.

Hay, W.W., Mohler, H.P., Roth, P.H., Schmidt, R.R. and Boudreaux, J.E., 1967. Calcareous nannoplankton zonation of the Cenozoic of the Gulf Coast and Caribbean-Antillean area, and transoceanic correlation. *Gulf Coast Assoc. Geol. Socs., Trans.*, **17**: 428–80.

Hay, W.W., Mohler, H.P. and Wade, M.E., 1966. Calcareous nannofossils from Nal'chik (northwest Caucasus). *Eclog. Geol. Helv.*, **59**: 379–99.

Heimdal, B.R., 1973. Two new taxa of Recent coccolithophorids. *'Meteor' Forsch-Ergebnisse*, Ser.D, **13**: 70–75.

Heimdal, B.R., 1982. Validation of the names of some species of *Zygosphaera* Kamptner. *I.N.A. Newsletter*, **4**: 52–56.

Heimdal, B.R. and Gaarder, K.R., 1980. Coccolithophorids from the northern part of the eastern central Atlantic. I. Holococcolithophorids. *'Meteor' Forsch-Ergebnisse*, Ser.D, **32**: 1–14.

Heimdal, B.R. and Gaarder, K.R., 1981. Coccolithophorids from the northern part of the eastern central Atlantic. II. Heterococcolithophorids. *'Meteor' Forsch-Ergebnisse*, Ser.D, **33**: 37–69.

Hibberd, D.J., 1976. The ultrastructure and taxonomy of the Chrysophyceae and Prymnesiophyceae (Haptophyceae): a survey with some new observations on the ultrastructure of the Chrysophyceae. *Bot. J. Linn. Soc.*, **72**: 55–80.

Hibberd, D.J., 1980. Prymnesiophytes (= Haptophytes). In *Phytoflagellates*, ed. E.R. Cox, pp. 273–318, Elsevier, North Holland.

Hussain, A. and Boney, A.D., 1971. Plant growth substances associated with the motile and non-motile phases of two *Cricosphaera* species (Order Prymnesiales, Class Haptophyceae). *Bot. mar.*, **14**: 17–21.

Huxley, T.H., 1868. On some organisms living at great depths in the North Atlantic Ocean. *Quart. J. Microsc. Sci.*, **8**: 203–12.

Inouye, I. and Chihara, M., 1979. Life history and taxonomy of *Cricosphaera roscoffensis* var. *haptonemofera* var. nov. (Class Prymnesiophyceae) from the Pacific. *Bot. Mag. Tokyo*, **93**: 195–208.

Inouye, I. and Chihara, M., 1983. Ultrastructure and taxonomy of *Jomonlithus littoralis* gen. et sp. nov. (Class Prymnesiophyceae), a coccolithophorid from the northwest Pacific. *Bot. Mag. Tokyo*, **96**: 365–76.

Inouye, I. and Pienaar, R.N., 1983. Observations on the life cycle and microanatomy of *Thoracosphaera heimii* (Dinophyceae) with special reference to its systematic position. *S. Afr. J. Bot.*, **2**: 63–75.

Inouye, I. and Pienaar, R.N., 1984. New observations on the coccolithophorid *Umbilicosphaera sibogae* var. *foliosa* (Prymnesiophyceae) with reference to cell covering, cell structure and flagellar apparatus. *Br. phycol. J.*, **19**: 357–69.

Jafar, S.A. and Martini, E., 1975. On the validity of the calcareous nannoplankton genus *Helicosphaera*. *Senckenbergiana Lethaea*, **56**: 381–97.

Jerkovic, L., 1970. *Noelaerhabdus* nov. gen. type d'un nouvelle famille de Coccolithophoridés fossiles: Noelaerhabdaceae du Miocene superieur de Yougoslavie. *C. r. hebd. Seanc. Acad. Sci.*, **270**D: 468–70.

Johansen, J.R., Doucette, G.J., Barclay, W.R. and Bull, J.D., 1988. The morphology and ecology of *Pleurochrysis carterae* var. *dentata* var. nov. (Prymnesiophyceae), a new coccolithophorid from an inland saline pond in New Mexico, USA. *Phycologia*, **27**: 78–88.

Jordan, R.W., Knappertsbusch, M., Simpson, W.R. and Chamberlain, A.H.L., 1991. *Turrilithus latericioides* gen. et sp. nov. a new coccolithophorid from the deep photic zone. *Br. phycol. J.*, **26**: 175–183.

Jordan, R.W. and Young, J.R., 1990. Proposed changes to the classification system of living coccolithophorids. *I.N.A. Newsletter*, **12** (1): 15–8.

Jordan, R.W., 1991. Problems in the taxonomy and terminology of living coccolithophorids. *INA Newsletter*, **13**(2): 52–3

Kamptner, E., 1927. Beitrag zur Kenntnis adriatischer Coccolithophoriden. *Arch. Protistenk.*, **58**: 173–84.

Kamptner, E., 1928. Über das system und die Phylogenie der Kalkflagellaten. *Arch. Protistenk.*, **64**: 19–43.

Kamptner, E., 1936. Über die Coccolithineen der Südwestküste von Istrien. *Anz. Akad. Wiss. Wien, Math.-Naturw. Kl.*, **73**: 243–7.

Kamptner, E., 1937. Neue und bemerkenswerte Coccolithineen aus dem Mittelmeer. *Arch. Protistenk.*, **89**: 279–316.

Kamptner, E., 1941. Die Coccolithineen der Südwestküste von Istrien. *Naturh. Mus. Wien, Ann. Anz.*, **51**: 54–149.

Kamptner, E., 1943. Zur Revision der Coccolithineen-Spezies *Pontosphaera huxleyi* Lohmann. *Akad. Wiss. Wien, Anz.*, **80**: 43–9.

Kamptner, E., 1950. Über den submikroskopischen Aufbau der Coccolithen. *Österr. Akad. Wiss., Anz., Math.- Nat. Kl.*, **87**: 152–8.

Kamptner, E., 1954. Untersuchungen über den Feinbau der Coccolithen. *Arch. Protistenk.*, **100**: 1–90.

Kamptner, E., 1963. Coccolithineen-Skelettreste aus Tiefseeablagerungen des Pazifischen Ozeans. *Annln. Naturh. Mus. Wien*, **66**: 139–204.

Kleijne, A., 1990. Distribution and malformation of extant calcareous nannoplankton in the Indonesian Seas. *Marine Micropaleontology*, **16**: 293–316.

Kleijne, A., 1991. Holococcolithophorids from the Indian Ocean, Red Sea, Mediterranean Sea and North Atlantic Ocean. *Marine Micropaleontology*, **17**: 1–76.

Kleijne, A., 1992. Extant Rhabdosphaeraceae (coccolithophorids, class Prymnesiophyceae) from the Indian Ocean, Red Sea Mediterranean Sea and North Atlantic Ocean. *Scripta Geologica*, **100**: 1–63.

Kleijne, A. and Jordan, R.W., 1990. Proposed changes to the classification system of living coccolithophorids. II. *I.N.A. Newsletter*, **12**(2): 13.

Kleijne, A., Jordan, R.W. and Chamberlain, A.H.L., 1991. *Flosculosphaera calceolariopsis* gen. et sp. nov., and *F. sacculus* sp. nov., new coccolithophorids (Prymnesiophyceae) from the N.E. Atlantic. *Br. phycol. J.*, **26**: 185–94.

Kleijne, A., Kroon, D. and Zevenboom, W., 1989. Phytoplankton and foraminiferal frequencies in northern Indian Ocean and Red Sea surface waters. *Proc. Snellius II Symp., Neth. J. Sea Res.*, **24**: 531–9.

Leadbeater, B.S.C., 1971. Observations on the life history of the haptophycean alga *Pleurochrysis scherffelii* with special reference to the microanatomy of the different types of motile cells. *Ann. Bot.*, **35**: 429–39.

Leadbeater, B.S.C. and Morton, C., 1973. Ultrastructural observations on the external morphology of some members of the Haptophyceae from the coast of Jugoslavia. *Nova Hedwigia*, **24**: 207–33.

Lecal, J., 1965a. *Navisolenia aprili* nouveau genre de Calciosolenidés (Coccolithophoridés). *Bull. Soc. His. de Toulouse*, **100**: 422–6.

Lecal, J., 1965b. A propos des modalités d'élaboration des formations épineuses des Coccolithophoridés. *Protistologica*, **1**: 63–70.

Lecal, J., 1966. Coccolithophoridés littoraux de Banyuls. *Vie et milieu*, Ser.B, Océanographie, **16**: 251–70.

Lecal, J., 1967. Le nannoplankton des côtes d'Israël. *Hydrobiologia*, **29**: 305–87.

Lecal, J. and Bernheim, A., 1960. Microstructure du squelette de quelques Coccolithophoridés. *Bull. Soc. Hist. Nat. Afr. Nord*, **51**: 273–97.

Lecal-Schlauder, J., 1949. Sur un *Coccolithus* n. sp., épiphyte d'une bacillariale, *Coscinodiscus* n. sp. *Bull. Soc. Hist. Nat. Afr. Nord*, **39**: 15–21.

Lecal-Schlauder, J., 1950. Notes préliminaires sur les Coccolithophoridés d'Afrique du Nord. *Bull. Soc. Hist. Nat. Afr. Nord*, **40**: 160–7.

Lecal-Schlauder, J., 1951. Recherches morphologiques et

biologiques sur les coccolithophoridés nord-africains. *Monaco, Inst. Océanogr., Ann.*, Ser.2, **26**: 255–362.

Lemmermann, E., 1908. Flagellatae, Chlorophyceae, Coccosphaerales und Silicoflagellatae. *Nordisches Plankton*, ed. K. Brandt and C. Apstein, **21**: 1–40.

Loeblich, A.R. Jr. and Tappan, H., 1963. Type fixation and validation of certain calcareous nannoplankton genera. *Proc. Biol. Soc. Wash.*, **76**: 191–6.

Loeblich, A.R. Jr. and Tappan, H., 1968. Annotated index and bibliography of the calcareous nannoplankton. II. *J. Paleont.*, **42**: 584–98.

Loeblich, A.R. Jr. and Tappan, H., 1978. The coccolithophorid genus *Calcidiscus* Kamptner and its synonyms. *J. Paleont.*, **52**: 1390–2.

Lohmann, H., 1902. Die Coccolithophoridae, eine Monographie der Coccolithen bildenden flagellaten, zugleich ein Betrag zur Kenntnis des Mittelmeerauftriebs. *Arch. Protistenk.*, **1**: 89–165.

Lohmann, H., 1903. Neue Untersuchungen über den Reichthum des Meeres an Plankton und über die Brauchbarkeit der verschiedenen Fangmethoden. Zugleich auch ein Beitrag zur Kenntnis des Mittelmeerauftriebs. *Wiss. Meeresuntersuch. Abth. Kiel*, n. ser., **7**: 1–87.

Lohmann, H., 1912. Untersuchungen über das Pflanzen- und Tierleben der Hochsee, zugleich ein Bericht über die biologischen Arbeiten auf der Fahrt der 'Deutschland' von Bremerhaven nach Buenos-Aires in der Zeit vom 7 Mai bis 7 September 1911. *Veröffentl. Instit. Meereskunde. Univ. Berlin, Neue Folge., A. Geogr.-naturwiss.*, Ser.1: 1–92.

Lohmann, H., 1913a. Beiträge zur Charakterisierung des Tier- und Pflanzenlebens in den von der 'Deutschland' während ihrer Fahrt nach Buenos Aires Durchfahrenen Gebieten des Atlantischen Ozeans. II. Teil. *Int. Rev. Hydrobiol. u. Hydrogr.*, **5**: 343–72.

Lohmann, H., 1913b. Über Coccolithophoriden. *Verh. dt. zool. Ges.*, **23**: 143–64.

Lohmann, H., 1919. Die Bevölkerung des Ozeans mit Plankton nach den Ergebnisse der Zentrifugenfänge während der Ausreise der 'Deutschland' 1911. *Arch. Biont.*, **4**: 1–617.

McIntyre, A. and Bé, A.W.H., 1967a. *Coccolithus neohelis* sp. n., a coccolith fossil type in contemporary seas. *Deep-Sea Res.*, **14**: 369–71.

McIntyre, A. and Bé, A.W.H., 1967b. Modern Coccolithophoridae of the Atlantic Ocean. - I. Placoliths and cyrtoliths. *Deep-Sea Res.*, **14**: 561–97.

Magne, F., 1954. Les Chrysophycées Marines de la Station Biologique de Roscoff. *Revue gén. Bot.*, **61**: 389–415.

Manton, I., Bremer, G. and Oates, K., 1984. Nanoplankton from the Galapagos Islands: *Michaelsarsia elegans* Gran and *Halopappus adriaticus* Schiller (Coccolithophorids) with special reference to coccoliths and their unmineralised components. *Phil. Trans. R. Soc. Lond.*, Ser.B, **305**: 183–99.

Manton, I. and Leedale, G.F., 1963. Observations on the microanatomy of *Crystallolithus hyalinus* Gaarder and Markali. *Arch. Mikrobiol.*, **47**: 115–36.

Manton, I. and Leedale, G.F., 1969. Observations on the microanatomy of *Coccolithus pelagicus* and *Cricosphaera carterae*, with special reference to the origin and nature of coccoliths and scales. *J. mar. biol. Ass. U.K.*, **49**: 1–16.

Manton, I. and Oates, K., 1975. Fine structural observations on *Papposphaera* Tangen from the southern hemisphere and on *Pappomonas* gen. nov. from South Africa and Greenland. *Br. phycol. J.*, **10**: 93–109.

Manton, I. and Oates, K., 1980. *Polycrater galapagensis* gen. et sp. nov., a putative coccolithophorid from the Galapagos

Islands with an unusual aragonite periplast. *Br. phycol. J.*, **15**: 95–103.

Manton, I. and Oates, K., 1983. Nanoplankton from the Galapagos Islands: Two genera of spectacular coccolithophorids (*Ophiaster* and *Calciopappus*) with special emphasis on unmineralised periplast components. *Phil. Trans. R. Soc. Lond.*, Ser.B, **300**: 435–62.

Manton, I. and Oates, K., 1985. Calciosoleniaceae (Coccolithophorids) from the Galapagos Islands: Unmineralized components and coccolith morphology in *Anoplosolenia* and *Calciosolenia*, with a comparative analysis of equivalents in the unmineralized genus *Navisolenia* (Haptophyceae = Prymnesiophyceae). *Phil. Trans. R. Soc. Lond.*, Ser.B, **309**: 461–77.

Manton, I. and Sutherland, J., 1975. Further observations on the genus *Pappomonas* Manton et Oates with special reference to *P. virgulosa* sp. nov. from West Greenland. *Br. phycol. J.*, **10**: 377–85.

Manton, I., Sutherland, J. and McCully, M., 1976. Fine structural observations on coccolithophorids from South Alaska in the genera *Papposphaera* Tangen and *Pappomonas* Manton and Oates. *Br. phycol. J.*, **11**: 225–38.

Manton, I., Sutherland, J. and Oates, K., 1976. Arctic coccolithophorids: two species of *Turrisphaera* gen. nov. from West Greenland, Alaska and the North-West Passage. *Proc. Roy. Soc.*, Ser.B, **194**: 179–94.

Manton, I., Sutherland, J. and Oates, K., 1977. Arctic coccolithophorids: *Wigwamma arctica* gen. et sp. nov. from Greenland and Arctic Canada, *W. annulifera* sp. nov. from South Africa and South Alaska and *Calciarcus alaskensis* gen. et sp. nov. from S. Alaska. *Proc. Roy. Soc.*, Ser.B, **197**: 145–68.

Markali, J. and Paasche, E., 1955. On two species of *Umbellosphaera*, a new marine coccolithophorid genus. *Nytt Mag. Bot.*, **4**: 95–100.

Martini, E. and Müller, C., 1972. Nannoplankton aus dem nördlichen Arabischen Meer. *'Meteor' Forschungs Ergebnisse*, Ser. C, **10**: 63–74.

Mills, J.T., 1975. *Hymenomonas coronata* sp. nov., a new coccolithophorid from the Texas coast. *J. Phycol.*, **11**: 149–54.

Müller, C., Blanc-Vernet, L., Chamley, H. and Froget, C., 1974. Les Coccolithophorides d'une carotte méditerranéenne. Comparaison paléoclimatologique avec les Foraminifères, les Ptéropodes et les argiles. *Tethys*, **6**(4): 805–28.

Murray, G. and Blackman, V.H., 1898. On the nature of the Coccospheres and Rhabdospheres. *Phil. Trans. R. Soc.*, Ser.B, **190**: 427–41.

Nishida, S., 1971. Nannofossils from Japan IV. *Trans. Proc. Palaeontol. Soc. Japan*, new ser., **83**: 143–61.

Nishida, S., 1979. Atlas of Pacific nannoplanktons. *News Osaka micropaleont., Spec. Pap.*, **3**: 1–31.

Norris, R.E., 1965. Living cells of *Ceratolithus cristatus* (Coccolithophorinae). *Arch. Protistenk.*, **108**: 19–24.

Norris, R.E., 1983. The family position of *Papposphaera* Tangen and *Pappomonas* Manton and Oates (Prymnesiophyceae) with records from the Indian Ocean. *Phycologia*, **22**: 161–9.

Norris, R.E., 1984. Indian Ocean nanoplankton. I. Rhabdosphaeraceae (Prymnesiophyceae) with a review of extant taxa. *J. Phycol.*, **20**: 27–41.

Norris, R.E., 1985. Indian Ocean nannoplankton. II. Holococcolithophorids (Calyptrosphaeraceae, Prymnesiophyceae) with a review of extant genera. *J. Phycol.*, **21**: 619–41.

Okada, H. and Honjo, S., 1973. The distribution of oceanic coccolithophorids in the Pacific. *Deep-Sea Res.*, **20**: 355–74.

Okada, H. and Honjo, S., 1975. Distribution of coccolithophores in marginal seas along the western Pacific Ocean and in the Red Sea. *Mar. Biol.*, **31**: 271–85.

Okada, H. and McIntyre, A., 1977. Modern coccolithophores of the Pacific and North Atlantic Oceans. *Micropaleontology*, **23**: 1–55.

Ostenfeld, C.H., 1899. Über *Coccosphaera* und einige neue Tintinniden im Plankton des nördlichen Atlantischen Oceans. *Zool. Anz.*, **22**: 433–9.

Ostenfeld, C.H., 1900. Über *Coccosphaera*. *Zool. Anz.*, **23**: 198–200.

Ostenfeld, C.H., 1910. *Thorosphaera*, eine neue Gattung der Coccolithophoriden. *Ber. dt. bot. Ges.*, **28**: 397–400.

Outka, D.E. and Williams, D.C., 1971. Sequential coccolith morphogenesis in *Hymenomonas carterae*. *J. Protozool.*, **18**: 285–97.

Papenfuss, G.F., 1955. Classification of the Algae. In *A Century of progress in the Natural Sciences*, 1853–1953, pp. 115–224. California Academy of Sciences: San Francisco.

Parke, M. and Adams, I., 1960. The motile (*Crystallolithus hyalinus* Gaarder and Markali) and the non-motile phases in the life history of *Coccolithus pelagicus* (Wallich) Schiller. *J. mar. biol. Ass. U.K.*, **39**: 263–74.

Parke, M. and Green, J.C., 1976. Haptophyta. In *Check-list of British Marine Algae*. third revision, ed. M. Parke and P.S. Dixon. *J. mar. biol. Ass. U.K.*, **56**: 551–5.

Parke, M., Manton, I. and Clarke, B., 1955. Studies on marine flagellates. II. Three new species of *Chrysochromulina*. *J. mar. biol. Ass. U.K.*, **35**: 387–414.

Pascher, A., 1910. Chrysomonaden aus dem Herschberger Grossteiche. *Monogr. Abh. Intern. Rev. Gesammt. Hydrobiol. Hydrogr.*, **1**: 1–66.

Pascher, A., 1914. Über Flagellaten und Algen. *Ber. dt. bot. Ges.*, **32**: 136–160.

Perch-Nielsen, K., 1985. Cenozoic calcareous nannofossils. In *Plankton Stratigraphy*, ed. H.M. Bolli, J.B. Saunders and K. Perch-Nielsen, vol.1, pp. 427–554. Cambridge University Press, Cambridge.

Pienaar, R.N., 1976. The microanatomy of *Hymenomonas lacuna* sp. nov. (Haptophyceae). *J. mar. biol. Ass. U.K.*, **56**: 1–11.

Poche, F., 1913. Das System der Proterozoa. *Arch. Protistenk.*, **30**: 125–321.

Pringsheim, E.G., 1955. Kleine Mitteilungen über Flagellaten und Algen. I. Algenartige Chrysophyceen in Reinkultur. *Arch. Mikrobiol.*, **21**: 401–10.

Reid, F.M.H., 1980. Coccolithophorids of the North Pacific Central Gyre with notes on their vertical and seasonal distribution. *Micropalaeontology*, **26**: 151–76.

Reinhardt, P., 1972. Coccolithen. Kalkiges Plankton seit Jahrmillionen. *Die neue Brehm-Bücherei*, 1–99. A. Ziemsen Verlag.

Rothmaler, W., 1951. Die Abteilungen und Klassen der Pflanzen. *Repert. Sp. Nov.*, **54**: 256–66.

Round, F.E., 1973. *The biology of the algae*. 2nd edn, Edward Arnold, London.

Rowson, J.D., Leadbeater, B.S.C. and Green, J.C., 1986. Calcium carbonate deposition in the motile (*Crystallolithus*) phase of *Coccolithus pelagicus*. *Br. phycol. J.*, **21**: 359–70.

Samtleben, C., 1980. Die Evolution der Coccolithophoriden-Gattung *Gephyrocapsa* nach Befunden im Atlantik. *Paläont. Z.*, **54**: 91–127.

Schiller, J., 1913. Vorläufige Ergebnisse der Phytoplankton-Untersuchungen auf den Fahrten S.M.S. 'Najade' in der Adria 1911–12. I. Die Coccolithophoriden. *K. Akad. Wiss. Wien, Sitzber., Math.-Naturw. Kl.*, **122**: 597–617.

Schiller, J., 1914. Bericht über Ergebnisse der Nannoplankton-untersuchungen anlässlich der Kreuzungen S.M.S. Najade in der Adria. *Internat. Rev. Gesamt. Hydrobiol. Hydrogr.*, **6** (Biol. Suppl.): 1–15.

Schiller, J., 1925. Die planktonischen Vegetationen des adriatischen Meeres. A. Die Coccolithophoriden-Vegetation in den Jahren 1911-14. *Arch. Protistenk.*, **51**: 1–130.

Schiller, J., 1926. Über Fortpflanzung, geissellose Gattungen und die Nomenklatur der Coccolithophoraceen nebst Mitteilung über Copulation bei *Dinobryon. Arch. Protistenk.*, **53**: 326–42.

Schiller, J., 1930. Coccolithineae. In *Kryptogamen-Flora von Deutschland, Österreich und der Schweiz*, ed. L. Rabenhorst, vol. **10**: 89–267. Akademische Verlagsgesellschaft: Leipzig.

Schlauder, J., 1945. Recherches sur les flagellés calcaires de la Baie d'Alger. *Univ. d'Alger, Dipl. Fac. Sci.*, pp. 1–51.

Schussnig, B., 1930. *Ochrosphaera neapolitana* nov. gen., nov. spec., eine neue Chrysomonade mit Kalkhülle. *Öst. bot. Z.*, **79**: 164–70.

Schwarz, E.H.L., 1894. Coccoliths. *Ann. Mag. nat. Hist.*, Ser.6, **14**: 341–6.

Senn, G., 1900. Flagellata. In: *Die Naturlichen Pflanzenfamilien*, ed. A. Engler and K. Prantl, Leipzig: Engelmann, vol. **1**: pp. 93–192.

Sorby, H.C., 1861. On the organic origin of the so-called 'crystalloids' of the chalk. *Ann. Mag. nat. Hist.*, Ser.3, **8**: 193–200.

Stein, F.R. von, 1878. *Der organismus der Infusionsthiere nach eigenen Forschungen in systematischer Reihenfolge bearbeitet.* Abt. 3, Flagellaten u. Geisselinfusorien. 1. Hälfte.

Steinmetz, J.C., 1991. Calcareous nannoplankton biocoenosis: sediment trap studies in the Equatorial Atlantic, Central Pacific, and Panama Basin. In *Ocean Biocoensis Series No.1*, ed. S. Honjo, pp.1–85. W.H.O.I., USA.

Stoermer, E.F. and Sicko-Goad, L., 1977. A new distribution record for *Hymenomonas roseola* Stein (Prymnesiophyceae, Coccolithophoraceae) and *Spiniferomonas trioralis* Takahashi (Chrysophyceae, Synuraceae) in the Laurentian Great Lakes. *Phycologia*, **16**: 355–8.

Takayama, T., 1967. First report on nannoplankton of the Upper Tertiary and Quaternary of the southern Kwanto region, Japan. *Austria, Geol. Bundesanst., Jahrb.*, **110**: 169–98.

Tangen, K., 1972. *Papposphaera lepida*, gen. nov., n. sp., a new marine coccolithophorid from Norwegian coastal waters. *Norw. J. Bot.*, **19**: 171–8.

Tappan, H., 1980. *The Paleobiology of Plant Protists.* Freeman, W.H. and Co., San Francisco.

Theodoridis, S., 1984. Calcareous nannofossil biostratigraphy of the Miocene and revision of the helicoliths and discoasters. *Utrecht Micropaleont. Bull.*, **32**: 1–271.

Thomsen, H.A., 1980a. *Wigwamma scenozonion* sp. nov.

(Prymnesiophyceae) from West Greenland. *Br. phycol. J.*, **15**: 335–42.

Thomsen, H.A., 1980b. *Quaternariella obscura* gen. et sp. nov. (Prymnesiophyceae) from West Greenland. *Phycologia*, **19**: 260–5.

Thomsen, H.A., 1980c. Two species of *Trigonaspis* gen. nov. (Prymnesiophyceae) from West Greenland. *Phycologia*, **19**: 218–29.

Thomsen, H.A., 1980d. *Turrisphaera polybotrys* sp. nov. (Prymnesiophyceae) from West Greenland. *J. mar. biol. Ass. U.K.*, **60**: 529–37.

Thomsen, H.A., 1981. Identification by electron microscopy of nanoplanktonic coccolithophorids (Prymnesiophyceae) from West Greenland, including the description of *Papposphaera sarion* sp. nov. *Br. phycol. J.*, **16**: 77–94.

Thomsen, H.A., Buck, K.R., Coale, S.L., Garrison, D.L., and Gowing, M.M., 1988. Nanoplanktonic coccolithophorids (Prymnesiophyceae, Haptophyceae) from the Weddell Sea, Antarctica. *Nord. J. Bot.*, **8**(4): 419–36.

Thomsen, H.A. and Oates, K., 1978. *Balaniger balticus* gen. et sp. nov. (Prymnesiophyceae) from Danish coastal waters. *J. mar. biol. Ass. U.K.*, **58**: 773–9.

Thomsen, H.A., Østergaard, J.B. and Hansen, L.E., 1991. Heteromorphic life histories in Arctic coccolithophorids (Prymnesiophyceae). *J. Phycol.*, **27**: 634–42.

Thomson, C.W., 1874. Preliminary notes on the nature of the seabottom procured by the soundings of HMS Challenger during her cruises in the 'Southern Sea' in the early part of the year 1874. *Proc. Roy. Soc. Lond.*, Ser.B, Biol. Sci., **23**: 32–49.

Wallich, G.C., 1861. Remarks on some novel phases of organic life, and on the boring powers of minute annelids, at great depths in the sea. *Ann. Mag. nat. Hist.*, Ser.3, **8**: 52–8.

Wallich, G.C., 1877. Observations on the coccosphere. *Ann. Mag. nat. Hist.*, Ser.4, **19**: 342–50.

Weaver, P.P.E. and Pujol, C., 1988. History of the last deglaciation in the Alboran Sea (Western Mediterranean) and adjacent North Atlantic as revealed by coccolith floras. *Palaeogeog., Palaeoclimatol., Palaeoecol.*, **64**: 35–42.

Weber-Van Bosse, A., 1901. Études sur les algues de l'Archipel Malaisien III. Note préliminaire sur les résultats algologiques de l'expédition du Siboga. *Jard. Bot. Buitenz., Ann.*, (Ser.2), **17**(2): 126–41.

Wilbur, K.M. and Watabe, N., 1963. Experimental studies on calcification in molluscs and the alga *Coccolithus huxleyi. Ann. N.Y. Acad. Sci.*, **109**: 82–112.

Winter, A., Reiss, Z. and Luz, B., 1978. Living *Gephyrocapsa protohuxleyi* in the Gulf of Elat. *Mar. Micropaleont.*, **3**: 295–8.

Winter, A., Reiss, Z. and Luz, B., 1979. Distribution of living coccolithophore assemblages in the Gulf of Elat ('Aqaba). *Mar. Micropaleont.*, **4**: 197–223.

7 Atlas of living coccolithophores

AMOS WINTER AND WILLIAM G. SIESSER

Introduction

The coccolithophore atlas owes much to the efforts of colleagues who contributed their SEM micrographs. We believe about 70% of all known living species are illustrated here. As those of us who take SEM micrographs know, a single good micrograph may represent many hours, if not days, of SEM viewing. The atlas therefore represents thousands of hours of devoted SEM work by the contributors. The name of each contributor appears above their micrograph. Taxonomy follows that of Jordan and Kleijne (see chapter 6). Bibliographic references for the taxa may also be found in chapter 6.

The known geological age-range and modern biogeographical distribution of each species are given below each micrograph. This information is presented with caution, as comparatively little is yet known about the range/distribution of many species. Biogeographical distributions are noted as follows: Pac = Pacific Ocean; Atl = Atlantic Ocean; Ind = Indian Ocean; Car = Caribbean Sea; Med = Mediterranean Sea; Red = Red Sea; Wed = Weddell Sea (N = north; NE = northeast; NW = northwest; C = central, S = south). 'Living' indicates no known fossil record. Information on the biogeographical distribution of each species is based on references in chapter 8 of this book. A scale bar under the micrographs is equal to 1 µm. Tables 1, 2 and 3 list each species by order of listing in the classification of Jordan and Kleijne (chapter 6), alphabetically by species (trivial) name, and alphabetically by genus, respectively. The collection location of each specimen illustrated is listed in the appendix to this chapter.

SEM micrograph contributors

Juan Alcober
Departmento de Biología Vegetal,
Facultad de Ciencias Biológicas,
Universidad de Valencia,
46100 Burjasot,
Valencia,
Spain

Jackie Bonilla
Department of Geology,
University of Puerto Rico,
P.O.Box 5000,
Mayaguez,
Puerto Rico 00681

Mara Y. Cortés
Geological Institute,
Swiss Federal Institute of Technology,
Sonneggstr. 5,
CH-8092 Zürich,
Switzerland

Jacqueline Fresnel
Laboratoire de Biologie et Biotechnologies Marines
   (Phycologie),
39 rue Desmoueux,
1400-Caen,
France

Peter Friedinger
Marine Geoscience Unit,
University of Cape Town,
Rondebosch 7700,
Cape Town,
South Africa

Ali T. Haidar
Geological Institute,
Swiss Federal Institute of Technology,
Sonneggstr. 5,
CH-8092 Zurich,
Switzerland

Ric W. Jordan
Institute of Oceanographical Sciences,
Deacon Laboratory,
Brook Road,
Wormley,
Surrey GU8 5UB,
U.K.

Annelies Kleijne
Duinoord 11,
2224 CA,
Katwijk,
The Netherlands

Stan A. Kling
Micropaleo Consultants Inc.,
681 Encinitas Blvd,
Suite 312,
Encinitas,
CA 92024,
U.S.A.

Michael Knappertsbusch
Vrije Universiteit,
Institute of Earth Sciences,
De Boelelaan 1085,
1007 MC Amsterdam,
The Netherlands

Shiro Nishida
Department of Earth Sciences,
Nara University of Education,
Nara,
630 Japan

Christian Samtleben,
Geologist Paläontologisches
   Institut der Universität Kiel.,
Olshausen Str. 40,
24098 Kiel,
Germany

Helge A. Thomsen
Institute for Sporeplanter,
University of Copenhagen,
Øster Farimagsgade 2D,
Copenhagen K,
Denmark 1353.

Amos Winter
Department of Marine Sciences,
University of Puerto Rico,
P.O.Box 5000,
Mayagüez,
Puerto Rico 00681

Table 1. *Living coccolithophores, listed following the classification of Jordan and Kleijne (chapter 6). Species illustrated in this atlas are denoted by an asterisk.*

Table 2. *Living coccolithophores, listed alphabetically by species (trivial) name. Species illustrated in this atlas are denoted by an asterisk.*

Heterococcolithophores

				page
68	*	*Acanthoica acanthifera*	acanthifera	127
77	*	*Anacanthoica acanthos*	acanthos	129
79	*	*Cyrtosphaera aculeate*	aculeate	130
97	*	*Michaelsarsia adriaticus*	adriaticus	133
124	*	*Calciarcus alaskensis*	alaskensis	140
104		*Syracosphaera ampliora*	ampliora	
16		*Umbilicosphaera angustiforamen*	angustiforamen	
133		*Wigwamma annulifera*	annulifera	
134		*Wigwamma antarctica*	antarctica	
105	*	*Syracosphaera anthos*	anthos	134
66	*	*Scyphosphaera apsteinii* f. *apsteinii*	apsteinii	127
135		*Wigwamma arctica*	arctica	
1	*	*Braarudosphaera bigelowii*	bigelowii	117
91	*	*Alveosphaera bimurata*	bimurata	132
94		*Coronosphaera binodata*	binodata	
69		*Acanthoica biscayensis*	biscayensis	
43		*Pappomonas flabellifera* var. *borealis*	borealis	
106	*	*Syracosphaera borealis*	borealis	135
9		*Coccolithus pelagicus* f. *braarudii*	braarudii	
2		*Anoplosolenia brasiliensis*	brasiliensis	
17		*Umbilicosphaera calvata*	calvata	
87		*Alisphaera capulata*	capulata	
56		*Pleurochrysis carterae* var. *carterae*	carterae	
23	*	*Helicosphaera carteri* var. *carteri*	carteri	121–2
92	*	*Calciopappus caudatus*	caudatus	132
14	*	*Oolithotus fragilis* var. *cavum*	cavum	120
78	*	*Anacanthoica cidaris*	cidaris	129
84	*	*Rhabdosphaera clavigera* var. *clavigera*	clavigera	130
13	*	*Neosphaera coccolithomorpha*	coccolithomorpha	120
107	*	*Syracosphaera corolla*	corolla	135
33	*	*Emiliania huxleyi* var. *corona*	corona	123
27		*Hymenomonas coronata*	coronata	
108	*	*Syracosphaera corrugis*	corrugis	135
34		*Gephyrocapsa crassipons*	crassipons	
4	*	*Ceratolithus cristatus* var. *cristatus*	cristatus	117–18
80	*	*Cyrtosphaera cucullata*	cucullata	130
76		*Anacanthoica deflandre*	deflandre	
57		*Pleurochrysis carterae* var. *dentata*	dentata	
67		*Scyphosphaera apsteinii* f. *dilatata*	dilatata	
111		*Syracosphaera halldalii* f. *dilatata*	dilatata	
62	*	*Pontosphaera discopora*	discopora	126
98	*	*Michaelsarsia elegans*	elegans	133
53		*Cricosphaera elongata*	elongata	
125	*	*Florisphaera profunda* var. *elongata*	elongata	140
109	*	*Syracosphaera epigrosa*	epigrosa	136
35	*	*Gephyrocapsa ericsonii*	ericsonii	123
110	*	*Syracosphaera exigua*	exigua	136
129	*	*Thorosphaera flabellata*	flabellata	141
44	*	*Pappomonas flabellifera* var. *flabellifera*	flabellifera	125

21	*	*Umbilicosphaera sibogae* var. *foliosa*	*foliosa*	121
99		*Ophiaster formosus* var. *formosus*	*formosus*	
15		*Oolithotus fragilis* var. *fragilis*	*fragilis*	
128	*	*Polycrater galapagensis*	*galapagensis*	141
54		*Cricosphaera gayraliae*	*gayraliae*	
28		*Hymenomonas globosa*	*globosa*	
112	*	*Syracosphaera halldalii* f. *halldalii*	*halldalii*	136
113	*	*Syracosphaera histrica*	*histrica*	136
18	*	*Umbilicosphaera hulburtiana*	*hulburtiana*	121
32	*	*Emiliania huxleyi* var. *huxleyi*	*huxleyi*	122
24	*	*Helicosphaera carteri* var. *hyalina*	*hyalina*	122
10	*	*Coccolithus pelagicus* f. *hyalinus*	*hyalinus*	119
101	*	*Ophiaster hydroideus*	*hydroideus*	134
100		*Ophiaster formosus* var. *inversus*	*inversus*	
131	*	*Umbellosphaera irregularis*	*irregularis*	142
70		*Acanthoica jancheni*	*jancheni*	
63	*	*Pontosphaera japonica*	*japonica*	126
29		*Hymenomonas lacuna*	*lacuna*	
114	*	*Syracosphaera lamina*	*lamina*	137
130	*	*Turrilithus latericioides*	*latericioides*	141
81		*Cyrtosphaera lecaliae*	*lecaliae*	
47	*	*Papposphaera lepida*	*lepida*	125
6	*	*Calcidiscus leptoporus* f. *leptoporus*	*leptoporus*	118
127		*Jomonlithus littoralis*	*littoralis*	
19		*Umbilicosphaera maceria*	*maceria*	
71	*	*Acanthoica maxima*	*maxima*	128
95		*Coronosphaera maxima*	*maxima*	
96	*	*Coronosphaera mediterranea*	*mediterranea*	133
102		*Ophiaster minimus*	*minimus*	
115	*	*Syracosphaera molischii*	*molischii*	137
36	*	*Gephyrocapsa muellerae*	*muellerae*	123
3	*	*Calciosolenia murrayi*	*murrayi*	117
116	*	*Syracosphaera nana*	*nana*	137
31		*Ochrosphaera neapolitana*	*neapolitana*	
8	*	*Coccolithus neohelis*	*neohelis*	119
117	*	*Syracosphaera nodosa*	*nodosa*	138
48		*Papposphaera obpyramidalis*	*obpyramidalis*	
37	*	*Gephyrocapsa oceanica*	*oceanica*	123
118	*	*Syracosphaera orbiculus*	*orbiculus*	138
88	*	*Alisphaera ordinata*	*ordinata*	131
38	*	*Gephyrocapsa ornata*	*ornata*	124
73	*	*Algirosphaera oryza*	*oryza*	128–9
119	*	*Syracosphaera ossa*	*ossa*	138
39	*	*Reticulofenestra parvula* var. *parvula*	*parvula*	124
26	*	*Helicosphaera pavimentum*	*pavimentum*	122
11	*	*Coccolithus pelagicus* f. *pelagicus*	*pelagicus*	119
12	*	*Hayaster perplexus*	*perplexus*	120
120	*	*Syracosphaera pirus*	*pirus*	139
58		*Pleurochrysis placolithoides*	*placolithoides*	
126	*	*Florisphaera profunda* var. *profunda*	*profunda*	140
121	*	*Syracosphaera prolongata*	*prolongata*	139
59		*Pleurochrysis pseudoroscoffensis*	*pseudoroscoffensis*	
122	*	*Syracosphaera pulchra*	*pulchra*	139
41		*Reticulofenestra punctata*	*punctata*	

74		*Algirosphaera quadricornu*	*quadricornu*	
55	*	*Cricosphaera quadrilaminata*	*quadrilaminata*	126
72	*	*Acanthoica quattrospina*	*quattrospina*	128
103	*	*Ophiaster reductus*	*reductus*	134
7	*	*Calcidiscus leptoporus* f. *rigidus*	*rigidus*	118
93	*	*Calciopappus rigidus*	*rigidus*	133
75	*	*Algirosphaera robusta*	*robusta*	129
60		*Pleurochrysis roscoffensis*	*roscoffensis*	
30		*Hymenomonas roseola*	*roseola*	
123	*	*Syracosphaera rotula*	*rotula*	140
49	*	*Papposphaera sagittifera*	*sagittifera*	125
50	*	*Papposphaera sarion*	*sarion*	125
136		*Wigwamma scenozonion*	*scenozonion*	
61		*Pleurochrysis scherffelii*	*scherffelii*	
20		*Umbilicosphaera scituloma*	*scituloma*	
42	*	*Reticulofenestra sessilis*	*sessilis*	124
22	*	*Umbilicosphaera sibogae* var. *sibogae*	*sibogae*	121
51		*Papposphaera simplicissima*	*simplicissima*	
89		*Alisphaera spatula*	*spatula*	
85	*	*Rhabdosphaera clavigera* var. *stylifera*	*stylifera*	131
64	*	*Pontosphaera syracusana*	*syracusana*	127
40		*Reticulofenestra parvula* var. *tecticentrum*	*tecticentrum*	
5	*	*Ceratolithus cristatus* var. *telesmus*	*telesmus*	118
132	*	*Umbellosphaera tenuis*	*tenuis*	142
52		*Papposphaera thomsenii*	*thomsenii*	
137		*Wigwamma triradiata*	*triradiata*	
82	*	*Discosphaera tubifera*	*tubifera*	130
65		*Pontosphaera turgida*	*turgida*	
90	*	*Alisphaera unicornis*	*unicornis*	131–2
83		*Palusphaera vandeli*	*vandeli*	
45		*Pappomonas virgulosa*	*virgulosa*	
25		*Helicosphaera carteri* var. *wallichii*	*wallichii*	
46		*Pappomonas weddellensis*	*weddellensis*	
86	*	*Rhabdosphaera xiphos*	*xiphos*	131
		Holococcolithophores		
186	*	*Sphaerocalyptra adenensis*	*adenensis*	154
150	*	*Gliscolithus amitakarenae*	*amitakarenae*	145
188	*	*Zygosphaera amoena*	*amoena*	154
151	*	*Homozygosphaera arethusae*	*arethusae*	145
179	*	*Poricalyptra aurisinae*	*aurisinae*	152
189	*	*Zygosphaera bannockii*	*bannockii*	154
157	*	*Syracolithus bicorium*	*bicorium*	147
138	*	*Calicasphaera blokii*	*blokii*	142
148	*	*Flosculosphaera calceolariopsis*	*calceolariopsis*	144
158		*Syracolithus catilliferus*	*catilliferus*	
141		*Calyptrosphaera cialdii*	*cialdii*	
139	*	*Calicasphaera concava*	*concava*	142
159	*	*Syracolithus confusus*	*confusus*	147
177	*	*Helladosphaera cornifera*	*cornifera*	151
160	*	*Syracolithus dalmaticus*	*dalmaticus*	147
142	*	*Calyptrosphaera dentata*	*dentata*	143
140	*	*Calicasphaera diconstricta*	*diconstricta*	143
167	*	*Calyptrolithina divergens*	*divergens*	149

164	*	*Anthosphaera fragaria*	*fragaria*	148
180	*	*Poricalyptra gaarderii*	*gaarderii*	152
143		*Calyptrosphaera galea*	*galea*	
174	*	*Corisphaera gracilis*	*gracilis*	151
171	*	*Calyptrolithophora gracillima*	*gracillima*	150
172	*	*Calyptrolithophora hasleana*	*hasleana*	150
144	*	*Calyptrosphaera heimdalae*	*heimdalae*	143
190	*	*Zygosphaera hellenica*	*hellenica*	155
181	*	*Poricalyptra isselii*	*isselii*	152
165	*	*Anthosphaera lafourcadii*	*lafourcadii*	148
182	*	*Poricalyptra magnaghii*	*magnaghii*	153
191	*	*Zygosphaera marsilii*	*marsilii*	155
183	*	*Poritectolithus maximus*	*maximus*	153
156	*	*Periphyllophora mirabilis*	*mirabilis*	146
169	*	*Calyptrolithina multipora*	*multipora*	149
145	*	*Calyptrosphaera oblonga*	*oblonga*	143–4
173	*	*Calyptrolithophora papillifera*	*papillifera*	150
166	*	*Anthosphaera periperforata*	*periperforata*	149
178	*	*Helladosphaera pienaarii*	*pienaarii*	152
147	*	*Daktylethra pirus*	*pirus*	144
161	*	*Syracolithus ponticuliferus*	*ponticuliferus*	147
184	*	*Poritectolithus poritectus*	*poritectus*	153
187	*	*Sphaerocalyptra quadridentata*	*quadridentata*	154
162	*	*Syracolithus quadriperforatus*	*quadriperforatus*	148
149	*	*Flosculosphaera sacculus*	*sacculus*	145
163	*	*Syracolithus schilleri*	*schilleri*	148
146	*	*Calyptrosphaera sphaeroidea*	*sphaeroidea*	144
152	*	*Homozygosphaera spinosa*	*spinosa*	146
175	*	*Corisphaera strigilis*	*strigilis*	151
153	*	*Homozygosphaera triarcha*	*triarcha*	146
168	*	*Calyptrolithina divergens* f. *tuberosa*	*tuberosa*	149
185	*	*Poritectolithus tyronus*	*tyronus*	153
176	*	*Corisphaera tyrrheniensis*	*tyrrheniensis*	151
154		*Homozygosphaera vavilovii*	*vavilovii*	
155	*	*Homozygosphaera vercellii*	*vercellii*	146
170	*	*Calyptrolithina wettsteinii*	*wettsteinii*	150

Partially calcified coccolithophores

197		*Turrisphaera arctica*	*arctica*	
192	*	*Balaniger balticus*	*balticus*	155
198	*	*Turrisphaera borealis*	*borealis*	156
194	*	*Trigonaspis diskoensis*	*diskoensis*	156
195	*	*Trigonaspis melvillea*	*melvillea*	156
196	*	*Trigonaspis minutissima*	*minutissima*	156
193	*	*Quaternariella obscura*	*obscura*	155
199		*Turrisphaera polybotrys*	*polybotrys*	

Table 3. *Living coccolithophores, listed alphabetically by genus. Species illustrated in this atlas are denoted by an asterisk.*

115

S. Nishida

A. Kleijne

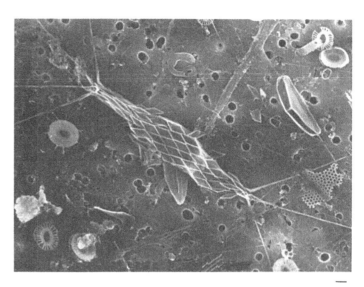

1 *Braarudosphaera bigelowii*
known range: Jurassic - Recent
known distribution: Pac (NE); Atl (N,C); Med

3 *Calciosolenia murrayi*
known range: Living
known distribution: Pac (NE,C); Atl (N) Car

A. Kleijne

M. Knappertsbusch

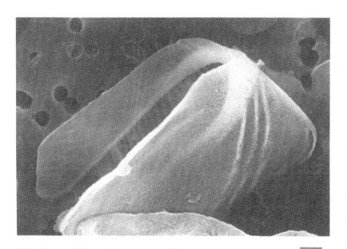

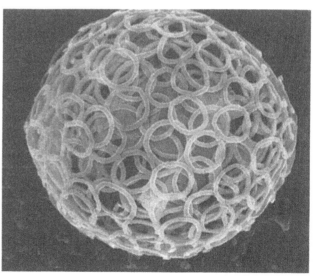

4A *Ceratolithus cristatus* var. *telesmus*
known range: Pleistocene - Recent
known distribution: Pac (NE,C); Atl (N,C); Med

4B *Ceratolithus cristatus* var. *cristatus*
known range: Pleistocene - Recent
known distribution: Pac (NE,C); Atl (N,C); Med

C. Samtleben

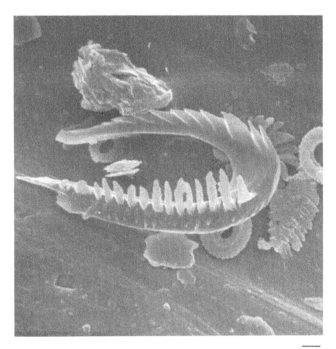

4C *Ceratolithus cristatus* var. *cristatus*
known range: Pleistocene - Recent
known distribution: Pac (NE,C); Atl (N,C); Med

A. Winter/P. Friedinger

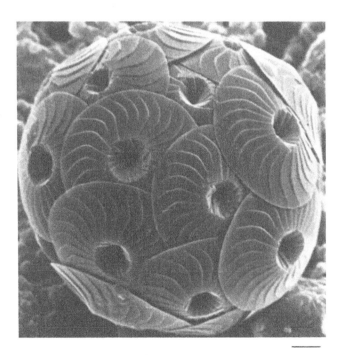

6 *Calcidiscus leptoporus* f. *leptoporus*
known range: Early Miocence - Recent
known distribution: Pac (C), Atl (N, C); Ind; Car

A. Kleijne

5 *Ceratolithus cristatus* var. *telesmus*
known range: Pleistocene - Recent
known distribution: Pac (C); Atl (N); Ind; Med

A. Kleijne

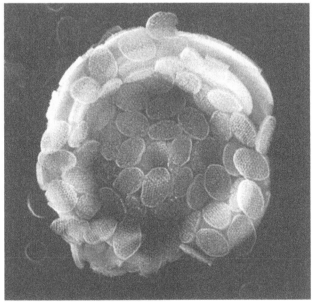

7 *Calcidiscus leptoporus* f. *rigidus*
known range: Living
known distribution: Atl (N,C); Ind; Med

J. Fresnel

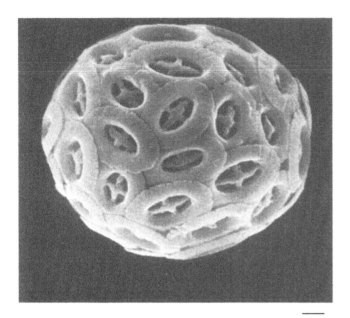

8 *Coccolithus neohelis*
known range: Living
known distribution: coastal Atl; Atl (N)

C. Samtleben

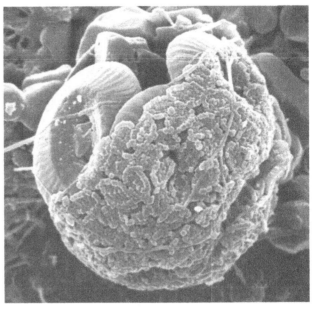

11A *Coccolithus pelagicus* f. *pelagicus* covered by cells of
Coccolithus pelagicus f. *hyalinus*
known range: Early Paleocene - Recent
known distribution: Pac (NE,C); Atl (N)

S. Nishida

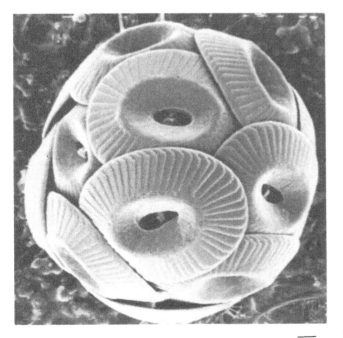

11B *Coccolithus pelagicus* f. *pelagicus*
known range: Early Paleocene - Recent
known distribution: Pac (NE,C); Atl (N)

C. Samtleben

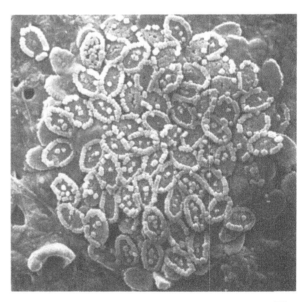

11C *Coccolithus pelagicus* f. *hyalinus*
known range: Early Paleocene - Recent
known distribution: Pac (NE,C); Atl (N)

S. Nishida

A. Winter

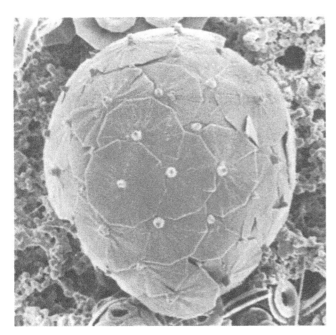

12 *Hayaster perplexus*
known range: Late Oligocene - Recent
known distribution: Pac (NE,C); Atl (N)

14 *Oolithotus fragilis* var. *cavum*
known range: Early Pliocence - Recent
known distribution: Pac (NE,C); Atl (N); Med

A. Winter/P. Friedinger

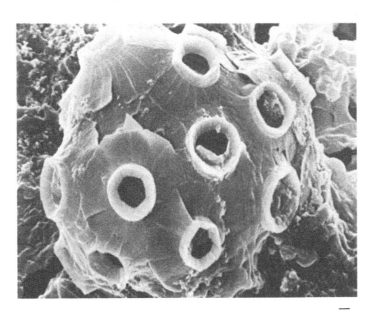

13 *Neosphaera coccolithomorpha*
known range: Late Miocene - Recent
known distribution: Pac (NE,C); Atl (N,C); Ind

A. Winter/P. Friedinger

J. Alcober

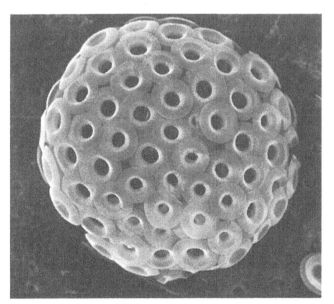

18 *Umbilicosphaera hulburtiana*
known range: Pleistocene – Recent
known distribution: Pac (NE,C); Atl (N,C); Ind; Med

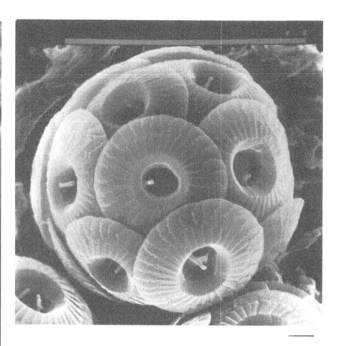

21 *Umbilicosphaera sibogae* var. *foliosa*
known range: Early Miocene - Recent
known distribution: Pac (NE,C); Atl (N)

A. Kleijne

A. Winter/P. Friedinger

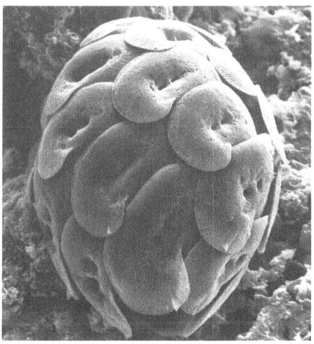

22 *Umbilicosphaera sibogae* var. *sibogae*
known range: Early Miocene - Recent
known distribution: Pac (NE,C); Atl (N,C); Ind; Med; Car

23A *Helicosphaera carteri* var. *carteri*
known range: Early Miocene - Recent
known distribution: Pac (NE,C); Atl (N,C); Ind; Car; Med

121

A. Winter/P. Friedinger

A. Kleijne

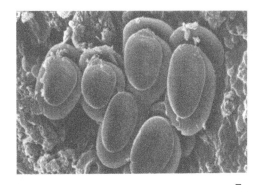

23B *Helicosphaera carteri* var. *carteri*
known range: Early Miocene - Recent
known distribution: Pac (NE,C); Atl (N,C); Ind; Car; Med

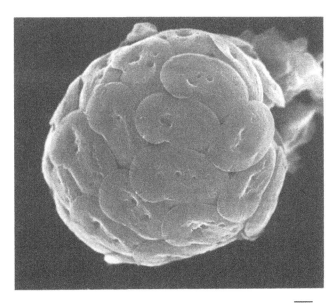

24 *Helicosphaera carteri* var. *hyalina*
known range: Late Pleistocene - Recent
known distribution: Pac (NE,C); Atl (N); Med

A. Winter/P. Friedinger

A. Kleijne

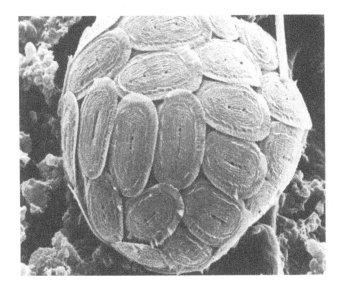

26 *Helicosphaera pavimentum*
known range: Late Pleistocene - Recent
known distribution: Pac (NE,C); Atl (N); Ind, Med

32 *Emiliania huxleyi* var. *huxleyi*
known range: Late Pleistocene - Recent
known distribution: Pac (NW,NE,C,S); Atl (N,C,S); Ind; Med;
Car; Red; Wed

S. Nishida

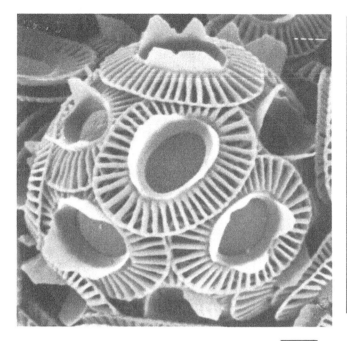

33 *Emiliania huxleyi* var. *corona*
known range: Late Pleistocene - Recent
known distribution: Pac (NE,C); Atl (N)

J. Alcober

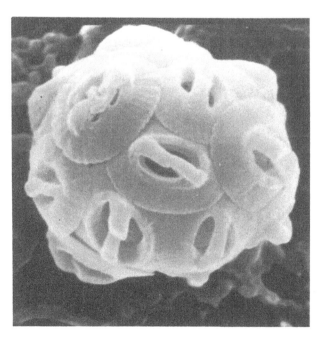

36 *Gephyrocapsa muellerae*
known range: Late Pleistocene - Recent
known distribution: Atl (N); Med

A. Kleijne

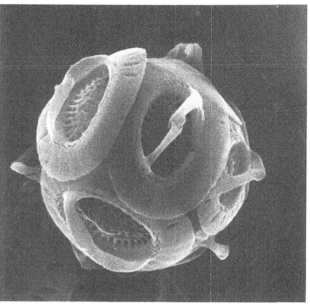

35 *Gephyrocapsa ericsonii*
known range: Pleistocene - Recent
known distribution: Pac (NE,C); Atl (N); Ind; Med; Red

S. Kling

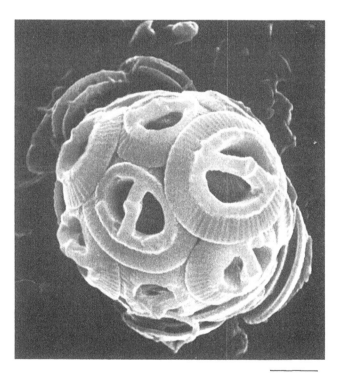

37 *Gephyrocapsa oceanica*
known range: Early Pleistocene - Recent
known distribution: Pac (NW,NE,C,S); Atl (N,C,S); Ind; Med;
Car; Red, Wed

123

S. Nishida

38 *Gephyrocapsa ornata*
known range: Late Pleistocene - Recent
known distribution: Pac (NE,C); Atl (N); Ind

J. Alcober

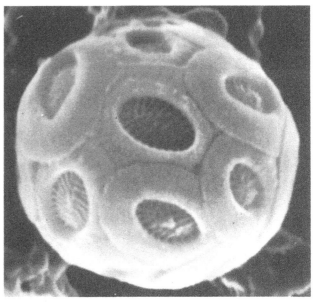

39 *Reticulofenestra parvula* var. *parvula*
known range: Living
known distribution: Pac (C); Atl (N)

S. Nishida

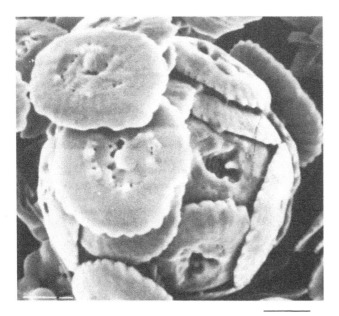

42A *Reticulofenestra sessilis*
known range: Living
known distribution: Pac (NE,C); Atl (N)

R. Jordan

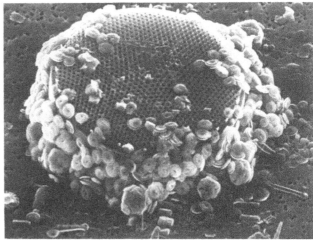

42B *Reticulofenestra sessilis*
known range: Living
known distribution: Pac (NE,C); Atl (N)

H. A. Thomsen

A. Kleijne

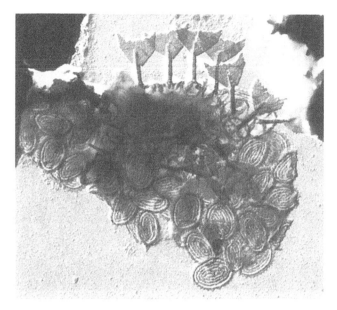

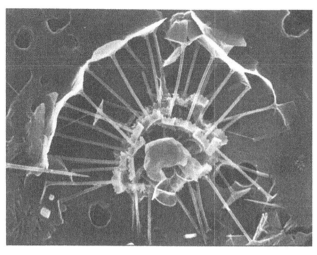

47 *Papposphaera lepida*
known range: Living
known distribution: Pac (NE,C); Atl (N)

44 *Pappomonas flabellifera* var. *flabellifera*
known range: Living
known distribution: Atl (N,S)

C. Samtleben

H. A. Thomsen

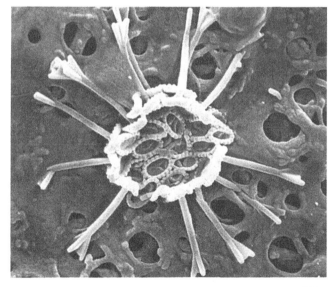

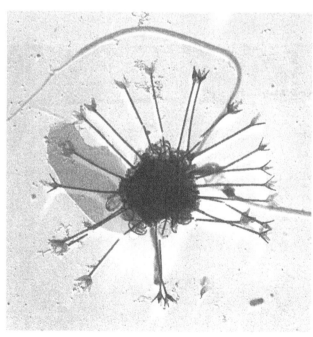

49 *Papposphaera sagittifera*
known range: Living
known distribution: Pac (NE); Atl (N); Wed

50 *Papposphaera sarion*
known range: Living
known distribution: Atl (N); Ind

R. Jordan

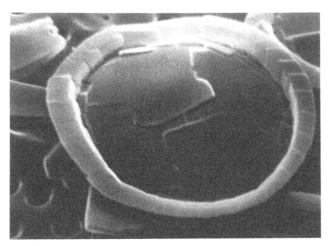

55 *Cricosphaera quadrilaminata*
known range: Living
known distribution: Pac (NE,C); Atl (N)

J. Fresnel

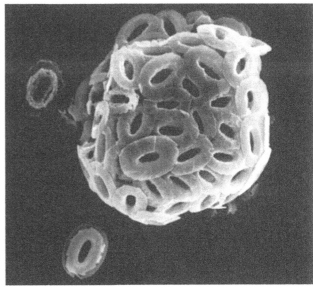

58 *Pleurochrysis placolithoide*s
known range: Living
known distribution: coastal

S. Nishida

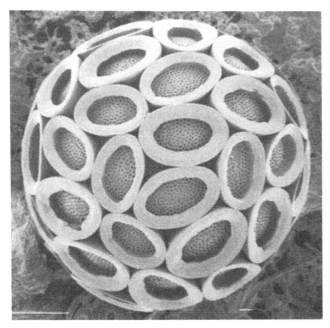

62 *Pontosphaera discopora*
known range: Living
known distribution: Pac (NE,C); Atl (N,C)

J. Alcober

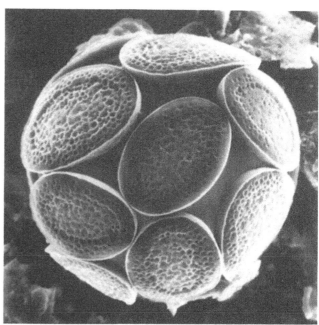

63 *Pontosphaera japonica*
known range: Living
known distribution: Pac (NE,C); Atl (N)

M. Y. Cortés / A. Winter / J. Bonilla

J. Alcober

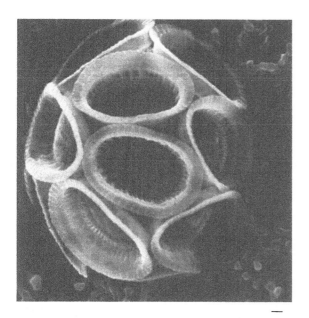

64 *Pontosphaera syracusana*
known range: Living
knwon distribution: Pac (NE,C); Atl (N); Ind

66 *Scyphosphaera apsteinii* f. *apsteinii*
known range: Early Eocence - Recent
known distribution: Pac (NE,C); Atl (N); Med

A. Kleijne

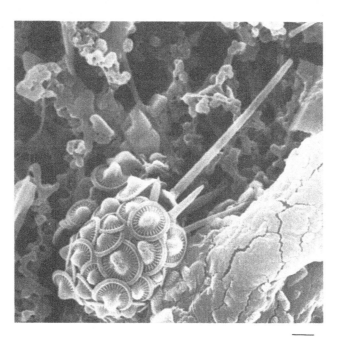

68 *Acanthoica acanthifera*
known range: Living
known distribution: Pac (NE,C); Atl (N); Med; Red

A. Kleijne

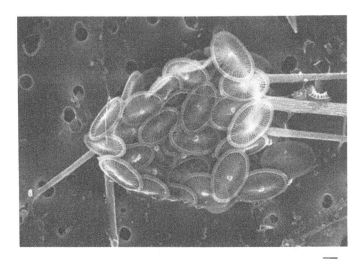

71 *Acanthoica maxima*
known range: Living
known distribution: Atl (C); Ind

S. Nishida

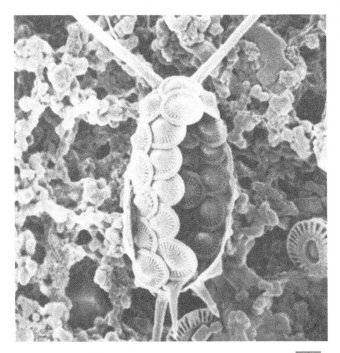

72 *Acanthoica quattrospina*
known range: Living
known distribution: Pac (NE,C); Atl (N); Ind; Med; Car

J. Alcober

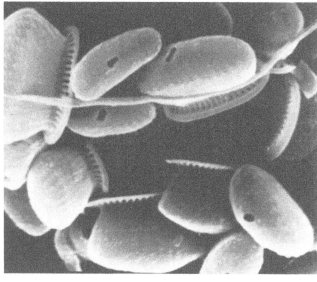

73A *Algirosphaera oryza*
known range: Living
known distribution: Pac (NE,,C); Atl (N); Car

J. Alcober

73B *Algirosphaera oryza*
known range: Living
known distribution: Pac (NE,,C); Atl (N); Car

M. Knappertsbusch

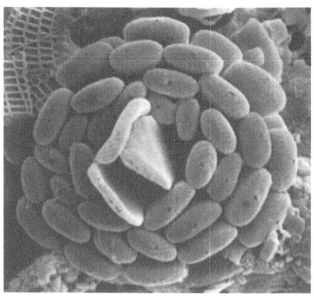

75 *Algirosphaera robusta*
known Living
known distribution: Pac (NE,,C); Atl (N); Ind; Car; Red

A. Kleijne

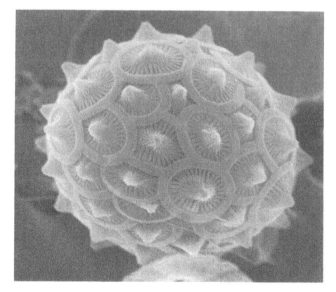

77 *Anacanthoica acanthos*
known range: Living
known distribution: Atl (N); Ind; Car; Med

A. Kleijne

78 *Anacanthoica cidaris*
known range: Living
known distribution: Ind; Med; Red

A. Winter / M. Y. Cortés / J. Bonilla

A. Kleijne

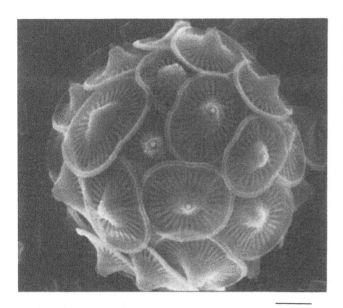

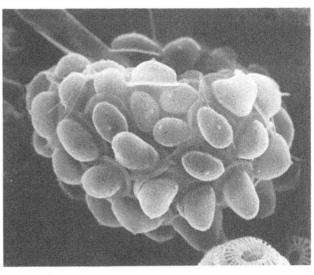

79 *Cyrtosphaera aculeata*
known range: Living
known distribution: Pac (NE,C); Atl (N); Car; Med; Red

80 *Cyrtosphaera cucullata*
known range: Living
known distribution: Atl (N); Ind; Med

A. Kleijne

A. Winter

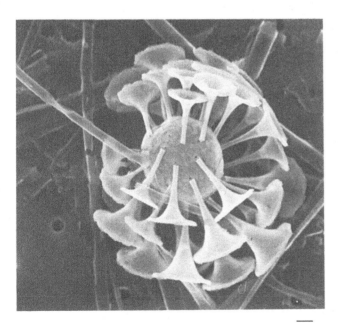

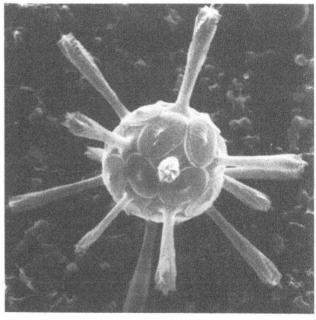

82 *Discosphaera tubifera*
known range: Pleistocene - Recent
known distribution: Pac (NE,C); Atl (N,C); Ind; Car; Med; Red

84 *Rhabdosphaera clavigera* var. *clavigera*
known range: Late Pliocene - Recent
known distribution: Pac (NE,C); Atl (N,C); Ind; Car; Med; Red

C. Samtleben

A. Kleijne

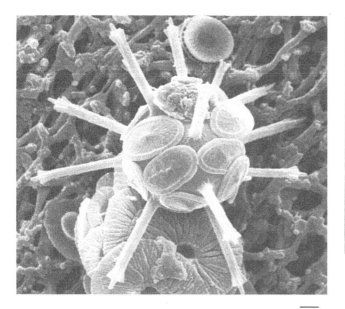

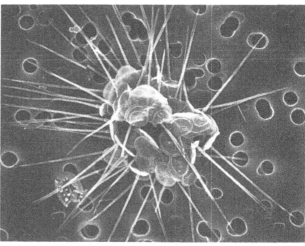

86 *Rhabdosphaera xiphos*
known range: Pliocene - Recent
known distribution: Pac (N); Atl (N); Ind; Med

85 *Rhabdosphaera clavigera var. stylifera*
known range: Living
known distribution: Pac (NE, C); Atl (N, C); Ind; Car; Med; Red

S. Nishida

A. Winter/P. Friedinger

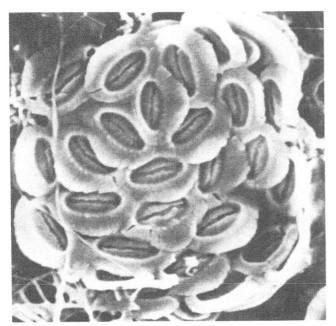

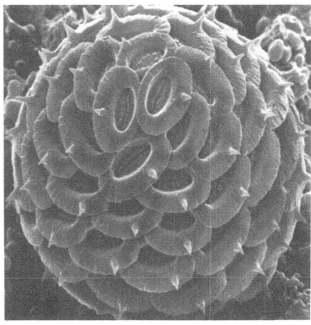

88 *Alisphaera ordinata*
known range: Living
known distribution: Pac (NE, C); Atl (N); Med

90A *Alisphaera unicornis*
known range: Living
known distribution: Pac (NE, C); Atl (N); Med

C. Samtleben

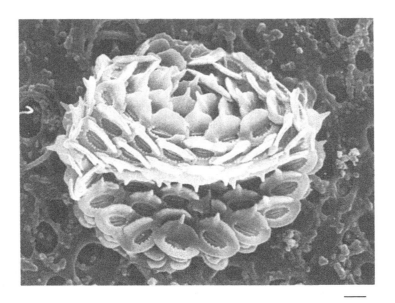

90B *Alisphaera unicornis*
known range: Living
known distribution: Pac (NE, C); Atl (N); Med

R. Jordan

C. Samtleben

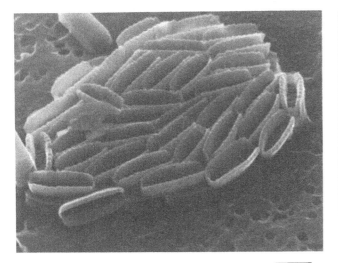

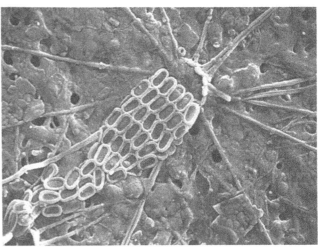

91 *Alveosphaera bimurata*
known range: Living
known distribution: Pac (NE, C); Atl (N)

92 *Calciopappus caudatus*
known range: Living
known distribution: Pac (NE,C); Atl (N); Car

A. Kleijne

J. Alcober

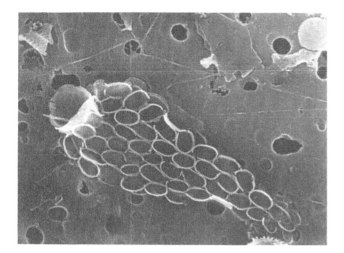

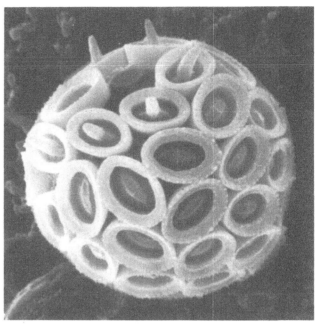

93 *Calciopappus rigidus*
known range: Living
known distribution: Atl (C); Car

96 *Coronosphaera mediterranea*
known range: Living
known distribution: Pac (NE, C); Atl (N,C)

S. Nishida

A. Winter

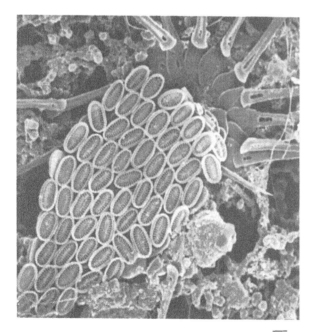

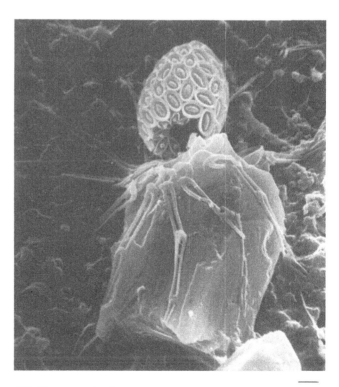

97 *Michaelsarsia adriaticus*
known range: Living
known distribution: Pac (NE,C); Atl (N);Car

98 *Michaelsarsia elegans*
known range: Living
known distribution: Atl (C); Car; Red

A. T. Haidar

A. Kleijne

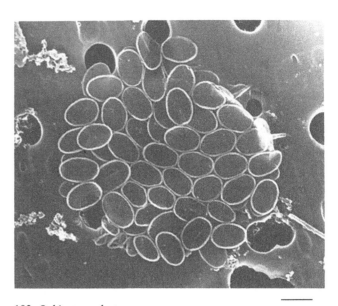

103 *Ophiaster reductus*
known range: Living
known distribution: Pac (C); Atl (N); Ind; Med; Red

101 *Ophiaster hydroideus*
known range: Living
known distribution: Pac (NE,C); Atl (N); Car

J. Alcober

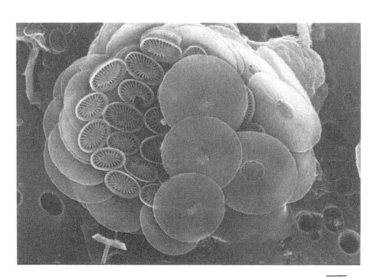

105 *Syracosphaera anthos*
known range: Living
known distribution: Pac (NE,C); Atl (N); Ind; Med; Red

C. Samtleben

A. Winter/P. Friedinger

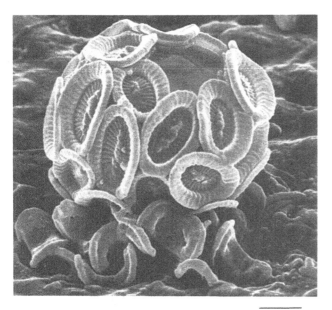

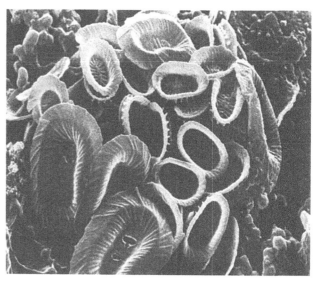

106 *Syracosphaera borealis*
known range: Living
known distribution: Pac (NE,C); Atl (N)

107 *Syracosphaera corolla*
known range: Living
known distribution: Pac (NE,C); Atl (N); Ind; Med

A. Winter

A. Winter/P. Friedinger

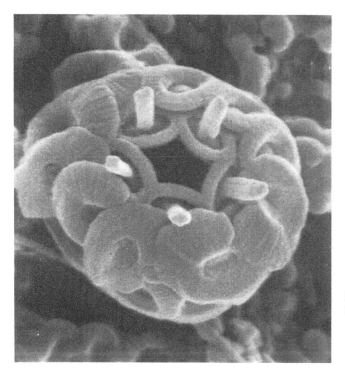

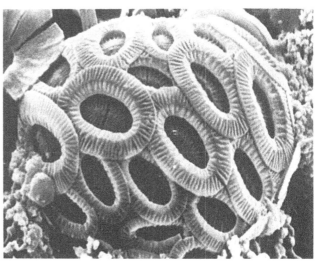

108B *Syracosphaera corrugis*
known range: Living
known distribution: Pac (NE,C); Atl (N); Red

108A *Syracosphaera corrugis*
known range: Living
known distribution: Pac (NE,C); Atl (N); Red

J. Alcober

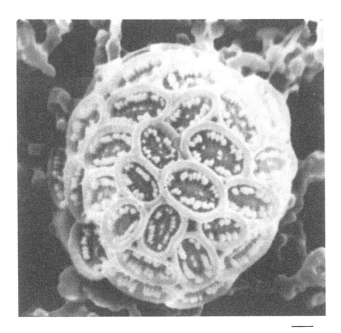

109 *Syracosphaera epigrosa*
known range: Living
known distribution: Pac (NE,C); Atl (N)

A. Winter/P. Friedinger

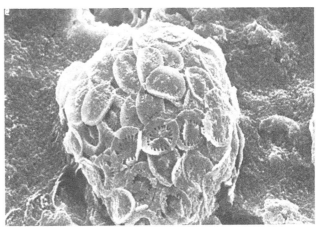

110 *Syracosphaera exigua*
known range: Living
known distribution: Pac (NE,C); Atl (N)

A. Winter/P. Friedinger

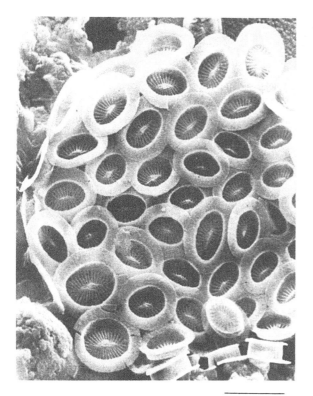

111 *Syracosphaera halldalii* f. *halldalii*
known range: Living
known distribution: Pac (NE,C); Atl (N); Med

M. Knappertsbusch

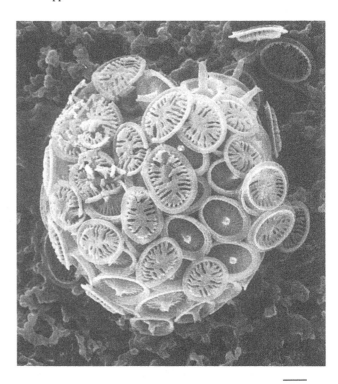

113 *Syracosphaera histrica*
known range: Living
known distribution: Pac (NE,C); Atl (N,C); Med; Car; Red

136

S. Nishida

A. Winter

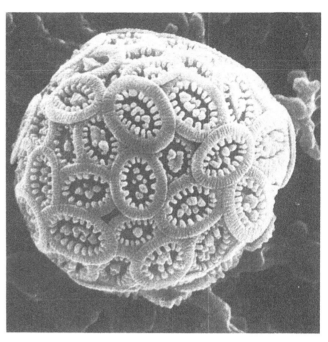

114 *Syracosphaera lamina*
known range: Living
known distribution: Pac (NE,C); Atl (N,C); Ind; Med

115A *Syracosphaera molischii*
known range: Living
known distribution: Pac (NE,C); Atl (N); Med; Car; Red

C. Samtleben

A. Kleijne

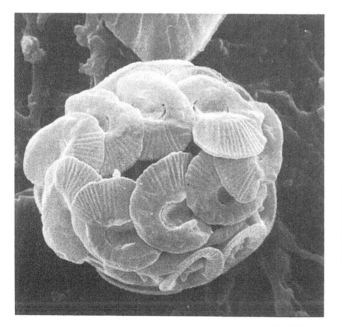

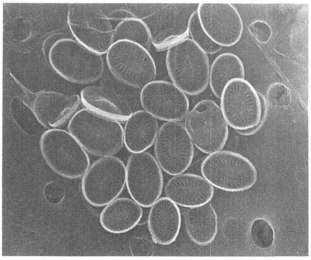

116 *Syracosphaera nana*
known range: Living
known distribution: Pac (NE,C); Atl (N,C); Red

115B *Syracosphaera molischii*
known range: Living
known distribution: Pac (NE,C); Atl (N); Med; Car; Red

S. Nishida

R. Jordan

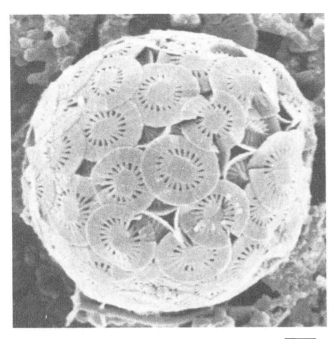

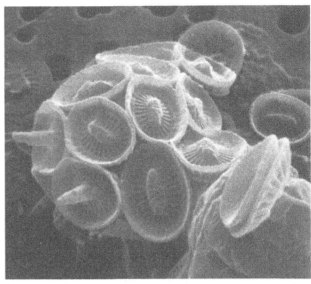

117B *Syracosphaera nodosa*
known range: Living
known distribution: Pac (NE,C); Atl (N,C); Ind; Car; Red

117A *Syracosphaera nodosa*
known range: Living
known distribution: Pac (NE,C); Atl (N,C); Ind; Car; Red

A. Kleijne

C. Samtleben

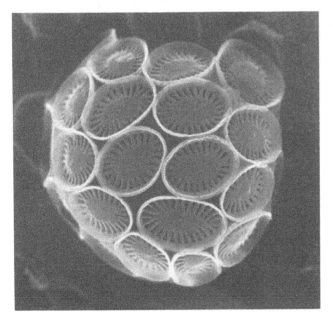

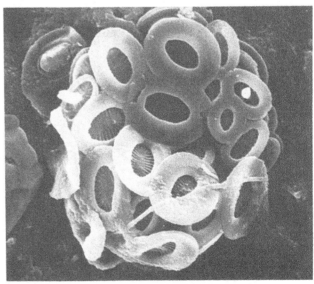

119 *Syracosphaera ossa*
known range: Living
known distribution: Pac (NE,C); Atl (N)

118 *Syracosphaera orbiculus*
known range: Living
known distribution: Pac (NE,C); Atl (N)

A. Winter

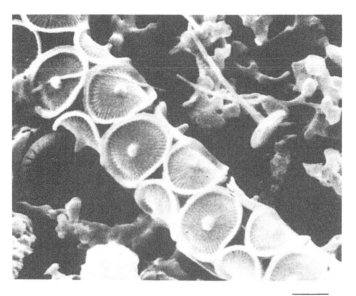

120 *Syracosphaera pirus*
known range: Living
known distribution: Pac (NE,C); Atl (N,C); Car; Red

M. Knappertsbusch

J. Alcober

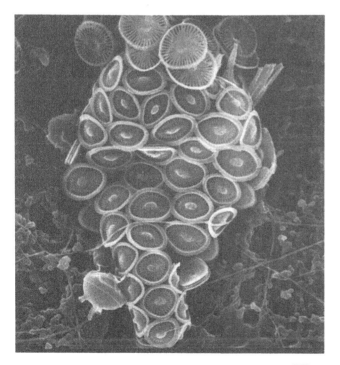

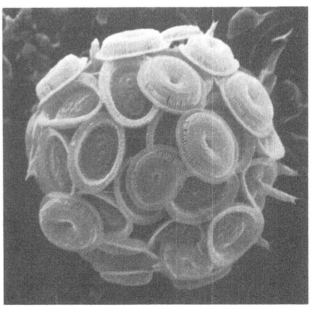

121 *Syracosphaera prolongata*
known range: Living
known distribution: Pac (NE,C); Atl (N,C); Car; Red

122 *Syracosphaera pulchra*
known range: Living
known distribution: Pac (NE,C); Atl (N,C); Ind; Med; Car; Red

J. Alcober

H. A. Thomsen

123 *Syracosphaera rotula*
known range: Living
known distribution: Pac (NE,C); Atl (N); Med; Red

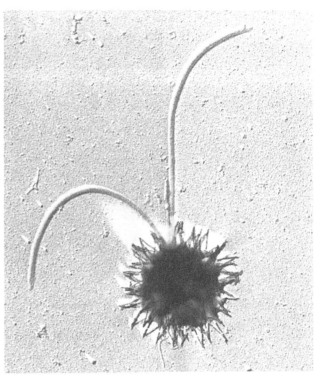

124 *Calciarcus alaskensis*
known range: Living
known distribution: Pac(NE); Atl(N); Red

S. Nishida

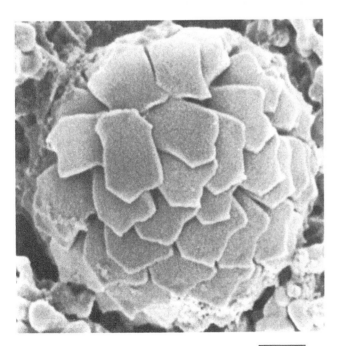

125 *Florisphaera profunda* var. *elongata*
known range: Late Miocene - Recent
known distribution: Pac (NE,C); Atl (N); Med

A. Winter/P. Friedinger

126 *Florisphaera profunda* var. *profunda*
known range: Late Miocene - Recent
known distribution: Pac (NE,C); Atl (N); Med

140

A. Winter/P. Friedinger

128 *Polycrater galapagensis*
known range: Living
known distribution: Pac (NE,C); Atl (N)

M. Knappertsbusch

129A *Thorosphaera flabellata*
known range: Living
known distribution: Pac (NE,C); Atl (N); Med; Car

R. Jordan

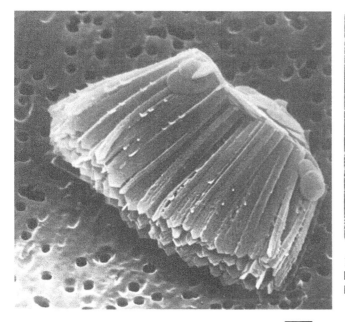

129B *Thorosphaera flabellata*
known range: Living
known distribution: Pac (NE,C); Atl (N); Med; Car

A. Winter/P. Friedinger

130 *Turrilithus latericioides*
known range: Living
known distribution: Pac (C); Atl (N); Ind; Med

141

A. Kleijne

A. Kleijne

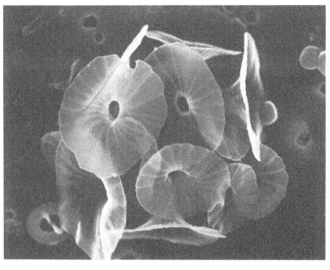

131 *Umbellosphaera irregularis*
known range: Late Pliocene - Recent
known distribution: Pac (NE,C); Atl (N,C); Ind; Car; Red

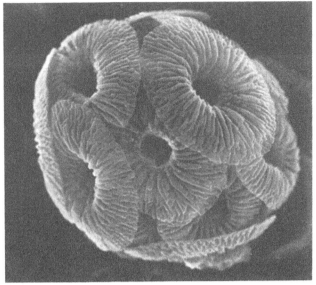

132 *Umbellosphaera tenuis*
known range: Recent
known distribution: Pac (NE,C); Atl (N,C); Ind; Car; Red

A. Kleijne

A. Kleijne

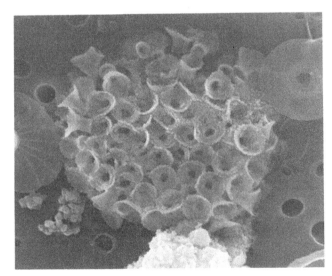

138 *Calicasphaera blokii*
known range: Living
known distribution: Med

139 *Calicasphaera concava*
known range: Living
known distribution: Atl (N)

A. Kleijne

A. Kleijne

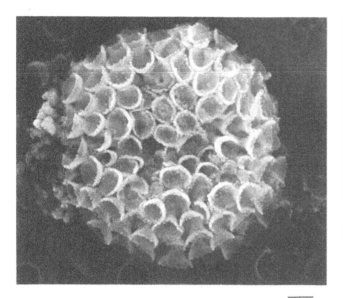

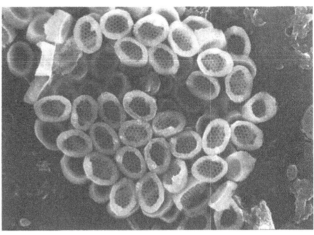

142 *Calicasphaera dentata*
known range: Living
known distribution: Pac (NE); Med

140 *Calicasphaera diconstricta*
known range: Living
known distribution: Med; Red

A. Kleijne

A. Winter

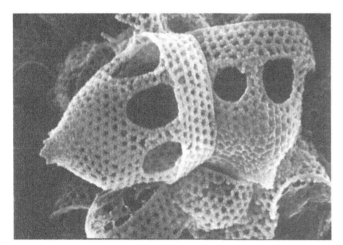

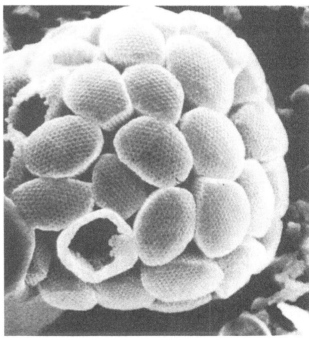

144 *Calyptrosphaera heimdalae*
known range: Living
known distribution: Med

145A *Calyptrosphaera oblonga*
known range: Living
known distribution: Pac (NW,NE,C); Atl (N,C); Ind; Med; Car;
Red

143

C. Samtleben

A. Kleijne

145B *Calyptrosphaera oblonga*
known range: Living
known distribution: Pac (NW,NE,C); Atl (N,C); Ind; Med; Car;
Red

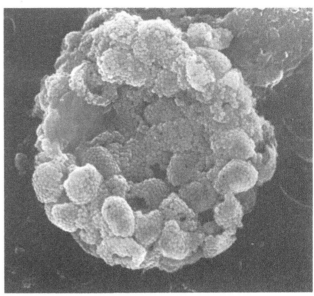

146 *Calyptrosphaera sphaeroidea*
known range: Living
known distribution: Pac (NE,C); Atl (N);Med

J. Alcober

A. Kleijne

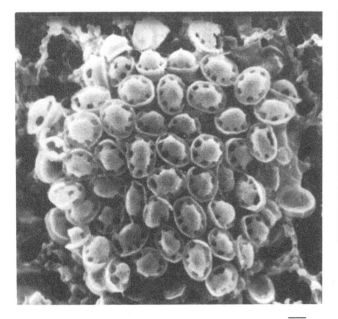

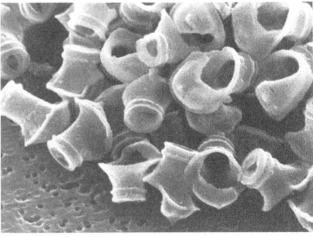

148 *Flosculosphaera calceolariopsis*
known range: Living
known distribution: Atl (N,NE)

147 *Daktylethra pirus*
known range: Living
known distribution: Pac (NE); Atl (N,C); Ind; Med; Car

A. Kleijne

J. Alcober

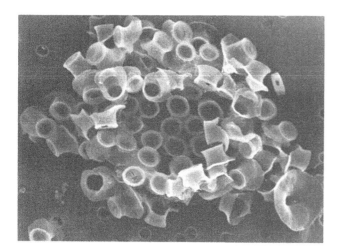

149 *Flosculosphaera sacculus*
known range: Living
known distribution: Atl (NE)

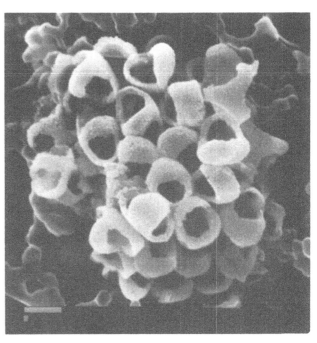

151 *Homozygosphaera arethusae*
known range: Living
known distribution: Atl (N,C); Ind; Med; Red

A. Kleijne

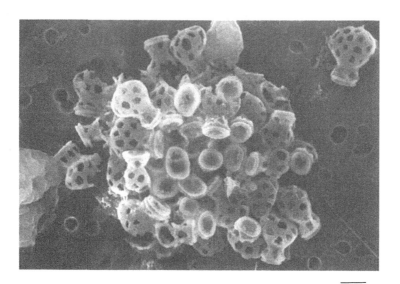

150 *Gliscolithus amitakarenae*
known range: Living
known distribution: Atl (N); Ind; Med

145

A. Kleijne

M. Knappertsbusch

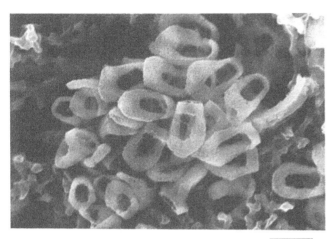

152 *Homozygosphaera spinosa*
known range: Living
known distribution: Atl (C); Ind; Med; Car

153 *Homozygosphaera triarcha*
known range: Living
known distribution: Pac (C); Atl (N); Ind; Med; Car; Red

J. Alcober

C. Samtleben

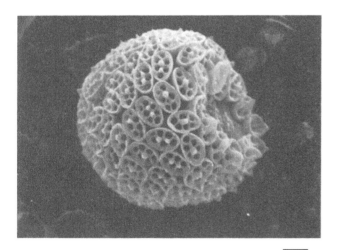

155 *Homozygosphaera vercellii*
known range: Living
known distribution: Pac (C); Atl (N); Ind; Med; Car; Red

156 *Periphyllophora mirabilis*
known range: Living
known distribution: Pac (NE,C,S); Atl (N,C); Ind; Med; Car

A. Kleijne

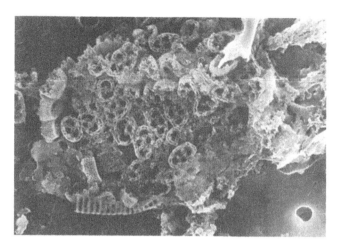

157 *Syracolithus bicorium*
known range: Living
known distribution: Atl (N); Med

A. Kleijne

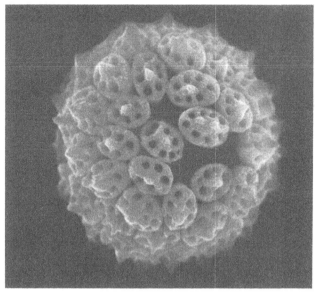

159 *Syracolithus confusus*
known range: Living
known distribution: Pac (NE); Med

J. Alcober

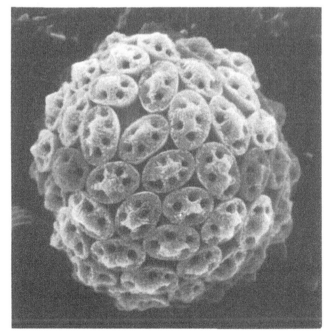

160 *Syracolithus dalmaticus*
known range: Living
known distribution: Pac (NE,C); Atl (N); Ind; Med

A. Kleijne

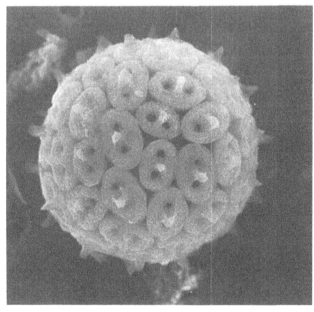

161 *Syracolithus ponticuliferus*
known range: Living
known distribution: Pac (NE,C); Atl (N); Car

147

A. Kleijne

M. Knappertsbusch

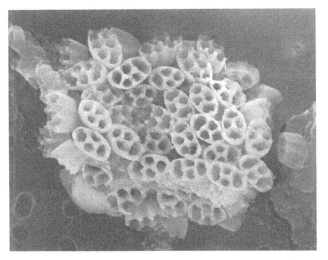

162 *Syracolithus quadriperforatus*
known range: Living
known distribution: Pac (NE,C); Atl (N,C); Ind; Med; Car

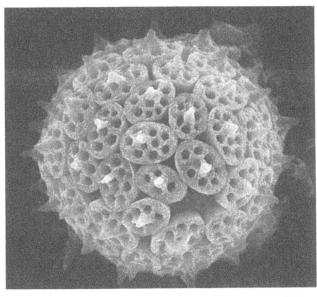

163 *Syracolithus schilleri*
known range: Living
known distribution: Pac (NE,C); Atl (N)

J. Alcober

A. Kleijne

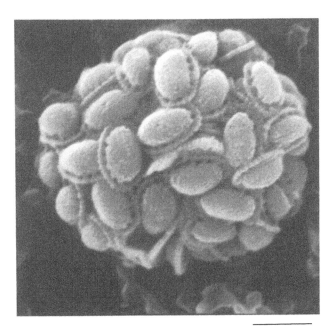

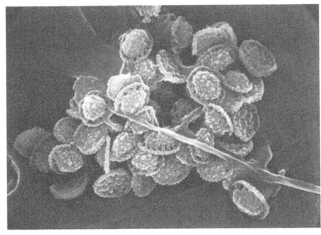

165 *Anthosphaera lafourcadii*
known range: Living
known distribution: Atl (N)

164 *Anthosphaera fragaria*
known range: Living
known distribution: Atl (N); Med

A. Kleijne

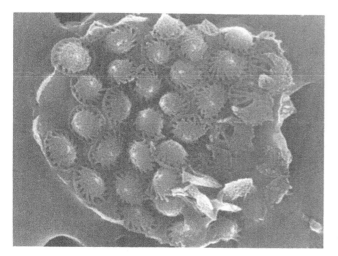

166 *Anthosphaera periperforata*
known range: Living
known distribution: Atl (N); Ind; Med

A. Kleijne

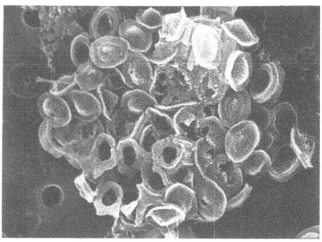

167 *Calyptrolithia divergens*
known range: Living
known distribution: Atl (N); Med

A. Kleijne

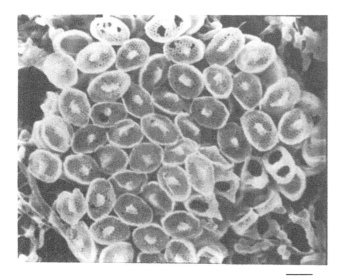

168 *Calyptrolithina divergens* f. *tuberosa*
known range: Living
known distribution: Pac (NE); Atl (N,C); Ind; Car

C. Samtleben

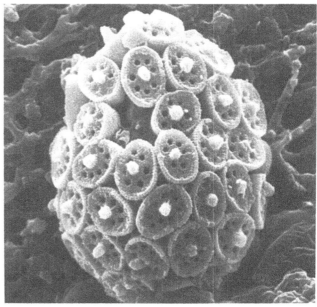

169 *Calyptrolithina multipora*
known range: Living
known distribution: Pac (NW,NE,C); Atl (N,C); Ind; Med; Car;
Red

A. Kleijne

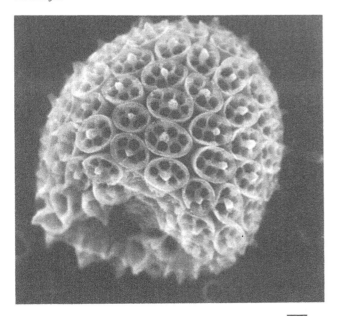

170 *Calyptrolithina wettsteinii*
known range: Living
known distribution: Pac (NE,C); Atl (N,C); Med; Car

S. Nishida

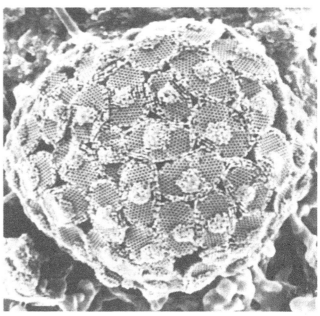

171 *Calyptrolithophora gracillima*
known range: Living
known distribution: Pac (NE,C); Atl (N,C); Med; Car

S. Nishida

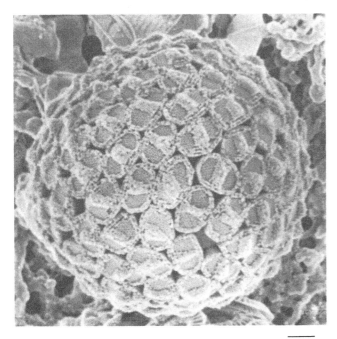

172 *Calyptrolithophora hasleana*
known range: Living
known distribution: Pac (NE,C); Atl (N,C); Med; Car

S. Nishida

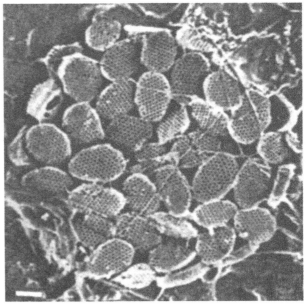

173 *Calyptrolithophora papillifera*
known range: Living
known distribution: Pac (NE,C); Atl (N,C); Ind; Med

A. Kleijne

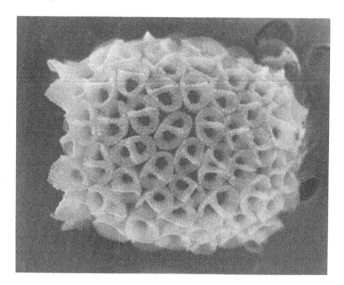

A. Kleijne

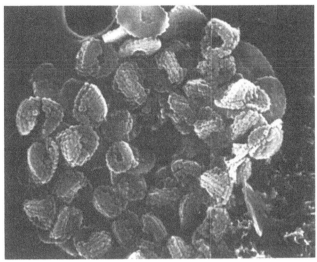

174 *Corisphaera gracilis*
known range: Living
known distribution: Pac (NE,C,S); Atl (N,C); Ind; Med; Red

175 *Corisphaera strigilis*
known range: Living
known distribution: Atl (N,C); Ind; Med; Red

A. Kleijne

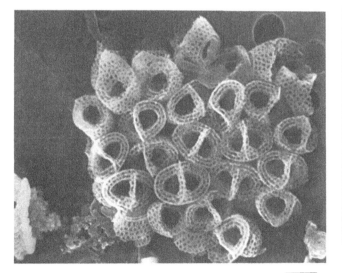

A. Kleijne

177 *Helladosphaera cornifera*
known range: Living
known distribution: Pac (C), Atl (N,C); Ind; Med; Car

176 *Corisphaera tyrrheniensis*
known range: Living
known distribution: Atl (N); Med

A. Kleijne

J. Alcober

178 *Helladosphaera pienaarii*
known range: Living
known distribution: Atl (N); Ind; Med

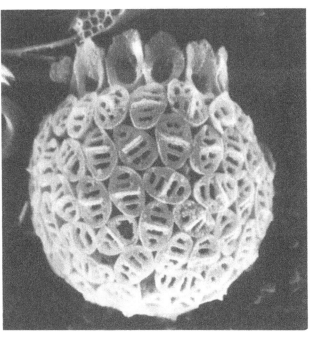

179 *Poricalyptra aurisinae*
known range: Living
known distribution: Pac (NE,C); Atl (N,C); Car

J. Alcober

C. Samtleben

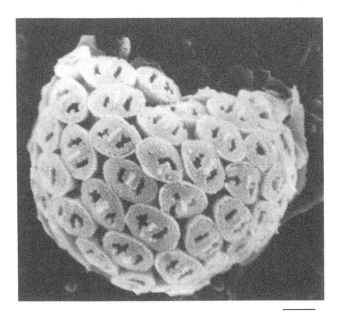

180 *Poricalyptra gaarderii*
known range: Living

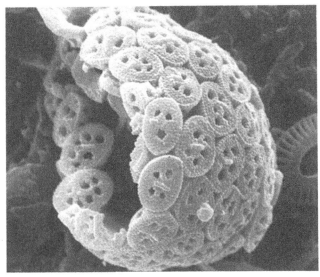

181 *Poricalyptra isselii*
known range: Living
known distribution: Atl (N); Med

C. Samtleben

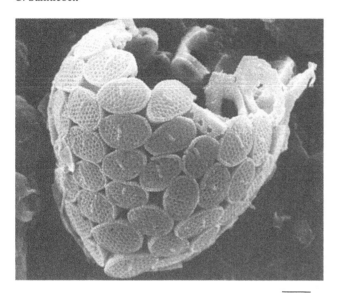

182 *Poricalyptra magnaghii*
known range: Living
known distribution: Pac (NE,C); Atl (N); Ind; Med; Car

A. Kleijne

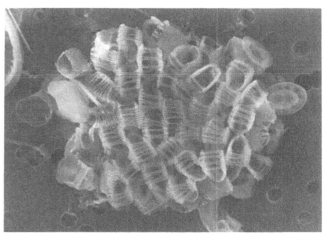

183 *Poritectolithus maximus*
known range: Living
known distribution: Pac (NE); Atl (N); Ind

A. Kleijne

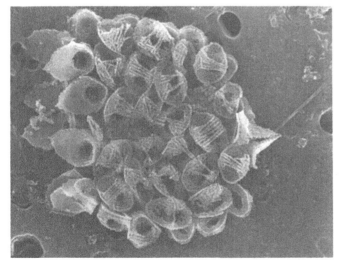

184 *Poritectolithus poritectus*
known range: Living
known distribution: Atl (N,C); Ind; Red

A. Kleijne

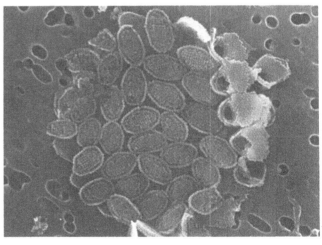

185 *Poritectolithus tyronus*
known range: Living
known distribution: Atl (N)

A. Kleijne

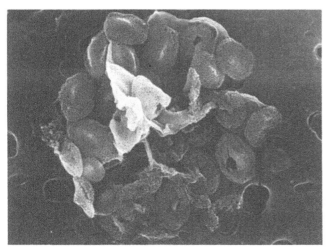

186 *Sphaerocalyptra adenensis*
known range: Living
known distribution: Ind; Red

A. Kleijne

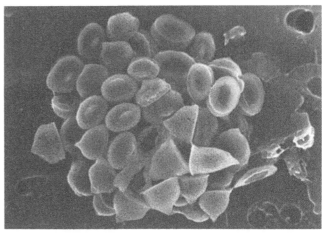

187 *Sphaerocalyptra quadridentata*
known range: Living
known distribution: Atl (N,C); Med

A. Kleijne

J. Alcober

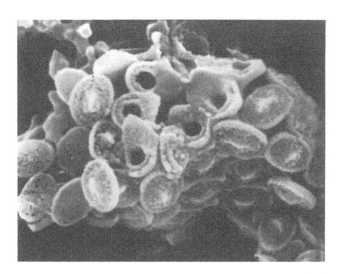

188 *Zygosphaera amoena*
known range: Living
known distribution: Atl (N); Ind; Med

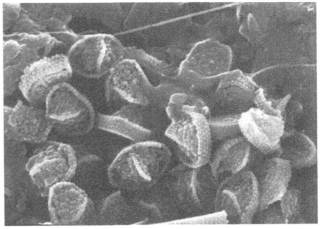

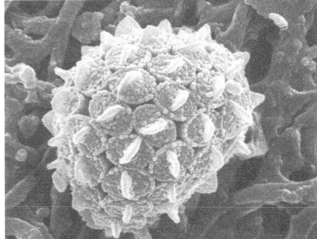

Top ⸻

Right
189 *Zygosphaera bannockii*
known range: Living
known distribution: Atl (N,C); Ind; Med

Bottom ⸻

J. Alcober

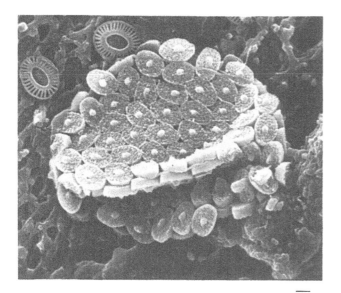

J. Alcober

190 *Zygosphaera hellenica*
known range: Living
known distribution: Pac (NE,C,S); Atl (N,C); Ind; Med

191 *Zygosphaera marsilii*
known range: Living
known distribution: Atl (N,C); Ind; Med; Red

H. A. Thomsen

H. A. Thomsen

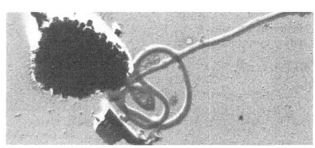

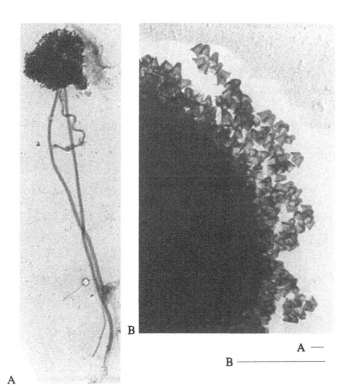

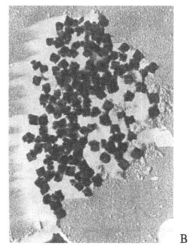

192 *Balaniger balticus*
known range: Living
known distribution: Atl(N)

193 *Quaternariella obscura*
known range: Living
known distribution: Atl (N)

H. A. Thomsen

H. A. Thomsen

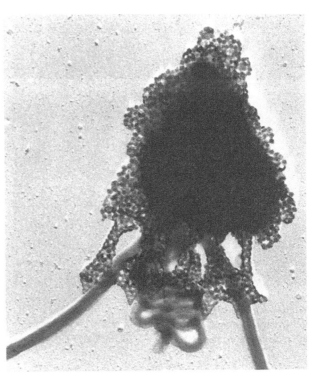

194 *Trigonaspis diskoensis*
known range: Living
known distribution: Atl (N)

195 *Trigonaspis melvillea*
known range: Living
known distribution: Wed

H. A. Thomsen

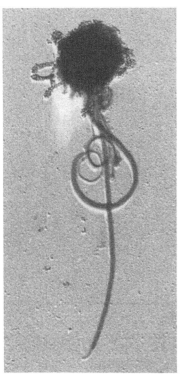

196 *Trigonaspis
minutissima*
known range: Living
known distribution: Atl (N)

H. A. Thomsen

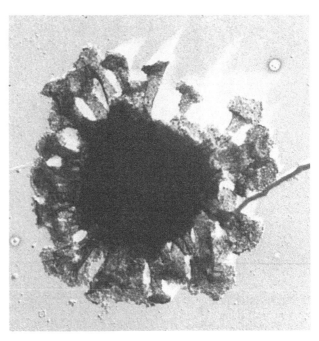

Right
198 *Turrisphaera borealis*
known range: Living
known distribution: Pac
(NE); Atl(N)

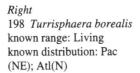

Appendix

	Species illustrated	Collection locality
1	*Braarudosphaera bigelowii*	northwestern Pacific
3	*Calciosolenia murrayi*	Arabian Sea
5	*Ceratolithus cristatus* var. *telesmus*	Flores Sea, Indonesia
4A	*Ceratolithus cristatus* var. *cristatus*	western Mediterranean
4B	*Ceratolithus cristatus* var. *cristatus*	western north-central Atlantic
4C	*Ceratolithus cristatus* var. *cristatus*	Vema channel (sediment)
5	*Ceratolithus cristatus* var. *telesmus*	Arabian Sea
6	*Calcidiscus leptoporus* f. *leptoporus*	southwestern Indian Ocean
7	*Calcidiscus leptoporus* f. *rigidus*	western Mediterranean
8	*Coccolithus neohelis*	Mediterranean
11A	*Coccolithus pelagicus* f. *pelagicus*	western equatorial Pacific
11B	*Coccolithus pelagicus* f. *pelagicus*	Norwegian Sea
11C	*Coccolithus pelagicus* f. *pelagicus*	Norwegian Sea
12	*Hayaster perplexus*	Kuroshio Current
13	*Neosphaera coccolithomorpha*	southwestern Indian Ocean
14	*Oolithotus fragilis* var. *cavum*	Gulf of Aqaba/Red Sea
18	*Umbilicosphaera hulburtiana*	southwestern Indian Ocean
21	*Umbilicosphaera sibogae* var. *foliosa*	western Mediterranean
22	*Umbilicosphaera sibogae* var. *sibogae*	northeastern Indian Ocean
23A	*Helicosphaera carteri* var. *carteri*	southwestern Indian Ocean
23B	*Helicosphaera carteri* var. *carteri*	southwestern Indian Ocean
24	*Helicosphaera carteri* var. *hyalina*	Arabian Sea
26	*Helicosphaera pavimentum*	southwestern Indian Ocean
32	*Emiliania huxleyi* var. *huxleyi*	Gulf of Suez
33	*Emiliania huxleyi* var. *corona*	Kuroshio Current
35	*Gephyrocapsa ericsonii*	Arabian Sea
36	*Gephyrocapsa muellerae*	western Mediterranean
37	*Gephyrocapsa oceanica*	Caribbean Sea
38	*Gephyrocapsa ornata*	Kuroshio Current
39	*Reticulofenestra parvula* var. *parvula*	western Mediterranean
42A	*Reticulofenestra sessilis*	northwestern Pacific
42B	*Reticulofenestra sessilis*	equatorial Atlantic
44	*Pappomonas flabellifera* var. *flabellifera*	west Greenland
47	*Papposphaera lepida*	central North Atlantic
49	*Papposphaera sagittifera*	Norwegian Sea
50	*Papposphaera sarion*	west Greenland
55	*Cricosphaera quadrilaminata*	northeast Atlantic
58	*Pleurochrysis placolithoides*	French Atlantic coast
62	*Pontosphaera discopora*	southwestern Pacific
63	*Pontosphaera japonica*	western Mediterranean
64	*Pontosphaera syracusana*	eastern Caribbean
66	*Scyphosphaera apsteinii* f. *apsteinii*	western Mediterranean
68	*Acanthoica acanthifera*	Arabian Sea
71	*Acanthoica maxima*	Arabian Sea
72	*Acanthoica quattrospina*	East China Sea
73A	*Algirosphaera oryza*	western Mediterranean
73B	*Algirosphaera oryza*	western Mediterranean
75	*Algirosphaera robusta*	western Mediterranean
77	*Anacanthoica acanthos*	western Mediterranean
78	*Anacanthoica cidaris*	Indian Ocean
79	*Cyrtosphaera aculeata*	eastern Caribbean
80	*Cyrtosphaera cucullata*	central North Atlantic
82	*Discosphaera tubifera*	Arabian Sea
84	*Rrabdosphaera clavigera* var. *clavigera*	southwestern Indian Ocean
85	*Rhabdosphaera clavigera* var. *stylifera*	southeast Atlantic
86	*Rhabdosphaera xiphos*	eastern Mediterranean Sea
88	*Alisphaera ordinata*	northwestern Pacific Ocean
90A	*Alisphaera unicornis*	southwestern Indian Ocean
90B	*Alisphaera unicornis*	equatorial Atlantic
91	*Alveosphaera bimurata*	northeastern Atlantic

	Species illustrated	Collection locality
92	*Calciopappus caudatus*	Norwegian Sea
93	*Calciopappus rigidus*	Arabian Sea
96	*Coronosphaera mediterranea*	western Mediterranean
97	*Michaelsarsia adriaticus*	northwestern Pacific Ocean
98	*Michaelsarsia elegans*	Gulf of Aqaba/Red Sea
101	*Ophiaster hydroideus*	Wester North Atlantic/ south east Bermuda
103	*Ophiaster reductus*	south of India
105	*Syracosphaera anthos*	central North Atlantic
106	*Syracosphaera borealis*	Norwegian Sea
107	*Syracosphaera corolla*	southwestern Indian Ocean
108A	*Syracosphaera corrugis*	Gulf of Aqaba/Red Sea
108B	*Syracosphaera corrugis*	southwestern Indian Ocean
109	*Syracosphaera epigrosa*	western Mediterranean
110	*Syracosphaera exigua*	southwestern Indian Ocean
111	*Syracosphaera halldalii* f. *halldalii*	southwestern Indian Ocean
113	*Syracosphaera histrica*	western Mediterranean
114	*Syracosphaera lamina*	East China Sea
115A	*Syracosphaera molischii*	Gulf of Aqaba/Red Sea
115B	*Syracosphaera molischii*	south Atlantic
116	*Syracosphaera nana*	eastern Mediterranean
117A	*Syracosphaera nodosa*	northeast Atlantic
117B	*Syracosphaera nodosa*	East China Sea
118	*Syracosphaera orbiculus*	south of India
119	*Syracosphaera ossa*	equatorial Atlantic
120	*Syracosphaera pirus*	southwestern Indian Ocean
121	*Syracosphaera prolongata*	western Mediterranean
122	*Syracosphaera pulchra*	western Mediterranean
123	*Syracosphaera rotula*	western Mediterranean
124	*Calciarcus alaskensis*	Weddell Sea
125	*Florisphaera profunda v. elongata*	Phillipine Sea
126	*Florisphaera profunda v. profunda*	southwestern Indian Ocean
128	*Polycrater galapagensis*	southwestern Indian Ocean
129A	*Thorosphaera flabellata*	western Mediterranean
129B	*Thorosphaera flabellata*	northeast Atlantic
130	*Turrilithus latericioides*	southwestern Indian Ocean
131	*Umbellosphaera irregularis*	Arabian
132	*Umbellosphaera enuis*	Gulf of Suez
138	*Calicasphaera blokii*	eastern Mediterranean
139	*Calicasphaera concava*	central North Atlantic
140	*Calicasphaera diconstricta*	Gulf of Suez
142	*Calyptrosphaera dentata*	eastern Mediterranean
144	*Calyptrosphaera heimdalae*	central North Atlantic
145A	*Calyptrosphaera oblonga*	southwestern Indian Ocean
145B	*Calyptrosphaera oblonga*	south Atlantic
146	*Calyptrosphaera sphaeroidea*	eastern Mediterannean
147	*Daktylethra pirus*	western Mediterranean
148	*Flosculosphaera calceolariopsis*	eastern Mediterannean
149	*Flosculosphaera sacculus*	eastern Mediterannean
150	*Gliscolithus amitakarenae*	central North Atlantic
151	*Homozygosphaera arethusae*	western Mediterranean
152	*Homozygosphaera spinosa*	central North Atlantic
153	*Homozygosphaera triarcha*	western Mediterranean
155	*Homozygosphaera vercellii*	western Mediterranean
156	*Periphyllophora mirabilis*	southern Atlantic
157	*Syracolithus bicorium*	eastern Mediterrranean
159	*Syracolithus confusus*	western Mediterranean
160	*Syracolithus dalmaticus*	western Mediterranean
161	*Syracolithus ponticuliferus*	central North Atlantic
162	*Syracolithus quadriperforatus*	central North Atlantic
163	*Syracolithus schilleri*	western Mediterranean
164	*Anthosphaera fragaria*	western Mediterranean
165	*Anthosphaera lafourcadii*	central North Atlantic

	Species illustrated	Collection locality
166	*Anthosphaera periperforata*	central North Atlantic
167	*Calyptrolithia divergens*	Arabian Sea
168	*Calyptrolithina divergens f. tuberosa*	eastern Mediterannean
169	*Calyptrolithina multipora*	southern Atlantic
170	*Calyptrolithina wettsteinii*	western Mediterranean
171	*Calyptrolithophora gracillima*	northwestern Pacific Ocean
172	*Calyptrolithophora hasleana*	East China Sea
173	*Calyptrolithophora papillifera*	northwestern Pacific Ocean
174	*Corisphaera gracilis*	central North Atlantic
175	*Corisphaera strigilis*	northeastern Indian Ocean
176	*Corisphaera tyrrheniensis*	central North Atlantic
177	*Helladosphaera cornifera*	central North Atlantic
178	*Helladosphaera pienaarii*	central North Atlantic
179	*Poricalyptra aurisinae*	western Mediterranean
180	*Poricalyptra gaarderii*	western Mediterranean
181	*Poricalyptra isselii*	southern Atlantic
182	*Poricalyptra magnaghii*	equatorial Atlantic
183	*Poritectolithus maximus*	central North Atlantic
184	*Poritectolithus poritectus*	northeastern Indian Ocean
185	*Poritectolithus tyronus*	central North Atlantic
186	*Sphaerocalyptra adenensis*	Arabian Sea
187	*Sphaerocalyptra quadridentata*	eastern Mediterranean
188	*Zygosphaera amoena*	western Mediterranean
189	*Zygosphaera bannockii*	eastern Mediterranean
190	*Zygosphaera hellenica*	western Mediterranean
191	*Zygosphaera marsilii*	western Mediterranean
192	*Balaniger balticus*	Danish Coast
193	*Quaternariella obscura*	Weddell Sea
194	*Trigonaspis diskoensis*	West Greenland
195	*Trigonaspis melvillea*	Weddell Sea
196	*Trigonaspis minutissima*	West Greenland
198	*Turrisphaera borealis*	West Greenland

8 Biogeography of living coccolithophores in ocean waters

AMOS WINTER, RIC W. JORDAN AND PETER H. ROTH

Introduction

There is still much discussion, primarily because of lack of information, as to what controls the vertical and horizontal distribution of living coccolithophores. Latitude, ocean currents, water masses (Okada and Honjo, 1973; McIntyre and Bé, 1967), nutrients, salinity, temperature, available light, vitamins and minerals (such as iron; Martin *et al.*, 1989) are all thought to control the distribution of coccolithophores, vertically and horizontally in the oceans. Unlike most other phytoplankton groups, coccolithophore diversity is highest in warm, low productivity 'blue water' regions such as gyre centers and in restricted areas of circulation such as the Red Sea and Gulf of California. However, under certain conditions coccolithophores (e.g., *Emiliania huxleyi*) bloom in cooler, nutrient rich waters such as the Norwegian Fjords where 10^7–10^8 coccolithophores have been counted in one liter of water (Birkenes and Braarud, 1952; Berge, 1962). These frequent blooms can often be seen clearly from satellites using the Coastal Zone Color Scanner (CZCS; Fig. 1). The amount of $CaCO_3$ in the top 60 m at one of these bloom sites has been conservatively calculated at 7.2×10^4 tonnes (Holligan *et al.*, 1983). Such great quantities of calcite, produced by organisms reproducing at rates as high as 2.6 divisions/day (Brand, 1981), must have an impact on the global CO_2 cycle and the chemistry of the oceans. In this chapter we will attempt to describe the biogeography of coccolithophores.

Methods

The contribution and presence of coccolithophores within oceanic waters has been recognized in a number of ways using a wide range of techniques. The discovery of long-chain alkenones and alkenoates in *Emiliania huxleyi* and some non-calcifying members of the Isochrysidales, has proved to be a powerful biomarker (Marlowe *et al.*, 1984), although one must assume that the signal is predominantly produced by the coccolithophore species and its nearest (untested) relatives, *Gephyrocapsa* and *Reticulofenestra* (Marlowe *et al.*, 1990). These compounds are absent in other phytoplankton groups and in the other coccolithophore species tested to date, and thus remain very selective taxonomic tools. Over the past 40 years significant progress has been made in the analysis of phytoplankton pigments and its use in chemotaxonomy. The continued development of thin layer chromatography (TLC) and high performance liquid chromatography (HPLC) has allowed workers to recognize and separate structurally distinct derivatives from previously grouped pigment classes. For instance, in the case of *E. huxleyi* the commonest pigments found were chlorophyll *a* and *c* and the carotenoid fucoxanthin (Jeffrey and Allen, 1964). The fucoxanthin pigment, on closer analysis, was found to be a new, but related, derivative called 19'Hexanoyloxyfucoxanthin (Arpin *et al.*, 1976). This compound has now been detected in a few species of prymnesiophytes, chysophytes and dinoflagellates, and hence its use as a chemotaxonomic tool is rather limited. The presence of chlorophyll *a* and *c* does not separate the coccolithophores from many of the other algal groups, but a closer look at the chlorophyll *c* unit of *E. huxleyi* has yielded some new, and as yet unique, pigments (Jeffrey and Wright, 1987; Nelson and Wakeham, 1989). Detection instruments such as flurometers and satellites cannot as yet separate out these new pigments as their reflectance spectra overlap with those of other pigments. Satellite imagery offers another way of studying the surface distribution of coccolithophores. The scattering of light caused by the presence of large quantities of suspended birefringent coccoliths has been well documented (Holligan *et al.*, 1983; Groom and Holligan, 1987). However, this technique cannot determine the composition of the reflecting particles and is limited in its penetration by the clarity of the upper water column. In most oceanic waters the satellite can 'see' about 15–30 m, but in turbid or high pigment waters this may be less than 15 m, and in some coastal regions less than 1 m (Holligan *et al.*, 1989). Furthermore, satellite imagery provides no information about the taxonomy, condition or quantity of photosynthetically active coccolithophore cells within this depth range or deeper. Most knowledge on the distribution of living coccolithophores comes from the collection of surface water samples (0–10 m) by a slowly moving research vessel. Many new coccolithophore species are described in this way, but little information is obtained about their vertical distribution or of species which inhabit deeper waters (80–220 m).

Fig. 1. Satellite image of a coccolithophore bloom in the North Sea. Image shows an extensive coccolithophore bloom in the northern North Sea extending as a plume into the southern Skagerrak between the Norwegian Channel and Danish coastal waters. Photo courtesy of P.M. Holligan, Plymouth, UK.

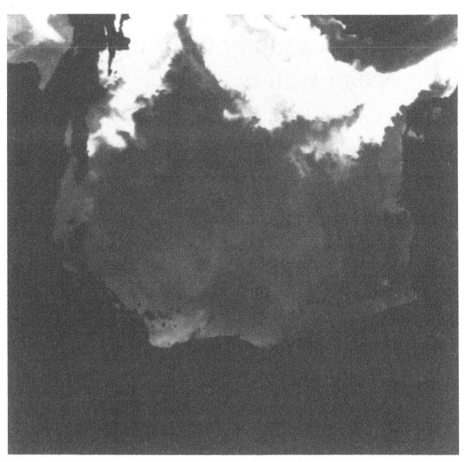

A

Fig. 2. Equipment for collecting coccolithophores in the water column and surface sediments. A. a neuston net; B. a large filtration pump system; C. a CTD rig carrying 2.5 liter water bottles; D. a multicorer.

B

C

Furthermore, few of these studies have provided quantitative data or noted the contribution of coccolithophores to phytoplankton assemblages.

The method used must be an important criterion when considering any review of past biogeographic studies. One must be aware that over the years different collection, preservation and preparation techniques have played a vital role in deciding what species may be present in a water sample. The early workers (e.g., Murray and Blackman, 1898; Gran, 1912) pumped seawater through silk meshes, whose capture efficiency increased with pump time. The samples were then fixed in Osmic acid, washed and preserved in camphor water or formaldehyde. They used light microscopy and for the most part under normal, not polarized, illumination. However, polarizing microscopes were available and used by geologists studying the coccoliths in chalk (Sorby, 1861). Tow-nets were later replaced by water bottles, which had been available since before the HMS *Challenger* expedition. From 1930 onwards the most commonly used were the Nansen, van Dorn and Niskin bottles (Fig. 2). The water from these bottles was either centrifuged and preserved in formalin or glutaraldehyde, or vacuum

pumped through filters (notably Millipore and then later Nuclepore), air or oven dried and stored. For surface water samples the shipboard seawater pump system is still a useful method of collection, especially when the ship is steaming or it is too rough for water bottle deployment. With the advent of the transmission electron microscope (TEM), carbon replicas of coccoliths and coccospheres were photographed to illustrate papers in the 1950s and 1960s. It was not until around 1970 that coccolithophores were published as scanning electron micrographs (SEM) and coccoliths could be viewed directly without forming replicas.

It is important to remember that initial studies by investigators (Huxley, 1868; Wallich, 1877), curious about the origins of the coccoliths previously seen within chalk deposits, started in the late nineteenth century (see chapter 1). These pioneer researchers used only simple light microscopes, often with lenses they had ground themselves, to observe the coccolithophores. Many of their line drawings were extremely detailed and have been subsequently confirmed in the SEM nearly 100 years later. For instance com-

D

Table 1. *Biogeographic coccolithophore zones*

Subarctic

In coastal regions of Arctic countries where low temperatures and salinities prevail, the only living coccolithophores are those belonging to the Papposphaeraceae and the partially calcified genera.

In open ocean, mainly during summer months, *Coccolithus pelagicus* (and its motile form *"Crystallolithus hyalinus"*) and *Calciopappus caudatus* are the only coccolithophores present. In waters of Atlantic origin *Emiliania huxleyi* and *Algirosphaera robusta* may be common.

Temperate

Dominated for most or all of the year by *E. huxleyi*. *Gephyrocapsa muellerae* common to abundant only in this zone (or upwelling waters), especially during the summer months. Flora also characterized by other placolith-bearing species.

Subtropical

High diversity with vertical zonation (see Fig. 5). Characterized by holococcolithophores, *Discosphaera tubifera*, *Rhabdosphaera clavigera*, *Umbellosphaera* spp., *Florisphaera profunda*, *Thorosphaera flabellata* and *Syracosphaera* spp.

Tropical

Dominated by placolith-bearing species, especially *E. huxleyi*, *Calcidiscus leptoporus* and *Gephyrocapsa oceanica*. *Umbellosphaera* spp., *F. profunda* and *T. flabellata* also present. *Reticulofenestra sessilis* only found in this zone.

Subantarctic

The only coccolithophores living in these subzero polar waters are those belonging to the Papposphaeraceae or the partially calcified genera.

pare the drawing of a cell of *Coccolithus pelagicus* by Murray and Blackman (1898), reproduced by Siesser (chapter 1) with the micrograph shown in Winter and Siesser (chapter 7).

Biogeography of coccolithophores

Coccolithophores, like most phytoplankton and microzooplankton groups, can be separated into five major latitudinal zones; Subarctic, Temperate (or Transitional), Subtropical (or Central), Tropical (or Equatorial) and Subantarctic (Fig. 3). These zones have been shown by various workers to be associated with the movements of the major water masses. For instance, the Subtropical Zones in both hemispheres of the Pacific and Atlantic Oceans are synonymous with the positions of the mid-ocean gyres. However, the boundaries of these zones are by no means static or simple straight lines. The frontal systems at all these boundaries are continually in motion, meandering back and forth, and with highly convoluted edges. Furthermore, there is a region of transition between two converging water masses, where the properties

of each are mixed to varying degrees. One must be also aware that the simplistic diagram in Fig. 3 does not account for such things as coastal currents, upwelling or isolated eddies. In open ocean away from frontal boundaries these problems are lessened.

From the extensive works in the Atlantic (McIntyre and Bé, 1967) and Pacific (Okada and Honjo, 1973), the rough positions of these zones have been determined. Distinct species assemblages can be assigned to each of these zones, and in most cases are similar to their counterparts in the opposing hemisphere (Table 1). The exceptions to this are the Subarctic–Subantarctic Zones, which have similar genera but few bipolar species. Coccolithophores generally live in the photic zone where they require sunlight for photosynthesis. The extent of this zone tends to deepen as one approaches the subtropics and is a direct result of the amount of suspended particles in the water. In high latitudes the depth at which 1% of the surface irradiance is found is about 30 m, whereas in the subtropics it may be as deep as 120 m.

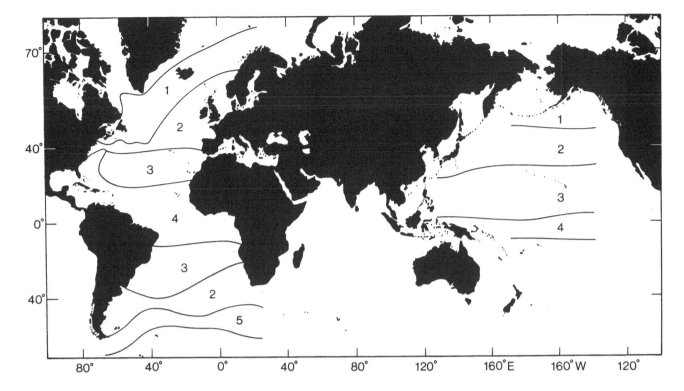

Fig. 3. Biogeographic coccolithophore zones from the Atlantic and Pacific Oceans redrawn from McIntyre and Bé (1967) and Okada and Honjo (1973) respectively. 1 = Subarctic, 2 = Temperate (Transitional), 3 = Subtropical (Central), 4 = Tropical (Equatorial) and 5 = Subantarctic.

In Temperate and Subpolar Zones, nutrients are seasonally replenished in the upper water column and phytoplankton (especially diatoms) can reach bloom proportions. This productivity increases the amount of particles within the upper water column and thus reduces the penetration depth which light can reach. As a consequence of this, species zonation within the photic zone is rather limited. In the subtropics nutrient recycling from deep water does not occur and so productivity within the surface waters is markedly lower. Relatively constant weather conditions allow the upper waters to remain permanently stratified. In low nutrient and particle conditions, light is not restricted to those cells at the surface. The Subtropical Zone has therefore developed a natural vertical zonation, with most species confined to either an upper photic (0–80 m) or lower photic (120–220 m) zone (Fig. 4; Table 2). The middle of the photic zone (80–120 m) is less distinctive and may incorporate species from either zone.

Dominant taxa and environmental preferences

Of the 195 coccolithophore taxa listed in this book, 130 are heterococcolithophores and 65 holococcolithophores (Jordan and Kleijne, chapter 6). However, only one of these species is found in nearly every sample of ocean water and late Quaternary sediment. *Emiliania huxleyi* is the most abundant and ubiquitous coccolithophore living in today's oceans, often occurring at a relative abundance of 60–80%.

It is one of the most euryhaline and eurythermal coccolithophore species. It thrives in the Red Sea at 41 ppt (Winter *et al.*, 1979) and can easily withstand the low salinities of 11 ppt and 17–18 ppt in the Sea of Azov and the Black Sea respectively (Bukry, 1974). *Emiliania huxleyi* probably has the largest temperature range (1 °C–30 °C) exhibited by any coccolithophore species (Okada and McIntyre, 1979), although this is not always reflected in laboratory cultures (e.g. Mjaaland, 1956). Its ability to grow in both eutrophic (e.g. Norwegian fjords) and oligotrophic (subtropical gyres) conditions suggest that this species can be successful over a wide range of nutrient levels. Equally its presence throughout the top 200 m in coccolithophore communities indicates that it can tolerate fluctuating light levels and still grow in less than 1% of the surface irradiance. As a direct result of its tolerance to changing environments, *E. huxleyi* can be cultured with relative ease. This has meant that more is known about this species than almost any other marine alga.

Evidence from the sedimentary record has indicated that *E. huxleyi* has dominated coccolithophore assemblages for approximately 73,000 years and probably evolved from the *Gephyrocapsa* species complex about 268,000 years ago (Thierstein *et al.*, 1977). The genus *Gephyrocapsa* may also dominate coccolithophore populations. *Gephyrocapsa*

Table 2. *The subtropical photic zones*

These zones are best defined in subtropical waters. Sometimes vertical zonation can be seen in the Tropical Zone but at shallower depths.

Upper photic zone (approx. 0–80 m)
Characterized by holococcolithophores, *Rhabdosphaera clavigera*, *Discosphaera tubifera*, *Neosphaera coccolithomorpha*, *Umbellosphaera* spp., *Acanthoica quattrospina*, *Ceratolithus cristatus*.

Middle photic zone (approx. 80–120 m)
This zone is not easily distinguished by a characteristic flora, but usually contains high abundances of a few species also found in the other zones. In particular, *Umbellosphaera tenuis*, *Syracosphaera* spp. and placolith-bearing species.

Lower photic zone (approx. 120–220 m)
Characterized by *Florisphaera profunda*, *Thorosphaera flabellata*, *Hayaster perplexus*, *Algirosphaera quadricornu*, *Turrilithus latericioides* and *Syracosphaera anthos*.

No depth preference
Most placolith-bearing species, *Helicosphaera* spp. and *Syracosphaera* spp.

Winter, 1987). *Gephyrocapsa muellerae* (Jordan, 1988) and *Coccolithus pelagicus* (Milliman, 1980) have also been recorded in quantities over 10^6 cells/liter, and the two species, *Oolithotus fragilis* (Bernard, 1948) and *Umbilicosphaera sibogae* (Honjo, 1982), may also produce large populations.

Many coccolithophore species prefer to live within specific depth ranges (Table 2). For example, *Discosphaera tubifera* and *Rhabdosphaera clavigera* are species that prefer to live in the upper photic layers (0–80 m) of the Pacific and Atlantic Oceans (Okada and Honjo, 1973 and Okada and McIntyre, 1979 respectively). Others such as *O. fragilis* and *U. tenuis* prefer the environmental conditions of the middle photic layer (80–120 m), whereas some species such as *Thorosphaera flabellata* and *Florisphaera profunda* appear adapted to life in the lower photic zone (120–220 m).

Some coccolithophore species thrive in a wide range of water temperatures (Okada and McIntyre, 1979). In a study of the distribution of coccolithophores in the western North Atlantic and Pacific Ocean, Okada and McIntyre (1979) found that the optimal temperatures for most coccolithophores lie between 12 °C and 27 °C. They also found that five species, *C. pelagicus*, *E. huxleyi*, *Algirosphaera robusta*, *Calciopappus caudatus* and *Syracosphaera borealis* are able to survive in temperatures as low as 1 °C. In fact

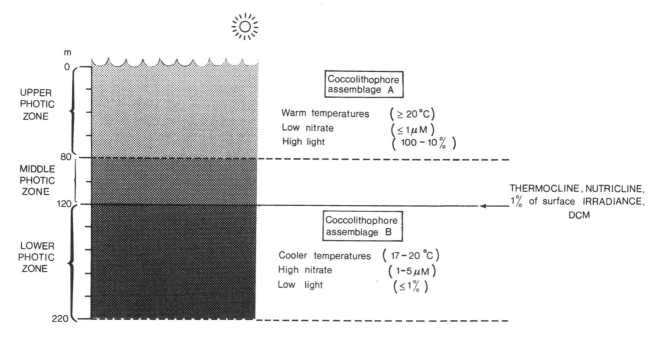

Fig. 4. Schematic diagram of the subtropical photic zone, adapted from data collected during the summer months in the NE Atlantic (R. Jordan, unpub. data), and with light irradiance levels from Jerlov (1976).

oceanica, is abundant in warm marginal seas (Okada and Honjo, 1975), is also found in upwelling areas (Mitchell-Innes and Winter, 1987; Kleijne *et al.*, 1989). *Gephyrocapsa oceanica* caused discoloration of the surface waters off the Cape Peninsula (Grindley and Taylor, 1970), where cell densities of 2.3×10^6 cells/liter (nearly exclusively *G. oceanica*) have since been recorded (Mitchell-Innes and

Coccolithus pelagicus may survive temperatures as low as −1.7 °C (Braarud, 1979). On the other hand, 11 other species live in temperatures up to 30 °C. One species in particular, *U. irregularis*, thrives in temperature from 25 °C to 30 °C.

Distribution of modern coccolithophores

This section attempts to summarize some of the most important contributions on the distribution of modern coccolithophores. The oceans and seas are reviewed in a geographic order to facilitate reading.

Atlantic Ocean

Research on coccolithophores in the surface waters probably originated in the Atlantic Ocean with the HMS *Challenger* expedition (Murray, 1885). Although few species were recorded on early expeditions (Murray, 1885; Murray and Blackman, 1898), the Atlantic is now generally regarded as having the richest coccolithophore flora (Gaarder, 1971). This diversity was recognized on subsequent expeditions to the central part of the North Atlantic (Lohmann, 1920) and Sargasso Sea (Gran, 1912). However, it was not until much later that information on the biogeography of coccolithophores improved as a result of more rigorous sampling strategies and the advent of the electron microscope.

A large quantitative study of the distribution of coccolithophores in Atlantic Ocean waters was conducted by McIntyre and Bé (1967). They also compared their results with the distribution of coccolithophores in surface sediments (see chapter 10). Surface water samples were collected by four research vessels from eight different cruises, each lasting a few months during 1964 to 1966. More than 70 coccolithophore species were identified. Of these, McIntyre and Bé (1967) described relative abundances of 13 placolith and cyrtolith species because of their importance in sediments. McIntyre and Bé (1967) erected four floral zones based on percent abundances of these species in surface waters as well as in sediments: (1) Tropical, (2) Subtropical, (3) Transitional (4) Subarctic and Subantarctic (Fig. 3). These floral zones coincide with major water mass boundaries. The most important species in each zone is given in Table 1.

Okada and McIntyre (1979) studied the seasonal and vertical distributions of coccolithophores from 540 water samples from five stations in the North Atlantic Ocean from 1969 to 1972. From these data they were also able to revise the classification system and describe a number of new species (Okada and McIntyre, 1977). Coccolithophore abundance decreased in the northern stations during winter months and in the southern stations during summer months. Seasonal abundance patterns for most of the major species were the same at all stations throughout the three year observation period. *Emiliania huxleyi* dominated all five collection stations. At station Bravo and to a lesser extent at station Charlie (Subarctic Zone of McIntyre and Bé, 1967) *Coccolithus pelagicus* monopolized the flora during spring and summer at all water depths (0 m, 50 m and 100 m) studied. At station Delta (Transitional Zone) *E. huxleyi* was replaced during summer months at the surface and 20 m primarily by *U. tenuis* and *U. hulburtiana* and at 100 m by *Florisphaera profunda*. Stations Echo and Hotel in the Subtropical and Tropical Zones respectively, showed similar seasonal distributions during most parts of the year. The assemblage was dominated by *E. huxleyi,* and although its numbers declined in the summer months, it still remained common. In the summer the percentage of *U. irregularis* increased at the surface, and that of *F. profunda* at 100 m.

Jordan (1988) studied the coccolithophore community structure from four cruises in the northeast Atlantic, including three visits to one site at Great Meteor East (around 31° N, 25° W). The cruises were conducted between April and July (1982–86), taking water samples from the whole water column, but concentrated between the surface and 200 m. Diatoms were the major phytoplankton component recorded from the April cruise off the coasts of Portugal and Ireland, although coccolithophores were abundant in the top 50 m. At the stations off Ireland (around 50° N) *E. huxleyi* was the most abundant coccolithophore, whereas at 42° N *Gephyrocapsa muellerae* and *G. ericsonii* were more numerous. Holococcolithophores (especially *Calyptrolithophora papillifera*) were absent during most of this cruise but became relatively abundant (35% of the coccolithophore count) over and near the Mid-Atlantic Ridge. A north–south transect from the UK to Madeira was conducted during June and July. At the northern stations coccolithophores and dinoflagellates were dominant in the top 40 m with pennate diatoms representing about 80% of the flora from 40–200 m. In the vicinity of the Azores Front coccolithophores became dominant over a greater depth range and reached their highest cell concentrations of nearly 10^6/liter. From north to south both the species diversity and the depth of the chlorophyll maximum increased. The coccolithophores were dominated by *G. muellerae* at almost all depths. South of the Front, coccolithophores dominated the phytoplankton and the chlorophyll maximum dropped to 118m. The coccolithophores were now characterized by the subtropical flora: *U. tenuis* in the top 60 m and the deep photic zone species between 120–160 m. *Gephyrocapsa muellerae* was absent or in low numbers. On three of the four cruises, from May to July, visits were made to the locality known as Great Meteor East. In early summer the thermocline, nutricline, and chlorophyll maximum lie between 80–110 m. The coccolithophores dominated the phytoplankton and showed maximum cell densities below 120 m. The species assemblage for all visits were characterized by *U. tenuis* in the top 100 m and the deep photic flora between 120–200 m. However, during the May visit *E. huxleyi* and *G. ericsonii* were very abundant at most depths, but were restricted to depths below 100 m on the other two visits. The holococcolithophores and the deep photic zone flora both had their highest percentages in July. Whether these visits truly represent a seasonal cycle has yet to be determined. An east–west transect along the 26° N line of latitude was conducted during May. Coccolithophores dominated at all depths representing 40–80% of the flora. Chlorophyll measurements were not taken during this cruise, but the nutricline was present between 120–140 m. The species assemblage was typically subtropical with *U. tenuis, Rhabdosphaera clavigera, Discosphaera tubifera,* and holococcolithophore spp. dominating the upper 100 m and *Florisphaera profunda* and *Thorosphaera flabellata* the deep photic zone. As the ship moved westwards the flora changed dramatically in both the upper and lower photic zones. The holococcolithophores, and to a lesser extent *Ceratolithus cristatus,* increased in percentage near and over the Mid-Atlantic Ridge. A phenomenon earlier recorded at 42° N. *Umbellosphaera tenuis* was more restricted to the middle photic waters (80–140 m), whereas *F. profunda* and *T. flabellata* occupied deeper waters than before and at higher percentages. Only four stations were analysed from a south–north transect in June from the equator to 31° N. Apart from the equatorial station, the transect was characterized by typical subtropical floras with *U. tenuis* dominating the upper 100 m and *F. profunda* the lower photic zone. At the equator, placolith-bearing species, notably *E. huxleyi* and *U. sibogae,* were the most common in the upper 25m, with *F. profunda* present at 50 m. The occurrence of the deep photic zone flora at this depth may have been caused by the equally shallow depth of both the thermocline and nutricline. Of ecological interest was the presence of a large number of *Reticulofenestra sessilis* cells at 80 m. This coccolithophore has only been reported from the equatorial regions and nearly always in close association with a centric diatom of the genus *Thalassiosira* (Okada and McIntyre, 1977). Although there is no direct evidence as yet, this combination may represent a symbiotic relationship.

Other workers have made valuable contributions to regional biogeography, and these will now be considered from pole to pole.

Partially calcified coccolithophores have been recorded in the Subarctic/Arctic off the coasts of Greenland and Canada (Manton *et al.* 1976b, 1977; Thomsen, 1981) during the summer months, and Denmark (Thomsen and Oates, 1978) and Finland (Thomsen, 1979) during the autumn and winter months, occasionally reaching abundances as high as 10^4 cells/liter off the Norwegian coast (Tangen, 1972). Their ability to grow in low temperatures and low salinity meltwaters have allowed them to occupy a restricted niche free from the competition of the other coccolithophore species.

Coccolithophore research in the North and Norwegian Seas has been primarily carried out by Norwegians (Gran, 1929; Smayda, 1958; Paasche, 1960). They identified a number of floral zones based on phytoplankton composition and related these to particular water masses or regions of mixing (Braarud *et al.,* 1953). Coccolithophores, especially *E. huxleyi* and *Algirosphaera robusta,* were abundant in inflowing Atlantic water or mixed waters of Atlantic origin and were generally absent in coastal assemblages or in truly polar waters (dominated by diatoms). In waters off northern Norway *Coccolithus pelagicus* (motile and non-motile forms) was the only coccolithophore found, whereas warm summer waters in the southern Norwegian Sea often had a high diversity, including some species normally associated with subtropical regions (Ramsfjell, 1960). The Norwegians have also gathered an immense amount of information on the phytoplankton in their fjords. It appears that *E. huxleyi* and *A. robusta* reach a maximum in the southern fjords in early summer, but do not increase in numbers until autumn in northern fjords (Braarud *et al.,* 1958). The occurrence of these coccolithophores in fjords is almost certainly linked to an inflow of Atlantic water, which is characterized by the two species above. On occasions the numbers of *E. huxleyi* within these fjords have reached bloom proportions on a scale unequalled in open ocean conditions (Birkenes and Braarud, 1952; Berge, 1962). Seasonal data collected from a weather ship (66° N, 4° E) showed two coccolithophore maxima, one in early summer and a larger one in autumn. Few coccolithophores were recorded between January and March (Halldal, 1953). The report included a new holococcolithophore species, *Calyptrosphaera papillifera* (now in *Calyptrolithophora*), which was found in September at 2.5×10^5 cells/liter.

In the English Channel phytoplankton populations are regularly monitored, but appear dominated by diatoms and dinoflagellates. Many neritic species have been described from the northern coast of France, and *Braarudosphaera bigelowii* has been recorded on a number of occasions from the Channel (Lefort, 1972; Boalch, 1987). However, coccolithophores bloom periodically around the coast of the United Kingdom. *Emiliania huxleyi* has been recorded in large numbers from the English Channel, Celtic and Amorican Shelfs, where cell concentrations have reached 8.5×10^6 cells/liter (Holligan *et al.,* 1983; Groom and Holligan, 1987). Extensive blooms of *E. huxleyi* have also been found south of Iceland over a number of years (BOFS unpub. data), and the largest concentrations of *C. pelagicus* ($>10^6$ cells/liter) have been reported in the surface waters off the west coast of Ireland (Milliman, 1980). In all cases the coccolithophores bloomed in the early summer months and persisted for several weeks.

Mesoscale blooms of *E. huxleyi* have been recorded by satellite from the western Atlantic, in the Gulf of Maine (Ackleson *et al.,* 1988; Balch *et al.,* 1991), where they reached maximum cell concentrations of 2×10^6 cells/liter during early summer. This ubiquitous species can be found along the entire eastern coast of the United States (Marshall, 1976).

The Sargasso Sea, like the Norwegian Sea, has become a center for coccolithophore research. Surveys along the eastern coast of the United States and off Venezuela revealed that diatoms dominate the shelf assemblages, with coccolithophores increasing in numbers seaward (Hulburt, 1962, 1963a, 1966; Marshall, 1968, 1969a, b). This distribution may be closely linked to salinity, with neritic diatoms favoring lower salinities and coccolithophores higher values (Hulburt and Rodman, 1963). Diatoms were also found to be the major group north of the Sargasso Sea at latitudes above 38° N, where the Gulf Stream is replaced by colder currents (Hulburt, 1964). Within the Sea itself coccolithophores, especially *E. huxleyi* and *Syracosphaera pulchra,* have been reported as the dominant phytoplankton species during most months of the year (Hulburt, 1963b; Marshall, 1966, 1968), however, diatoms have been known to outnumber coccolithophores at some stations during spring (Hulburt *et al.,* 1960; Hulburt, 1990). Studies on the vertical distribution of coccolithophores revealed that most species preferred the top 100 m, and although others could be found down to 1500 m, no species were exclusive to the deep photic zone (Marshall, 1966). This would suggest that either the thermocline is too unstable to support a deep photic zone flora or the indicative species for this zone were absent or overlooked by the above workers. Stations Hotel and Echo are in or near the Sargasso Sea and possessed abundant *F. profunda* at 100 m (Okada and McIntyre, 1979).

Not much work has been undertaken in the South Atlantic. Hentschel (1932, 1936) on board the RV *Meteor* studied the general phytoplankton community structure including coccolithophores. Mitchell-Innes and Winter (1987) sampled coccolithophores as well as other phytoplankton at 10 stations along an upwelling plume off the Cape Peninsula in the southeastern Atlantic. Thirteen coccolithophore species were identified with *E. huxleyi* dominating the assemblages. Coccolithophores were the most abundant phytoplankton except at one station.

The Southern Ocean is largely dominated by diatoms, although coccolithophores have been recorded from the Weddell Sea (see later) and are present north of the Antarctic Convergence (R. Jordan, unpub. obs).

Caribbean and Gulf of Mexico

Gaarder and Hasle (1971) examined coccolithophores from 10 water samples and eight net hauls on two cruises (October, 1966 and October–November, 1968) from the Gulf of Mexico. Thirty-two species were observed. *Syracosphaera* (now *Coronosphaera*) *mediterranea* was the most abundant followed by *Ceratolithus cristatus, Emiliania huxleyi, Discosphaera tubifera* and

Gephyrocapsa oceanica. A sample of water taken at 100 m contained nearly the same assemblage as at the surface indicating that there was little vertical zonation of species at the time and location the sample was taken. Hulburt (1968) studied the phytoplankton in the western Caribbean Sea and found *Emiliania huxleyi* to be the dominant species in 26 water samples. *E. huxleyi* was more abundant in coastal stations than stations lying far offshore.

Mediterranean Sea

Many of the early studies on coccolithophores were carried out in the Mediterranean Sea. These studies resulted in classical works on coccolithophore biology, taxonomy and distribution. Lohmann (1902) published the first monograph on coccolithophores. His research was conducted off Sicily and became an important source for all later work. Several years later, Schiller (1930) published an extensive monograph which summarized primarily his studies of coccolithophores in the Adriatic Sea. The Adriatic Sea studies were continued by Schiller's colleague Kamptner (1941) and extended into other parts of the Mediterranean (Kamptner 1937, 1944). Other studies of coccolithophore distribution in the northern Mediterranean occurred in the Ligurian (Bernard, 1939; Bernhard and Rampi,1965), Tyrrhenian (Borsetti and Cati, 1972, 1976, 1979) and Adriatic Seas (Vilicic and Fanuko, 1984), and around Sardinia and Corsica (Saugestad, 1967). In the southern Mediterranean sampling has been limited to the coastlines of Israel (Lecal, 1967; Kimor and Wood, 1975) and Algeria (Lecal, 1954, 1965). Lecal (1967) collected 10 water samples, five from the surface and five at 3.5 m water depth. Of the 35 species recognized *Emiliania huxleyi* was the most abundant except for the southernmost sample which contained abundant *Gephyrocapsa oceanica.* Kimor and Wood (1975) showed that coccolithophores were the most abundant phytoplankton of the deep chlorophyll maximum (DCM) and near surface layer of the Mediterranean Sea near Israel.

Quantitative studies on the distribution of coccolithophores in the entire Mediterranean were recently conducted by Knappertsbusch (1990) and Kleijne (1991). The results of Knappertsbusch were based on two cruises along the entire length of the Mediterranean in the fall of 1986 and the late winter of 1988. Distribution of coccolithophores from 21 stations at six different water depths from 0 to 200 m were compared to physical and chemical parameters. Seventy two living species were identified, seven of which dominated the assemblages. Fluctuations in species abundances were attributed to the seasonal hydrological changes of the Mediterranean. In summer the water is highly stratified, whereas in winter the water column is well mixed due to increased winds. Assemblages from the summer period seem to reflect strong hydrological gradients between the eastern and western basins whereas assemblages during the winter period were dominated by blooms of *E. huxleyi*. Cell abundances in winter were 10^4 to 3×10^5 per liter which is 10 times greater than in summer. Kleijne (1991) collected surface water samples along an east–west transect of the Mediterranean during the summer of 1985. Her results concerned only the contribution made by the holococcolithophores, which at times reached 50–70% of the total coccolithophore count. However, even at these high percentage stations the abundance of coccolithophores was low ($<10^4$ per liter), corroborating the earlier findings of Knappertsbusch (1990). Kleijne also noted a strong east–west gradient, with higher absolute frequencies of holococcolithophores increasing westward with decreasing salinity.

Restricted seas

Coccolithophores seem equally adapted to living in restricted seas, especially the Red, Black, and Weddell Seas discussed below. Okada and Honjo (1975) sampled the middle and southern parts of the Red Sea in November, and found that the distribution of coccolithophores was consistently high, apart from in the middle section and the channel between the Red Sea and the Gulf of Aden. The southern part was dominated by *Gephyrocapsa oceanica* in a low diversity assemblage exhibiting a high degree of malformation and with no vertical zonation. The assemblage at the middle section, although dominated by *G. oceanica* and *Emiliania huxleyi,* had a higher diversity with vertical zonation and showed little sign of malformation. Kleijne (1991) collected summer (June to July) surface water samples from the Gulf of Aden, Red Sea and the Gulf of Suez. The assemblage in the Gulf of Aden and Red Sea was dominated by *Umbellosphaera irregularis* and *U. tenuis* in warm, oligotrophic conditions and in the Gulf of Suez by *E. huxleyi* in relatively lower temperatures and higher salinities. Winter *et al.* (1979) studied the bathymetrical and seasonal distribution of coccolithophores between the surface and 400 m depth in the Gulf of Aqaba, the northeast arm of the Red Sea. Fifty-two species and two different coccolithophore assemblages were observed throughout the year from the Gulf of Aqaba. Assemblages were dominated by *E. huxleyi* and *G. ericsonii* and abundant *Algirosphaera quadricornu*, *Syracosphaera* spp. and scapholith-bearing coccolithophores. A profusion of holococcolithophores, *U. tenuis* and *Rhabdosphaera xiphos* characterized the summer assemblage. High salinities in the Gulf of Aqaba (>41 ppt) seem to prevent the entrance of many common pelagic species. Seasonal morphological changes in the central area of *E. huxleyi* and *G. ericsonii* were observed and may prove useful tools in the paleoecological reconstruction of the Gulf of Aqaba. Coccolithophores in the Gulf of Aqaba constitute a major component of the phytoplankton and contribute greatly to primary production. Heimdal *et al.* (1977) only

recorded three holococcolithophore species from a phytoplankton assemblage dominated by diatoms and dinoflagellates, during net hauls in January from the Great Bitter Lake, Suez Canal.

Coccolithophores have been reported from the Black Sea (Morozova-Vodyanitskaya and Belogorskaya, 1957; Valkanov, 1962; Mikhailova, 1965) where 18 species have been recorded in low salinities of 17–18 ppt. The Black Sea gradually changed from freshwater to brackish over 5000 years ago and the first invasion by *E. huxleyi* took place about 1500 years BP. *Emiliania huxleyi* still dominates the coccolithophore assemblages and blooms during the summer and autumn months (Hay and Honjo, 1989).

Coccolithophores are also found in cool waters near the polar regions where they often occur in large numbers near the ice edge (Goes and Devassy, 1983). Recently, Thomsen *et al.* (1988) described 13 taxa of coccolithophores from the Weddell Sea. Eight of these were described for the first time, although two of them remain unnamed. Three others, *Wigwamma annulifera, W. arctica* and *Papposphaera sagittifera* are now considered bipolar, whereas specimens resembling the arctic species, *Calciarcus alaskensis* and *Turrisphaera arctica* were also found.

Indian Ocean

The Indian Ocean had until a few years ago received little attention. In two early investigations, Silva (1960) found that coccolithophores were as abundant as diatoms off the Mozambique Coast, and Bernard and Lecal (1960) described a number of taxa from the western Indian Ocean. Since then several taxonomic studies (Norris, 1971, 1983, 1984, 1985; Kleijne, 1991) have been conducted in the Indian Ocean from either surface water samples or from the contents of salp guts. Friedinger and Winter (1987) recognized 59 coccolithophore species from 35 surface water samples taken from the southwest Indian Ocean near southern Africa. Cluster analysis revealed four different species assemblages dominated by *Emiliania huxleyi, Umbellosphaera hulburtiana, U. tenuis* and *Gephyrocapsa oceanica*. The assemblages reflected the two major oceanographic features of the study area: the Agulhas Current and the Agulhas Return Current. Kleijne *et al.* (1989) described the distribution patterns of living coccolithophores from 77 out of 160 samples along an east–west transect in the northern Indian Ocean and Red Sea. The same assemblage of coccolithophore species which dominated in the southwest Indian Ocean (Friedinger and Winter, 1987) were also dominant in the northern Indian Ocean. The only exception was that *U. hulburtiana* was replaced by *U. irregularis*. Coccolithophore standing crops in the northern Indian Ocean samples ranged from 2 to 11×10^4 coccolithophores per liter. In the Indian Ocean, *E. huxleyi* and *G. oceanica* were the two most dominant species. Both species increase in frequency during upwelling conditions. In the Gulf of Aden *U. irregularis* and *U. tenuis* are the two most abundant species. Hallegraeff (1984) examined the coccolithophore flora from Australian tropical and subtropical waters. *Emiliania huxleyi* and *G. oceanica* were the most abundant of 42 species described. Coccolithophore standing crop reached 10^5 per liter, sometimes occurring in a sharp subsurface maximum at 75 m depth. Nishida (1986) described three nannoplankton assemblages between Australia and the Antarctic continent – subtropical, subantarctic, and antarctic. The subtropical flora was characterized by a diverse assemblage, dominated by *E. huxleyi* below southeast Australia and by *E. huxleyi* and *U. tenuis* in the southwest. The subantarctic flora was also dominated by *E. huxleyi,* but the assemblage was reduced to four or five species, including *Calciopappus caudatus* a known cold water form. The subzero temperatures of his Antarctic Zone leads one to suppose that the cells recorded at these latitudes were no longer viable. This is supported by the scarcity of coccolithophores and the occasional presence in this zone of *U. irregularis,* a species with a preference for high temperatures (25–30 °C).

Pacific Ocean

The study of the distribution of coccolithophores in the Pacific Ocean began much later than in the Atlantic Ocean. The initial studies of phytoplankton assemblages were mostly concerned with the dominance of the diatoms and dinoflagellates. Marshall (1933) found low numbers of coccolithophores (mostly *Discosphaera tubifera*) from the Great Barrier Reef region, and attributed this to unstable salinity conditions. When the salinity dropped below 35 ppt during the rainy season the coccolithophores disappeared, a phenomenon first noted by Gran (1929) from the Norwegian Sea. Phytoplankton studies were later conducted in the Okhotsk Sea (Smirnova, 1959), the equatorial and subantarctic Pacific (Hasle, 1959, 1960) and the South Pacific (Norris, 1961). However, information on coccolithophore biogeography did not improve until McIntyre *et al.* (1970) studied their distribution from the North and South Pacific. They examined the contents of 96 surface water samples collected by the RV *Vema* and RV *Conrad* over an eight-year period, and from their resulting data published distribution maps of paleontologically important species. Not surprisingly *Emiliania huxleyi* showed the widest biogeographical distribution. On the other hand *Umbellosphaera irregularis* was restricted to tropical latitudes between the 21 °C isotherms. McIntyre *et al.* (1970) reported that *Coccolithus pelagicus* occurred only in northern latitudes north of the winter position of the 14 °C isotherm, with a temperature range of 6–14 °C. Its unimodel distribution was said to be a result of this species losing its ecological niche after a warm water incursion approximately 8000 years BP. It is known however, that *C. pelagicus* occurs

in specific coastal currents off South Africa in both the Atlantic and Indian Oceans (Mitchell-Innes and Winter, 1987; A. Winter unpub. data), and in the Great Australian Bight off western Tasmania (Hallegraeff, 1984). The failure of McIntyre *et al.* (1970) to find *C. pelagicus* in the southern hemisphere of the Pacific may have been due to sparse sampling near the major continental coastlines. *Umbellosphaera tenuis* occurred in both hemispheres between the summer 15 °C isotherms and preferred subtropical gyre water of the North and South Pacific. Both *Discosphaera tubifera* and *Rhabdosphaera clavigera* var. *stylifera* occurred abundantly in cool subtropical waters. Their poleward boundary corresponded with subpolar waters. *Discosphaera tubifera* had an optimum temperature preference of about 1 °C higher than *R. clavigera* var. *stylifera*. McIntyre *et al.* (1970) did not find either species in equatorial waters.

The *Gephyrocapsa* species are a common component of Quaternary sediment. For this reason McIntyre *et al.* (1970) were especially interested in studying their distribution. Unfortunately, there is still no agreement concerning the taxonomic position of some of the *Gephyrocapsa* species. There is also debate about whether *G. caribbeanica* is a living species (see chapter 6). Nevertheless, it is useful to summarize the biogeography of *Gephyrocapsa* species according to McIntyre *et al.* (1970). *Gephyrocapsa caribbeanica* preferred tropical to subpolar waters between 14 °C and 21 °C with an optimum temperature of 19 °C. This "cosmopolitan" species lives in waters down to 6 °C. According to McIntyre *et al.* (1970), *G. ericsonii* exhibited two different biogeographical and temperature ranges and could be split into two forms. A cold-water form preferring subtropical waters between 14 °C and 21 °C with an optimum temperature around 19 °C, and a rarely occurring warm-water form living in water between 23 °C and 30 °C. Another species, *G. oceanica,* on the other hand, was limited to tropical and subtropical waters and was absent in cool subtropical waters.

McIntyre *et al.* (1970) recognized two morphotypes of *Calcidiscus leptoporus*, although recently a third has been reported by Kleijne (pers. comm.) in the water column and by Knappertsbusch (1990) in the sediments. The distinguishing characteristic is the number of elements forming the placolith shield. One is a eurythermal form whose poleward range extends up to the summer 8 °C isotherm and has an average of 30 elements. Although the other favors warm tropical and subtropical waters, has an average of 20 elements and is present between the summer 18 °C isotherms with its maximum abundance centered on the equator.

A large scale, quantitative study of coccolithophore distribution primarily from the North Pacific was undertaken by Okada and Honjo (1973) and Honjo and Okada (1974). They studied the community structure and distribution of coccolithophores from material gathered on two different cruises from November to January (1968–69) and from August to

November (1969). During three north–south sampling tra-
verses from 15° S to 50° N and three east–west transverses
across the northern Pacific they were able to collect 600
water samples from 232 stations. On three of these transects
samples were collected at different water depths down to
200 m. At least 200 coccolithophore individuals were
counted from each sample using electron microscopes. In
all, 90 species were identified. From the distribution patterns
of these species, six coccolithophore zones were erected:
Subarctic, Transitional, Central-North, Equatorial-North,
Equatorial-South and Central-South (Fig. 3). These zones
were in close association with the positions of the surface
water current boundaries and a previously published zooplankton zonation scheme (Brinton, 1962). The Subarctic
flora was composed almost exclusively of *E. huxleyi,* with
occasional *Calciopappus caudatus.* The Transitional Zone
possessed a high diversity, with *R. clavigera* and *D. tubifera*
in the surface waters, *U. tenuis* between 75–100 m, and
*Florisphaera profunda, Thorosphaera flabellata,
Algirosphaera quadricornu* and *Oolithotus fragilis* in the
deeper waters (150-200 m). The Central Zones were characterized by *Umbellosphaera* spp. in the top 125m, with *U.
tenuis* situated more in the middle portion of the photic zone
and to a lesser extent in the southern hemisphere. The deeper
waters were dominated by *F. profunda* and *T. flabellata.* The
Equatorial Zones were characterized by placolith-bearing
species in the top 125m, notably *G. oceanica, E. huxleyi,
Calcidiscus leptoporus* and *O. fragilis.* The deeper waters
were dominated by *T. flabellata* and to a lesser extent *F. profunda.* Based on the coccolithophore distribution in the top
200 m, Okada and Honjo (1973) subdivided the Transitional
and Central Zones into three vertical layers and the
Equatorial Zones into two. The Subarctic Zone only possessed one layer. In the lower latitudes Okada and Honjo
(1973) recognized key species which could be placed into
either the upper, middle, or lower photic zones, although
some species (e.g. *E. huxleyi*) were present at all depths.
The presence of a distinct deep photic zone flora, especially
in the subtropics, was revealed for the first time. It consisted
largely of *F. profunda, T. flabellata* and *A. quadricornu.* In
recent years *F. profunda* has been recognized as a major
contributor to Atlantic sediments underlying subtropical and
tropical waters, where the level of its abundance may be
used in paleoceanography to determine nutricline dynamics
(Molfino *et al.,* 1989; Molfino and McIntyre, 1990).

Other workers have concentrated on more regional studies
and these will be discussed from north to south.

A number of species have been collected from various
depths down to 20 m in the Gulf of Alaska using a hand-held
bottle (Manton *et al.,* 1976a, b, 1977). However, there is little
information at present on their distribution at high latitudes.

Nishida (1979) studied the distribution of coccolithophores in about 500 surface water samples collected during nine cruises, mostly from the northern and western

Pacific between 1972 and 1978. He also studied the vertical
distribution of coccolithophores from the surface to 400 m on
a cruise along the Kuroshio current. *Emiliania huxleyi* and *G.
oceanica* were respectively the two most dominant coccolithophore species except in the high diversity tropical–subtropical waters, and in subarctic waters where *Coccolithus
pelagicus* and *Calciopappus caudatus* were the most common coccolithophores. In the subtropics *U. irregularis* and
D. tubifera and were the most abundant. Coccolithophore
abundance was highest close to 40° N and 34° S near the subtropical convergence zones. Highest standing crop occurred
between 30° to 40° N and S where water temperature was
between 12 °C and 17 °C. In the water column coccolithophores were most abundant near 50 m water depth.

Okada and Honjo (1975) investigated the distribution of
coccolithophores in the marginal seas off the western Pacific.
They used over 1000 samples collected over three years during various months. The assemblages were low in diversity
and dominated by *G. oceanica,* regardless of the time of year,
with rare to common *E. huxleyi* and *Umbellosphaera* spp. The
assemblages were largely characterized by high degrees of
malformation. Kleijne (1990) recorded the distribution of 36
living coccolithophore species from surface and subsurface
waters of the Banda Sea and adjacent waters. Coccolithophores were nearly absent in the upper 5 m but were
abundant in waters from 20 to 300 m. During the Northwest
Monsoon (January–February), when water nutrients are generally depleted, Kleijne (1990) proposed that the coccolithophores would be found without their covering of coccoliths. However, in laboratory strains of two species, coccolith
production showed an inverse relationship with nitrogen concentration (Wilbur and Watabe, 1963; Baumann *et al.,* 1978),
whereas *E. huxleyi* has been known to produce coccoliths during bloom conditions in high nutrient levels (Birkenes and
Braarud, 1952; Berge, 1962). Loss of calcification may be due
to other environmental parameters like decreases in salinity
(Okada and Honjo, 1975; Kleijne, 1990).

Beers *et al.* (1975) reported 35 coccolithophore taxa from
the North Central Gyre (around 28° N 155° W), collected in
June from bottles taken every 20 m to a depth of 200 m. The
assemblages, identified using an inverted microscope, were
dominated by *Helladosphaera cornifera* in the surface samples and by the deep photic flora of Okada and Honjo (1973)
in deeper waters. Reid (1980) presented the seasonal and
vertical distributions of coccolithophores using the data of
Beers *et al.* (1975) and that of four other expeditions to the
same locality. The seasonal data from January to September
was compiled using months of two different years. Reid
(1980) found at least 53 species with *E. huxleyi* dominating
all depths from January to March, whereas from May
onwards its relative abundance declined and it was replaced
by the upper and lower photic zone floras. The upper photic
zone was dominated by *Helladosphaera cornifera,* with
contributions from *D. tubifera* and *U. tenuis.* In the lower

photic zone *F. profunda* was the most abundant, with common *A. oryza (A. quadricornu ?)*, *T. flabellata* and *Deutchlandia* spp. Later, Hoepffner and Haas (1990) collected phytoplankton from three depths (the surface, 60–70 m and 90–110 m) at a number of sites in this same region. Coccolithophores and other prymnesiophytes represented about 55% of the assemblage and dominated (especially *E. huxleyi*) the deep chlorophyll maximum. They identified 11 taxa using the TEM including a few species not previously reported by Reid (1980).

A study of the coastal flora off southern California by Reid *et al.* (1978) showed a March assemblage dominated by diatoms and dinoflagellates, with *Calciosolenia murrayi* being the only numerically important coccolithophore. An investigation of the coccolithophore community by Winter (1985) at approximately the same location some years later, revealed a much richer flora dominated by *E. huxleyi*, *U. sibogae*, *G. oceanica* and '*R. longistylis*'. Winter (1985) recognized four assemblages in March which appeared to reflect distinct water masses, but by June the assemblages were more uniform possibly indicating more stable conditions in the region. In the Gulf of California coccolithophores, especially *E. huxleyi*, were the most important numerical contributor to the chlorophyll maximum (Hernandez-Becerril, 1987).

Marshall (1970) studying the phytoplankton diversity along the coast of Ecuador and Panama, found that the coccolithophores and silicoflagellates dominated the pelagic stations. He found only three species of coccolithophore, *E. huxleyi*, *G. oceanica* and *Calcidiscus leptoporus*.

Norris (1961) studied the phytoplankton community off northern New Zealand and the Kermadec, Fiji and Tonga Islands. Coccolithophores were present at all stations but reached highest diversity (26 taxa) amongst the Kermadec Islands.

Hasle (1960) reported five species from the subantarctic between December and February, with only *E. huxleyi* being numerically abundant. It produced 2.05×10^5 cells/liter at 2.3 °C. No living cells were recorded at lower temperatures or south of the Antarctic Convergence.

Future research

Coccolithophores have recently gained much attention as important players in global climate change and carbon cycles. However, the environmental parameters that control the horizontal and vertical distribution of them are still poorly understood. A concerted effort needs to be made to collect seasonal samples not only from the surface but at different water depths over consecutive years. In the past some of these criteria have been met. The Continuous Plankton Recorder (CPR) has been operating in the North Atlantic since 1931, collecting plankton samples on a monthly basis along designated routes. However, the mesh size used during tows, limits the catch to zooplankton and very large phytoplankton species. To our knowledge this technique has not yielded any coccolithophore data. Weather ships operating in the North Atlantic have been used to conduct long-term sampling but at fixed positions (Halldal, 1953; Okada and McIntyre, 1979), whereas McIntyre *et al.* (1970) collated the data from water and surface sediment samples collected throughout the Pacific Ocean over an eight year period. However, the vertical sampling in these studies was rather limited, with few samples collected below 100 m. More recently Jordan (1988) has collected vertical profiles through the top 220 m at about 20 stations in the northeast Atlantic, but only during the months of April to July. Over the last few years (1989–92) the Biogeochemical Ocean Flux Study (BOFS) has spent some of its time monitoring *Emiliania huxleyi* blooms in the northeast Atlantic, with an intense inter-disciplinary research programme. This study directed most of its interest at this one coccolithophore species and was restricted geographically. What is really needed is a project on the scale of CLIMAP (1976), which involved an extremely detailed analysis of the Last Glacial Maximum using a vast number of cores throughout the Atlantic Ocean. The program took 15 years to complete and subsequent paleoceanographic reconstructions were based on foraminiferal data. A smaller project involving coccoliths from only 51 Pacific Ocean core tops has been used to reconstruct paleotemperatures (Geitzenauer *et al.*, 1976). To use this approach for coccolithophore sampling would be very time consuming and expensive and therefore must be organized as a multi-national operation, like the Ocean Drilling Program but with more ships. It would be equally important for ecological reasons for samples to be accompanied with good hydrographic data, on the sort of scale now being conducted by various ships as part of Joint Global Ocean Flux Study. Only in this way will we learn quickly enough to keep pace with the paleoceanographers and to provide the information for the ever-growing and ever-eager modelling community.

Appendix

Procedure for collecting, preserving and viewing living coccolithophores.

Filtration

1. Collect one or more liters of seawater.
2. Immediately after arrival on board pass the water through a Nuclepore filter (preferably 0.22 μm porosity). Use moderate suction of 10 to 45 dm.
3. Rinse filter with alkaline water (pH 8).
4. Dry filter in oven at 50 °C or air dry.
5. Store in plastic case or bag and label.

Preservation

1. Collect one or more liters of seawater.
2. Concentrate the material either by centrifugation or filtration.*
3. Wash the pellet or filter thoroughly with distilled water (pH 8). Discard the filter. Recentrifuge the resuspended material and reduce the supernatant to a smaller volume.
4. Preserve with an equal volume of 1% (or stronger) glutaraldehyde.
5. Store in a labelled bottle in a cool (4 °C), dark place.

* Alternatively, concentrate the material until 40 ml is left above the filter. Pipette some of this concentrate into a clean bottle and add the preservative.

Viewing

LIGHT MICROSCOPY

Using a filter:

1. Cut out a small strip from the filter and place on a clean microscope slide.
2. Place immersion oil on the filter (Nuclepore filters are best) to render filter transparent.
3. Observe with a polarizing microscope at high magnification using immersion oil and polarized light.
4. For quantitative analysis, extrapolate from the number of cells found per unit area to the total filter area and then divide by the number of liters passed through the filter.

Using a preserved water sample

1. Pipette several drops of material, taken from the bottom of the sample bottle, onto a clean glass slide.
2. Place a cover slip onto the material, carefully avoiding air bubbles.
3. View under normal or polarized light.
4. If a permanent mount is needed, dry the material onto a slide in an oven at 50 °C.
5. Add several drops of a mounting medium (e.g., Entellan) to the dried material and place a cover slip on top. Avoid air bubbles.
6. Let preparation dry for 24 hours. Then view with a polarizing microscope.

TRANSMISSION ELECTRON MICROSCOPY

1. Remove large organisms and debris by pouring the preserved water sample through a 20 μm net.
2. Concentrate the material by centrifugation.
3. Resuspend material onto a formvar/carbon coated grid, fix in osmic vapor for 30 seconds and let dry.
4. Remove salt crystals by thoroughly washing in distilled water for 10 mins.
5. Dry and store in gelatine capsules or grid boxes.
6. Shadow-cast with gold-palladium.
7. Place in the TEM.

SCANNING ELECTRON MICROSCOPY

1. Cut a small strip from the filter (e.g., 8 × 8 mm) and mount on an aluminium stub using double-sided adhesive tape.
2. Coat stub with gold-palladium using a vacuum evaporator.
3. Outline filter with a silver colloidal suspension to alleviate problems with 'charging'.
4. Secure stub onto the SEM stage.

References

Ackleson, S., Balch, W.M. and Holligan, P.M., 1988. White waters of the Gulf of Maine. *Oceanography*, **1**(2): 18–22.

Arpin, N., Svec, W.A. and Liaaen-Jensen, S., 1976. New fucoxanthin-related carotenoids from *Coccolithus huxleyi*. *Phytochemistry*, **15**: 529–32.

Balch, W.M., Holligan, P.M., Ackleson, S.G. and Voss, K.J., 1991. Biological and optical properties of mesoscale coccolithophore blooms in the Gulf of Maine. *Limnol. Oceanogr.*, **36**: 629–43.

Baumann, F.G., Isenberg, H.D. and Gennaro, J. Jr., 1978. The inverse relationship between nutrient nitrogen concentration and coccolith calcification in cultures of the coccolithophorid *Hymenomonas* sp. *J. Protozool.*, **25**(2): 253–6.

Beers, J.R., Reid, F.M.H. and Stewart, G.L., 1975. Microplankton of the North Pacific Central Gyre. Population structure and abundance, June 1973. *Int. Revue ges. Hydrobiol.* **60**: 607–38.

Berge, G., 1962. Discoloration of the sea due to *Coccolithus huxleyi* "bloom". *Sarsia*, **6**: 27–40.

Bernard, F., 1939. Recherches sur les Coccolithophorides. I. Principales espèces du plancton à Monaco. *Bull. Inst. Océanogr.* (Monaco), **767**: 1–19.

Bernard, F., 1948. Recherches sur le cycle du *Coccolithus fragilis* Lohm., flagellé dominant des mers chaudes. *J. Cons. Int. Explor. Mer.*, **15**: 177–88.

Bernard, F. and Lecal, J., 1960. Plankton unicellulaire récolté dans l'océan Indien par le *Charcot* (1950) et le *Norsel* (1955–56). *Bull. Inst. Océanogr.* (Monaco), **1166**: 1–59.

Bernhard, M. and Rampi, L., 1965. Horizontal micro-distribution of marine phytoplankton in the Ligurian Sea. *Botanica Gothoburgensia III. Proc. Fifth Mar. Biol. Symp. 1965*, 13–24.

Birkenes, E. and Braarud, T., 1952. Phytoplankton in the Oslo Fjord during a 'Coccolithus huxleyi-summer'. *Avh. Norske Vidensk Akad. Oslo I. Mat.-Nat. Kl.*, **2**: 1–23.

Boalch, G.T., 1987. Changes in the phytoplankton of the Western English Channel in recent years. *Br. Phycol. J.*, **22**: 225–35.

Borsetti, A.M. and Cati, F., 1972. Il nannoplancton calcareo vivente nel tirreno centro-meridionale. *Giorn. Geol.*, ser. 2a, **38**: 395–452.

Borsetti, A.M. and Cati, F., 1976. Il nannoplancton calcareo vivente nel tirreno centro-meridionale. Parte II. *Giorn. Geol.*, ser. 2a, **40**: 209–40.

Borsetti, A.M. and Cati, F., 1979. Il nannoplancton calcareo vivente nel tirreno centro-meridionale. Parte III. *Giorn. Geol.*, ser. 2a, **43**:157–74.

Braarud, T., 1979. The temperature range of the non-motile stage of *Coccolithus pelagicus* in the North Atlantic region. *Br. Phycol. J.*, **14**: 349–52.

Braarud, T., Gaarder, K.R. and Grøntved, J., 1953. The phytoplankton of the North Sea and adjacent waters in May, 1948. *Rapp. Cons. Explor. Mer.*, **133**: 1–87.

Braarud, T., Gaarder, K.R. and Nordli, O., 1958. Seasonal changes in the phyroplankton at various points off the Norwegian West Coast. *Fiskerdir. Skr. Havundersøk.*, **12**(3):1–77.

Brand, L.E., 1981. Genetic variability in reproduction rates in marine phytoplankton populations. *Evolution*, **35** (6): 1117–27.

Brinton, E., 1962. The distribution of Pacific Euphausiids. *Bull. Scripps Inst. Oceanogr., Univ. Calif.*, **8**: 51–270.

Bukry, D., 1974. Coccoliths as paleosalinity indicators - evidence from the Black Sea. In *The Black Sea - Geology, Chemistry and Biology*, ed. E.T. Degens and D.A. Ross, pp. 353-633. Amer. Assn. Petrol. Geol., Memoir 20.

CLIMAP Project Members, 1976. The surface of the Ice Age Earth. *Science*, **191**: 1131–7.

Friedinger, P.J.J. and Winter, A., 1987. Distribution of modern coccolithophore assemblages in the southwest Indian Ocean off southern Africa. *J. Micropalaeontol.*, **6**(1): 49–56.

Gaarder, K.R., 1971. Comments on the distribution of coccolithophorids in the ocean. In *The Micropalaeontology of Oceans.* Cambridge Univ. Press, ed. B.M. Funnell and W.R. Riedel, pp. 97-103. Cambridge.

Gaarder, K.R. and Hasle, G.R., 1971. Coccolithophorids of the Gulf of Mexico. *Bull Mar. Sci.*, **21**(2): 519–44.

Geitzenauer, K.R., Roche, M.B. and McIntyre, A., 1976. Modern Pacific coccolith assemblages: derivation and application to Late Pleistocene paleotemperature analysis. In *Investigation of Late Quaternary Paleoceanography and Paleoclimatology.* ed. R.M. Cline and J.D. Hays, Geol. Soc. Amer. Mem., **145**: 423–48.

Goes, J.I and Devassy, V.P., 1983. Phytoplankton organisms collected during the First Indian Antarctic Expedition. In: *Scientific Reports of First Indian Expedition to Antarctica, Technical Publication No. 1.*, 1983, pp. 198–201; Tech Publ. Sci. Rep. FIEA; No 1.

Gran, H.H., 1912. Pelagic plant life. In *The Depths of the Ocean*, ed. Murray, J. and Hjort, S. London, pp. 307–86.

Gran, H.H., 1929. Quantitative plankton investigations carried out during the expedition with the *'Michael Sars',* July–Sept., 1924. *Rapp. Cons. Explor. Mer.*, LVI (5): 1–50.

Grindley, J.R. and Taylor, F.J.R., 1970. Factors affecting plankton blooms in False Bay. *Trans. Roy. Soc. S. Afr.*, **39**: 201–10.

Groom, S.B. and Holligan, P.M., 1987. Remote sensing of coccolithophore blooms. *Adv. Space Res.*, **7** (2): 73–8.

Halldal, P., 1953. Phytoplankton investigations from weather ship M in the Norwegian Sea, 1948–49. Including observations during the *Armauer Hansen*–cruise, July, 1949. *Hvalrad. Skr. Norske Vidensk. Akad.* **38**: 5–91.

Hallegraeff, G.M., 1984. Coccolithophorids (Calcareous nanoplankton) from Australian waters. *Bot. Mar.*, **27**(6): 229–47.

Hasle, G.R., 1959. A quantitative study of phytoplankton from the equatorial Pacific. *Deep-Sea Res.*, **6**: 38–59.

Hasle, G.R., 1960. Plankton coccolithophorids from the subantarctic and equatorial Pacific. *Nytt Mag. Bot.*, **8**: 77–88.

Hay, B.J. and Honjo, S., 1989. Particle sources in the present and Holocene Black Sea. *Oceanography*, **2**: 26–31.

Heimdal, B.R., Taasen, J.P. and Elbrächter, M., 1977. Net phytoplankton of the Great Bitter Lake in the Suez Canal. *Sarsia*, **63**: 75–83.

Hentschel, E., 1932. Die biologische Methoden und das biologische Beobachtungsmaterial der *Meteor*-Expedition. *Wiss. Ergebn. Atlant. Exped. Meteor*, **10**: 1–274.

Hentschel, E., 1936. Allgemeine Biologie des Südatlantischen Ozeans. *Wiss. Ergebn. Atlant. Exped. Meteor*, **11**: 1–344.

Hernandez-Becerril, D.U., 1987. Vertical distribution of phytoplankton in the central and northern part of the Gulf of California (June 1982). P.S.Z.N. I: *Mae Ecol.*, **8**(3): 237–51.

Hoepffner, N. and Haas, L.W., 1990. Electron microscopy of nanoplankton from the North Pacific Central Gyre. *J. Phycol.*, **26**: 421–39.

Holligan, P.M., Aarup, T. and Groom, S.B., 1989. The North Sea: Satellite Colour Atlas. *Cont. Shelf Res.*, **9**(8): 667–765.

Holligan, P.M., Viollier, M., Harbour, D.S., Camus, P. and Champagne-Philippe, M., 1983. Satellite and ship studies of coccolithophore production along a continental shelf edge. *Nature*, **304** (5924): 339–42.

Honjo, S., 1982. Seasonality and interaction of biogenic and lithogenic particulate flux at the Panama Basin. *Science*, **218**: 883–4.

Honjo, S., and Okada, H., 1974. Community structure of coccolithophores in the photic layer of the mid-Pacific. *Micropaleontol.*, **29**: 209–30.

Hulburt, E.M., 1962. Phytoplankton in the southwestern Sargasso Sea and North Equatorial Current, February, 1961. *Limnol. Oceanogr.*, **7**(3): 307–15.

Hulburt, E.M., 1963a. Distribution of phytoplankton in coastal waters of Venezuela. *Ecology*, **44**(1): 169–71.

Hulburt, E.M., 1963b. The diversity of phytoplankton populations in oceanic, coastal, and estuarine regions. *J. Mar. Res.*, **21**(2): 81–93.

Hulburt, E.M., 1964. Succession and diversity in the plankton flora of the western North Atlantic. *Bull. Mar. Sci. Gulf Caribb.*, **14**: 33–44.

Hulburt, E.M., 1966. The distribution of phytoplankton and its relationship to hydrography, between southern New England and Venezuela. *J. Mar. Res.*, **24**(1): 67–81.

Hulburt, E.M., 1968. Phytoplankton observations in the western Caribbean Sea. *Bull. Mar. Sci. Gulf Caribb.*, **18**: 388–99.

Hulburt, E.M., 1990. Description of phytoplankton and nutrient in spring in the western North Atlantic Ocean. *J. Plankt. Res.*, **12**: 1–28.

Hulburt, E.M. and Rodman, J., 1963. Distribution of phytoplankton species with respect to salinity between the coasts of southern New England and Bermuda. *Limnol. Oceanogr.*, **8**(2): 263–9.

Hulburt, E.M., Ryther, J.H. and Guillard, R.R.L., 1960. The phytoplankton of the Sargasso Sea off Bermuda. *J. Cons.*, **25**(2): 115–28.

Huxley, T.H., 1868. On some organisms living at great depths in the North Atlantic Ocean. *Quart. J. Microscopical. Sci.*, new ser., **8**: 203–12.

Jeffrey, S.W. and Allen, M.B., 1964. Pigments, growth and photosynthesis in cultures of two chrysomonads, *Coccolithus huxleyi* and a *Hymenomonas* sp. *J. Gen. Microbiol.*, **36**: 277–88.

Jeffrey, S.W. and Wright, S.W., 1987. A new spectrally distinct component in preparations of chlorophyll *c* from preparations of *Emiliania huxleyi* (Prymnesiophyceae). *Biochem. Biophys. Acta*, **894**: 180–8.

Jerlov, N.G., 1976. *Marine Optics.* Elsevier Oceanographic Series **14**. Elsevier, New York.

Jordan, R.W., 1988. Coccolithophorid Communities in the North-East Atlantic. Unpub. PhD thesis, Univ. Surrey, Guildford.

Kamptner, E., 1937. Neue und bemerkenswerte Coccolithineen aus dem Mittelmeer. *Arch. Protistenk.*, **89**(3): 279–316.

Kamptner, E., 1941. Die Coccolithineen der Südwestküste von Istrien. *Ann. Naturh. Mus. Wien.*, **51**: 54–149.

Kamptner, E., 1944. Coccolithineen-Studien im Golf von Neapel. *Wien. Bot. Z.* **93**: 138–47.

Kimor, B. and Wood, E.J.F., 1975. A plankton study of the Eastern Mediterranean Sea. *Mar. Biol.*, **29**: 321–33.

Kleijne, A., Kroon, D. and Zevenboom, W., 1989. Phytoplankton and foraminiferal frequencies in northern Indian Ocean and Red Sea surface waters. *Neth. J. Sea Res.*, **24**(4): 531–9.

Kleijne, A., 1990. Distribution and malformation of extant calcareous nannoplankton in the Indonesian Seas. *Mar. Micropaleontol.*, **16**: 293–316.

Kleijne, A., 1991. Holococcolithophorids from the Indian Ocean, Red Sea, Mediterranean Sea and North Atlantic Ocean. *Mar. Micropaleontol*, **17**: 1–76.

Knappertsbusch, M.W., 1990. Geographic distribution of modern coccolithophores in the Mediterranean Sea and morphological evolution of *Calcidiscus leptoporus*. Unpub. Ph.D. dissertation, Swiss Federal Inst. of Tech., Zurich ETH, Nr. 9169.

Lecal, J., 1954. Richesse en microplancton estival des eaux méditerranéennes de Port-Vendres à Oran. *Vie et Milieu*, **3**: 13–95.

Lecal, J., 1965. Coccolithophorides littoraux de Banyuls. *Vie et Milieu*, **16**: 251–70.

Lecal, J., 1967. Le nannoplancton des Côtes d'Israel. *Hydrobiologia*, **29**(3/4): 305–87.

Lefort, F., 1972. Quelques caractères morphologiques de deux espèces actuelles de *Braarudosphaera* (Chrysophycées, Coccolithophoracées). *Le Botaniste*, **55**: 82–93.

Lohmann, H., 1902. Die Coccolithophoridae. *Arch. Protistenk.*, **1**: 89–165.

Lohmann, H., 1920. Die Bevölkerung des Ozeans mit plankton. Nach den Ergebnissen der Zentrifugenfängen während der Ausreise der *Deutschland* 1911. *Arch. Biontol. Berl.*, **4**(1916-19): 1–617.

McIntyre, A. and Bé, A.W.H., 1967. Modern coccolithophores of the Atlantic Ocean – I. Placolith and cyrtoliths. *Deep-Sea Res.*, **14**: 561–97.

McIntyre, A., Bé, A.W.H. and Roche, M.B., 1970. Modern Pacific Coccolithophorida: a paleontological thermometer. *Trans. N.Y. Acad. Sci.*, Ser.II, **32**(6): 720–31.

Manton, I., Sutherland, J. and McCully, M., 1976a. Fine structural observations on coccolithophorids from south Alaska in the genera *Papposphaera* Tangen and *Pappomonas* Manton and Oates. *Br. Phycol. J.*, **11**: 225–38.

Manton, I., Sutherland, J. and Oates, K., 1976b. Arctic coccolithophorids: two species of *Turrisphaera* gen. nov. from West Greenland, Alaska and the North-West Passage. *Proc. Roy. Soc.*, Ser.B, **194**: 179–94.

Manton, I., Sutherland, J. and Oates, K., 1977. Arctic coccolithophorids: *Wigwamma arctica* gen. et sp. nov. from Greenland and Arctic Canada, *W. annulifera* sp. nov. from South Africa and south Alaska and *Calciarcus alaskensis* gen. et sp. nov. from S. Alaska. *Proc. Roy. Soc.*, Ser.B, **197**: 145–68.

Marlowe, I.T., Brassell, S.C., Eglinton, G. and Green, J.C., 1990. Long-chain alkenones and alkyl alkenoates and the fossil coccolith record of marine sediments. *Chem. Geol.*, **88**: 349–75.

Marlowe, I.T., Green, J.C., Neal, A.C., Brassell, S.C., Eglinton, G. and Course, P.A., 1984. Long chain (n-C_{37}-C_{39}) alkenones in the Prymnesiophyceae. Distribution of alkenones and other lipids and their taxonomic significance. *Br. Phycol. J.*, **19**: 203–16.

Marshall, H.G., 1966. Observations on the vertical distribution of coccolithophores in the northwestern Sargasso Sea. *Limnol. Oceanogr.* **11**(3): 432–5.

Marshall, H.G., 1968. Coccolithophores in the northwest Sargasso Sea. *Limnol. Oceanogr.*, **13**: 370–6.

Marshall, H.G., 1969a. Observations on the spatial concentrations of phytoplankton. *Castanea*, **34**: 217–22.

Marshall, H.G., 1969b. Phytoplankton distribution off the North Carolina coast. *Amer. Mid. Nat.*, **82**: 241–57.

Marshall, H.G., 1970. Phytoplankton in tropical surface waters between the coast of Ecuador and the Gulf of Panama. *J. Wash. Acad. Sci.*, **60**: 18–21.

Marshall, H.G., 1976. Phytoplankton distribution along the eastern coast of the USA. I. Phytoplankton composition. *Mar. Biol.*, **38**: 81–9.

Marshall, S.M., 1933. The production of microplankton in the Great Barrier Reef region. *Sci. Rept. Great. Barrier Reef Exped.* 1928-1929. **2**: 112–57.

Martin, J.H., Gordon, R.M., Fitzwater, S. and Broenkow, W.W., 1989. VERTEX: Phytoplankton/iron studies in the Gulf of Alaska. *Deep-Sea Res.*, **36**: 649–80.

Mikhailova, N.F., 1965. Novyi dlya Chernogo morya vid kokkolitoforid *Calciosolenia granii* var. *cylindrothecae-formis* Schiller (A species of coccolithophorid new for the Black Sea). *Trud. Sevastopol. Biol. Stants.*, **15**: 50–2.

Milliman, J.D., 1980. Coccolithophorid production and sedimentation, Rockall Bank. *Deep-Sea Res.*, **27**: 959–63.

Mitchell-Innes, B.A. and Winter, A., 1987. Coccolithophores; a major phytoplankton component in mature upwelled waters off the Cape Peninsula, South Africa in March, 1983. *Mar. Biol.*, **95**: 25–30

Mjaaland, G., 1956. Some laboratory experiments on the coccolithophorid *Coccolithus huxleyi*. *Oikos*, **7**: 251–5.

Molfino, B. and McIntyre, A., 1990. Precessional forcing of nutricline dynamics in the Equatorial Atlantic. *Science*, **249**: 766–9.

Molfino, B., McIntyre, A. and Campbell, D., 1989. Equatorial Atlantic nutricline variations aligned with the Younger Dryas. *Trans. Amer. Geophys. Union.*, **70**(43): 1134.

Morozova-Vodyanitskaya N.V. and Belogorskaya, E.V., 1957. Oznachenii kokkolitoforid i osobenno pontosfery v planktone Chernogo morya (On the importance of the coccolithophorids and particularly *Pontosphaera* in the plankton of the Black Sea). *Trud. Sevastopol. Biol. Stants.*, **9**: 14–21.

Murray, G. and Blackman, V.H., 1898. On the nature of the coccospheres and rhabdospheres. *Phil. Trans. Roy. Soc. (London)*, Ser. B, **190**: 427–41.

Murray, J., 1885. Coccospheres and rhabdospheres. In *Report on the Scientific Results Of the Exploring Voyage of the H.M.S. Challenger during the Years 1873-76.* vol. 1, part 2. Neill and CO, Edinburgh, For HMSO.

Nelson, J.R. and Wakeham, S.G., 1989. A phytol-substituted chlorophyll *c* from *Emiliania huxleyi*. *J. Phycol.*, **25**: 761–6.

Nishida, S., 1979. Atlas of Pacific Nannoplanktons. *Micropaleontol. Soc. Osaka*, Spec. Paper, No. 3.

Nishida, S., 1986. Nannoplankton flora in the Southern Ocean, with special reference to siliceous varieties. In *Proceedings of the Seventh Symposium on Polar Biology. Mem. Natl. Inst. Polar Biol.* (Japan) (Spec. Issue) ed. T. Hoshiai, T. Nemoto and Y. Naito.; No. 40, pp. 56–68.

Norris, R.E., 1961. Observations on phytoplankton organisms collected on the N.Z.O.I. Pacific Cruise, September 1958. *N.Z. J. Sci.*, **4**: 162–88.

Norris, R.E., 1971. Extant calcareous nannoplankton from the Indian Ocean. In *Proc. Plankt. Conf.*, Rome 1970, ed. A. Farinacci, pp. 899–909.

Norris, R.E., 1983. The family position of *Papposphaera* Tangen and *Pappomonas* Manton and Oates (Prymnesiophyceae) with records from the Indian Ocean. *Phycologia*, **22**: 161–9.

Norris, R.E., 1984. Indian Ocean nanoplankton. I. Rhabdosphaeraceae (Prymnesiophyceae) with a review of extant taxa. *J. Phycol.*, **20**: 27–41.

Norris, R.E., 1985. Indian Ocean nannoplankton. II. Holococcolithophorids (Calyptrosphaeraceae, Prymnesiophyceae) with a review of extant genera. *J. Phycol.*, **21**: 619–41.

Okada, H. and Honjo, S., 1973. The distribution of oceanic coccolithophorids in the Pacific. *Deep-Sea Res.*, **20**: 355–74.

Okada, H. and Honjo, S., 1975. Distribution of coccolithophores in marginal seas along the western Pacific Ocean and in the Red Sea. *Mar. Biol.*, **31**: 271–85.

Okada, H. and McIntyre, A., 1977. Modern coccolithophores of the Pacific and North Atlantic Oceans. *Micropaleontol.*, **23**(1): 1–55.

Okada, H. and McIntyre, A., 1979. Seasonal distribution of modern coccolithophores in the western North Atlantic Ocean. *Mar. Biol.*, **54**: 319–28.

Paasche, E., 1960. Phytoplankton distribution in the Norwegian Sea in June, 1954, related to hydrography and compared with primary production data. *Fiskeridir. Skr. Havundersøk.*, **12**(11): 1–77.

Ramsfjell, E., 1960. Phytoplankton distribution in the Norwegian Sea in June, 1952 and 1953. *Fiskeridir. Skr. Havundersøk.*, **12**(10): 1–112.

Reid, F.M.H., 1980. Coccolithophorids of the North Pacific Central Gyre with notes on their vertical and seasonal distribution. *Micropaleont.*, **26**: 151–76.

Reid, F.M.H., Stewart, E., Eppley, R.W. and Goodman, D., 1978. Spatial distribution of phytoplankton species in chlorophyll maximum layers off southern California. *Limnol. Oceanogr.*, **23**: 219–26.

Saugestad, A., 1967. Planteplankton i veslige Midelhav mars–april 1961. Manuscript, University of Oslo.

Schiller, J., 1930. Coccolithineae. In Rabehnorst's Kryptgamenflora von Deutschland, Österreich und der Schweiz, Leipzig. **10**: 89–267.

Silva, E.S., 1960. O microplâncton de superfície nos meses de setembro e outubre na estaçáo de Inhaca (Moáçambique). *Mem. Junta. Invest. Ultram.* 2 Sér., **18**: 1–50.

Smayda, T.J., 1958. Phytoplankton studied around Jan Mayen Island March–April, 1955. *Nytt Mag. Bot.*, **6**: 75–96.

Smirnova L.T., 1959. Phytoplankton in the Okhotsk Sea and Kuril Island region. (in Russian). *Trudy Inst. Okeanol. Akad. Nauk SSSR*, **30**: 3–51.

Sorby, H.C., 1861. On the organic origin of the so-called 'crystalloids' of the chalk. *Ann. Mag. Nat. Hist.*, Ser.3, **8**: 193–200.

Tangen, K., 1972. *Papposphaera lepida,* gen. nov., n. sp., a new marine coccolithophorid from Norwegian coastal waters. *Norw. J. Bot.*, **19**: 171–8.

Thierstein, H.R., Geitzenauer, K.R., Molfino, B. and Shackleton, N.J., 1977. Global synchroneity of late Quaternary coccolith datum levels: validation by oxygen isotopes. *Geology*, **5**: 400–4.

Thomsen, H.A., 1979. Electron microscopical observations on brackish-water nannoplankton from the Tvärminne area, SW coast of Finland. *Acta Bot. Fennica*, **110**: 11–37.

Thomsen, H.A., 1981. Identification by electron microscopy of nanoplanktonic coccolithophorids (Prymnesiophyceae) from West Greenland, including the description of *Papposphaera sarion* sp. nov. *Br. Phycol. J.*, **16**: 77–94.

Thomsen, H.A, Buck, K.R., Coale, S.L., Garrison, D.L. and Gowing, M.M., 1988. Nanoplankton coccolithophorids (Prymnesiophyceae, Haptophyceae) from the Weddell Sea, Antarctica. *Nord. J. Bot.*, **8**(4): 419–36.

Thomsen, H.A. and Oates, K., 1978. *Balaniger balticus* gen. et sp. nov. (Prymnesiophyceae) from Danish coastal waters. *J. Mar. Biol. Assn. U.K.*, **58**: 773–9.

Valkanov, A., 1962. Uber die Entwicklung von *Hymenomonas coccolithophora* Conrad. *Rev. Algol. N.S.*, **6**: 220–6.

Vilicic, D. and Fanuko, N., 1984. A study of phytoplankton in offshore waters of southern Adriatic, January 1980. *Nova Thalassia*, **6**: 67–82.

Wallich, G.C., 1877. Observations on the coccosphere. *Ann. Mag. Nat. Hist.*, Ser. 4, **19**: 342–50.

Wilbur, K.M. and Watabe, N., 1963. Experimental studies on calcification in molluscs and the alga *Coccolithus huxleyi*. *Ann. N.Y. Acad. Sci.*, **109**: 82–104.

Winter, A., 1985. Distribution of living coccolithophores in the California Current system, southern California Borderland. *Mar. Micropaleontol.*, **9**: 385–93.

Winter, A., Reiss, Z. and Luz, B.,1979. Distribution of living coccolithophore assemblages in the Gulf of Elat ('Aqaba). *Mar. Micropaleontol.*, **4**: 197–223.

9 Sedimentation of coccolithophores

JOHN C. STEINMETZ

Introduction

The global carbon cycle is undergoing intense scrutiny in light of growing evidence that anthropogenic carbon dioxide is impacting it in unknown ways. It is well recognized that processes in the ocean are fundamental to this cycling, and that the oceans constitute the major global reservoir for carbon. Currently, much research is involved in understanding the dynamic exchange of carbon dioxide between the atmosphere and the ocean. A critical part of evaluating global environmental changes related to carbon dioxide input is a clearer understanding of the processes governing the production, fate, and cycling of biogenic materials in the sea (SCOR, 1987, 1990; Houghton *et al.*, 1990).

Coccolithophores are the major primary producers that convert dissolved carbon dioxide in the ocean to $CaCO_3$. The pathway between production of this mineral at the ocean's surface to storage on the sea floor is an important one, since it represents an essential variable in the global carbon cycle equation. The purpose of this paper is to review the current knowledge regarding the vertical transport of coccolithophores in the open ocean.

The sedimentation of coccolithophores to the deep-sea floor is not a simple phenomenon of minute particles settling slowly through the water column. The mass of a single coccolith is so small (about 8×10^{-6} μg) and its surface area is so large (about 32 μm²) that unassisted its settling velocity is extremely slow (about 1.6 μm/sec or 13.8 cm/day). Additionally, a slow descent rate coupled with a large surface area would contribute to the rapid dissolution of coccolithophores in water undersaturated with respect to $CaCO_3$ (Honjo, 1975). It is becoming increasingly apparent through laboratory experiments and sediment-trap studies that the vast majority of coccolithophores, either as intact coccospheres or as individual coccoliths, descend rapidly in fecal pellets, marine snow, or similar, relatively large aggregate particles. Data from sediment-trap studies also reveal the proportions of nannoplankton that survive the descent and those that dissolve *en route*.

Stokes' equation, relating the settling time of particles in a fluid, predicts that an individual coccolith from a coccosphere would take several tens of years to settle unassisted in the open ocean (Honjo, 1976). In a mere fraction of that time, however, a settling coccolith would be likely to drift in ocean currents far beyond its original latitudinal distribution zone and undergo marked, if not complete, dissolution. And yet, the sedimentary thanatocoenosis (death or sedimentary assemblage) closely resembles the biocoenosis (living assemblage): well-preserved coccolith ooze present on the sea-floor is a subset of the assemblage composition in the euphotic zone community directly overlying. Post-depositional processes, can, of course, further modify this.

Coccolithophores, together with diatoms and other phytoplankton or single-celled algae, form the base of the food pyramid in the open ocean, and as such, they are often likely to be consumed before completing their life cycles (Honjo, 1975). Grazing zooplankton commonly pass coccoliths through their guts and excrete aggregates of them in fecal pellets. The mechanical and chemical effects of digestion on coccoliths have been found to be minimal; in fact, often delicately preserved, intact coccospheres are found within zooplankton fecal pellets (Honjo, 1975, 1976; Honjo and Roman, 1978). Fecal pellets and other oceanic macroaggregates are largely responsible for the rapid vertical transport of the majority of coccospheres and coccoliths through undersaturated waters to the sea-floor (Honjo, 1975, 1976; Pilskaln and Honjo, 1987). This transport mechanism also explains why coccoliths may exhibit little or no evidence of progressive dissolution even below the calcite compensation depth (Roth *et al.*, 1975).

It is highly unlikely that individual coccoliths found at depth in the water column have descended very slowly to that point. Instead, they most probably were transported in larger, faster-descending aggregates to be released there (Honjo, 1976). Coccoliths found separate or free in the water column were termed 'free-coccoliths' by Steinmetz (1991) as it is likely that they had been liberated from fecal pellets or similar aggregate particles. Honjo (1976) observed that free (or fresh) coccoliths and coccospheres are replenished at all depths by descending (and spilling) fecal pellets. Of course, once the contents of the host fecal pellet are spilled, their rate of descent decreases a thousand-fold, and they are then fully exposed to undersaturated deep water (Honjo, 1975). In this manner, free-coccoliths are quickly remineralized (dissolved) in the water column; only those particles that descend rapidly in aggregates through undersaturated waters are likely to become part of the thanatocoenosis.

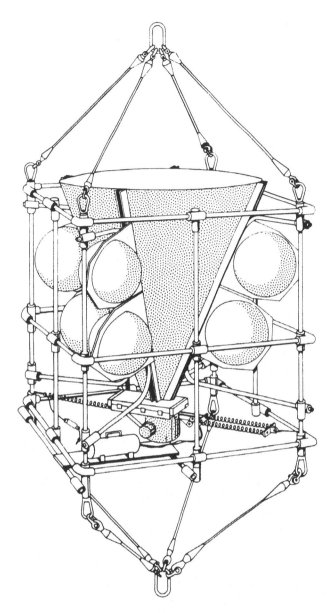

Fig. 1. A PARFLUX Mark II sediment trap (from Honjo et al., 1980)

Plankton-tow and waterbottle-casting studies have provided a wealth of qualitative and quantitative information regarding the biology and ecology of coccolithophores (see Winter *et al.*, Chapter 8): standing-stock estimates, geographic and vertical distribution of species, and seasonal variation in abundance and composition (e.g., McIntyre and Bé, 1967; Gaarder and Hasle, 1971; Okada and Honjo, 1973; Honjo and Okada, 1974; Nishida, 1979; Hara and Tanoue, 1985; Samtleben and Schröder, 1990). This information, however, does not translate into sedimentation rate, accumulation rate, or even preservational state on the sea-floor. Surface-sediment studies, utilizing samples collected from dredgehauls, box corers, or piston-core tops, provide infor-

mation about the spatial distribution of calcareous nannoplankton in the overlying water, and certainly something about their own state of preservation, but they reveal little about coccolith production, vertical transport, or the proportion that become part of the fossil, or sedimentary, record.

Sediment traps

Sediment traps are devices that passively capture particles settling through the water column (Fig. 1). This apparently simple function, however, should not imply that they are simple devices. Sediment-trap technology has evolved to include such considerations as the following: deployment of the traps without collecting before they reach the desired depth, collection without biasing the particle sizes captured, poisoning of collected particles to inhibit their microbial degradation, rotating collecting containers to capture time-series sampling, and self-closing containers to prohibit loss of sample during trap retrieval. Sediment traps can be deployed either in anchored, vertical arrays or as free-floating collecting devices, for hours, days, or months at a time. Once retrieved, the collected particles are identified and enumerated, and particle fluxes can be directly calculated (Wiebe *et al.*, 1976; Soutar *et al.*, 1977; Honjo, 1978, 1980; Spencer *et al.*, 1978; Knauer *et al.*, 1979; Rowe and Gardner, 1979; Thunell and Honjo, 1981; Betzer *et al.*, 1984). No one trap design is utilized by all scientists; in fact, traps of various designs have been developed and tested, and they have been calibrated against each other in various sediment-trap intercomparison experiments (Gardner, 1980a, 1980b; Spencer, 1981). The techniques of handling, filtering, and splitting of sediment samples were discussed by Honjo (1980).

Inasmuch as only two detailed studies of coccoliths caught in sediment traps have been made (Samtleben and Bickert, 1990; Steinmetz, 1991), the techniques of their enumeration have not as yet been standardized. As a result, methods of sample preparation, specimen counting, and flux calculations differ among investigators, often hampering comparisons. Nevertheless, adjustments can be made, and the correlation of results can be attempted. Samtleben and Bickert (1990) examined the fine fraction (<63 μm) from sample splits on a section of filter using a scanning electron microscope (SEM). Identifiable coccoliths were counted, and their total number was projected, based on the proportion of the surface area of the filter and the volume of sample split. Steinmetz (1991) also examined the fine fraction (<63 μm) of sediment-trap samples, but he employed an optical microscope to count coccoliths in a cell-counting chamber. He introduced an aliquot of coccoliths suspended in seawater into a hemocytometer with improved Neubauer ruling. The total number of coccoliths in the sediment-trap sample was calculated by taking the product of the frequency of coc-

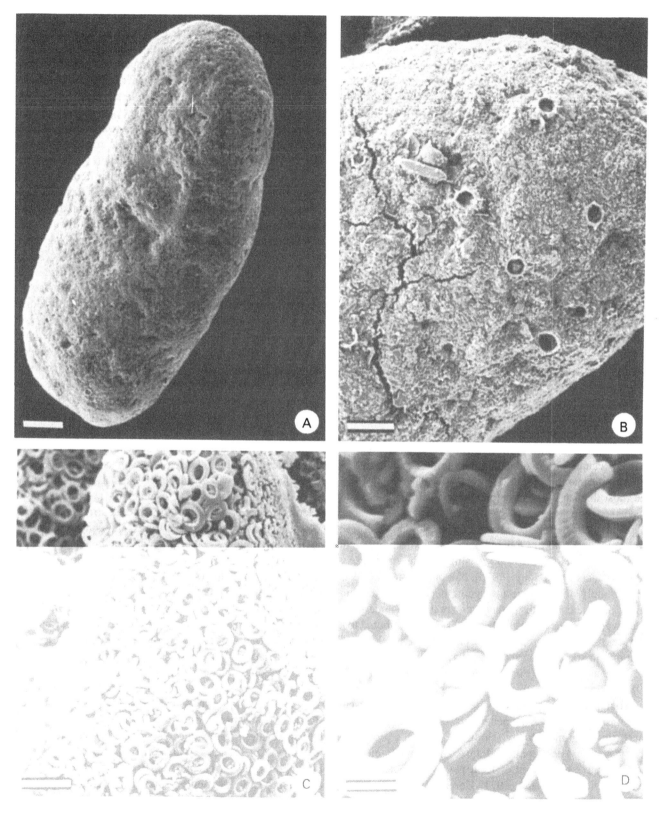

Fig. 2. Fecal Pellets. A. Typical fecal pellet. Scale bar is 100 μm. B. Close-up of the surface of a fecal pellet. Circular objects on the surface are silicoflagellates. Scale bar is 50 μm. C. and D. Close-up of the surface of a fecal pellet composed entirely of the coccolithophore *Umbilicosphaera sibogae* . Scale bars are 10 μm and 2 μm, respectively. SEM photos compliments of C. Pilskaln; C. and D. from Pilskaln (1985).

coliths per milliliter, the original volume of liquid sample in the aliquot fraction, and the inverse of the fraction aliquot. Steinmetz utilized the SEM to identify taxa, estimate their relative abundance, and to assess the preservation of the taxa. Both groups of investigators then estimated the coccolith fluxes from the calculated number of coccoliths, the surface area of the trap opening, and the length of time that the trap was open to descending particles.

Mechanisms of vertical transport

It is well established that the dominant mechanism for the vertical transport of particulate matter from the euphotic zone to the deep sea is the relatively rapid settling of large particles (McCave, 1975; Bishop et al., 1977; Lal, 1977; Knauer et al., 1984). Two particle types are commonly recognized: fecal pellets and 'marine snow'.

Fecal pellets (Fig. 2) are produced by zooplankton such as copepods, chaetognaths, siphonophores, euphausids, tunicates, and amphipods feeding in the euphotic zone. Lohmann (1902) was the first to suggest that fecal pellets may serve as a means of vertical transport of coccoliths and coccospheres to the deep-sea floor. Since then, numerous investigators have reported coccoliths in zooplankton fecal pellets (Marshall and Orr, 1956, 1962; Bernard, 1963; Roth et al., 1975; Honjo and Roman, 1978; Pilskaln and Honjo, 1987). Coccoliths and coccospheres pass through the alimentary canals of the zooplankton with no dissolution and often no mechanical effects (Honjo and Roman, 1978). Even the most delicate microstructures of a coccolith shield can survive copepod digestion (Roth et al., 1975). The size range of pellets collected in the Panama Basin (Station PB_1) (see Fig. 4) was 10–200 μm in width and 100–700 μm in length (Pilskaln and Honjo, 1987). A single pellet may carry 100,000 coccoliths (Honjo, 1976). Not only do fecal pellets act as rapidly sinking large aggregates, but they also serve to protect calcite coccoliths in their descent through undersaturated waters. Some of this protection has been attributed to a surface membrane or pellicle enclosing the fecal pellet, although in some instances pellets have been found devoid of these membranes (Pilskaln and Honjo, 1987; Pilskaln, 1988). Where present, the pellicle is important in protecting the contents of the pellet, provides a chemical barrier, and smooths the surface of the pellet, thereby reducing drag and increasing sinking velocity (Honjo, 1976). This protective role has also been confirmed through scanning electron microscopic examination of fecal pellets deposited on the deep-sea floor (Schrader, 1971; Honjo, 1975; Roth et al., 1975; Honjo and Roman, 1978).

Marine snow particles (Fig. 3) are fragile, amorphous particulates ranging in size from about 0.5 mm up to centimeters or even meters in diameter. The term marine snow is somewhat generic and can include various macroscopic aggregates, agglomerates, or flocculates of inorganic and organic matter loosely bonded together by intermolecular, intramolecular, or atomic forces, surface tension, organic cohesion, or electrostatic force (Knauer et al., 1984). These particles are composed of living and dead bacteria, phytoplankton, and zooplankton, fecal pellets and detritus, river- and wind-derived inorganic minerals, organic material such as free algal cells, pigmented granules, waxy particles, and nutrients aggregated together by organic mucus (Silver et al., 1978; Honjo et al., 1982a). They contain all the individual microplankton taxa found in the water column (although often in different ratios). Marine snow aggregates may contain as many as 2000 fecal pellets, suggesting that the aggregates may capture faster-sinking fecal pellets (Silver et al., 1978). Marine snow is a common component in pelagic surface and deep waters throughout the world's oceans (Silver et al., 1978; Alldredge, 1984). Their abundance in the water column ranges from several tens per liter near the surface to one per liter or less at depth (Shanks and Trent, 1980; Honjo et al., 1984; Alldredge and Gotschalk, 1989).

The sinking rates of fecal pellets and marine snow have been measured by various SCUBA or photographic means in the open ocean (Table 1). The rate of descent of fecal pellets averages about 200 m/day, twice that of marine snow aggregates. Studies have also revealed that these two kinds of particles are contributed in different proportions to the overall total reaching the sea-floor.

In shallower waters, fecal pellets predominate. Dunbar and Berger (1981) found at least 60% and as much as 90% of trapped material was fecal pellets at 341 m depth in the Santa Barbara Basin of California. They were not able to determine what proportion of the material arriving at the trap was marine snow. In deeper settings, and contrary to previous assumptions, the contribution of fecal pellets to the total mass flux of biogenic particulate material is no more than 5% (Pilskaln and Honjo, 1987). Asper (1987) reported that marine snow aggregates constituted essentially all the settling particles arriving at the sea-floor in the Panama Basin (Station PB_1). Fecal pellets are believed to be particularly efficient in transporting fine particles originating at the surface, and marine snow is thought to play the dominant role in transportation to the deeper layers (Shanks and Trent, 1980; Honjo et al., 1984). The depth where fecal pellets are largely replaced by marine snow particles is, as yet, unknown. Breakage of pellets, and the incorporation of their contents into aggregates is likely due to coprophagy (organisms feeding on fecal pellets) (Frankenberg and Smith, 1967; Paffenhöfer and Strickland, 1970) or bacterial degradation of the protective organic pellicle surrounding the pellet (Johannes and Satomi, 1966).

A fecal pellet settling undisturbed in 5 km of water at between 100 and 300 m/day would take between 17 and 50 days to reach the bottom. If such a pellet were subjected to a unidirectional advective current of 3 cm s^{-1}, it would be dis-

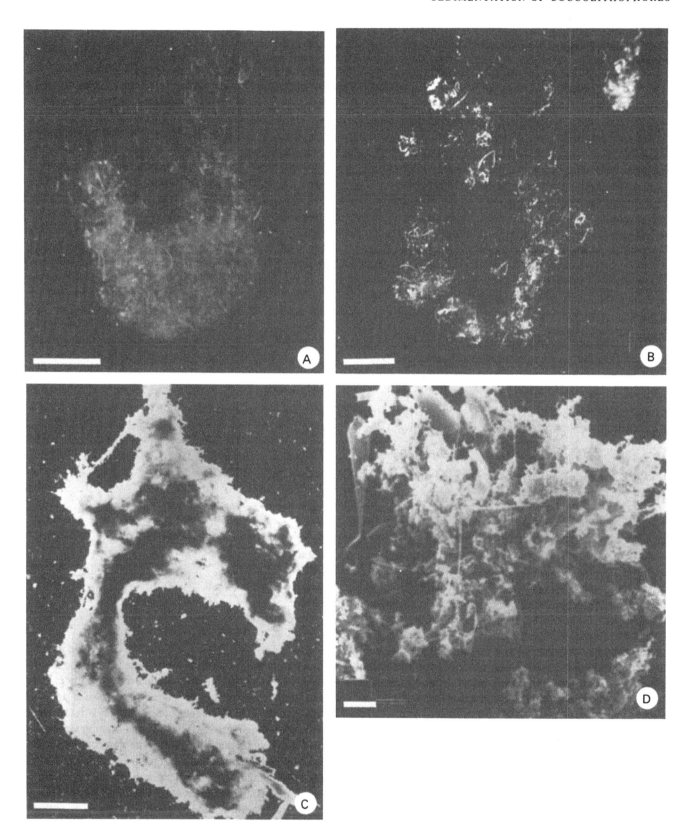

Fig. 3. Marine snow particles or macroaggregates. A. and B. From Alldredge and Gotschalk (1989); scale bars are 1 cm. C. From Honjo (1980); scale bar is 1 cm. D. From Asper (1987); scale bar is 100 μm. (A., B., and D. Reprinted with permission of Pergamon Press PLC; C. compliments of the *J. Marine Res.*)

Table 1. *The sinking rates of fecal pellets and marine snow measured in the open ocean.*

Particle type	Sinking rate range m/d	Sinking rate average m/d	Reference
fecal pellets	50–225	160	Wiebe *et al.*, 1976
fecal pellets	266–400	–	Bishop *et al.*, 1977
Acartia tonsa fecal pellets	80–150	120	Honjo and Roman, 1978
Calanus finmarchicus fecal pellets	180–220	–	Honjo and Roman, 1978
copepod fecal pellets	10–145*	75	Paffenhöfer and Knowles, 1979
various fecal pellets	40–3200**	350	Honjo *et al.*, 1982a
green fecal pellets	237–750	343	Pilskaln, 1982
marine snow	43–95	68	Shanks and Trent, 1980
marine snow	1–36	–	Asper, 1987
marine snow: 'diatom flocs'	50–200	117	Alldredge and Gotschalk, 1989

*based on copepod genera of various sizes.
**based on copepods, copepodites, and salps.

Table 2. *Location and logistics of PARFLUX sediment traps in Honjo et al. (1982a), Pilskaln and Honjo (1987), and Steinmetz (1991)*

	PARFLUX E	PARFLUX P_1	PARFLUX PB_1
Location	13°31' N 54°00' W	15°21' N 151°28' W	5°21' N 81°53' W
Term	11/77–2/78	9/78–11/78	7/79–11/79
Duration	98 days	61 days	112 days
Trap depths	389 m 988 m 3755 m 5068 m	378 m 978 m 2778 m 4280 m 5582 m	667 m 1268 m 2265 m 2869 m 3769 m 3791 m
Ocean depth	5288 m	5792 m	3856 m

placed laterally between 44 and 130 km. These distances, therefore, estimate the resolution of replication of the biocoenosis and the thanatocoenosis in the deep-sea sediment, and they agree with similar calculations of Honjo (1976) and Bishop *et al.* (1977).

Flux calculations

Numerous sediment-trap studies have shown that there exists a strong relationship between primary production and particle flux out of the euphotic zone (Deuser and Ross, 1980; Deuser *et al.*, 1981; Honjo, 1982; Betzer *et al.*, 1984; Knauer *et al.*, 1984). The exact relationship, however, is not simple, and it is still largely unresolved (Betzer *et al.*, 1984; Knauer *et al.*, 1984).

The total mass flux is the rate of arrival of oceanic materials to a certain depth. Since flux is a rate term, it is commonly expressed as mass per unit area per unit time, usually mg m^{-2} day^{-1} (Honjo, 1986). The lithogenic fraction, such as clay, quartz, and volcanic fragments, constitutes a relatively small fraction of sinking particles. The biogenic fraction represents most of the particles that sink through the water column in the ocean, and it includes carbonates, opaline silica, and particulate organic matter (cellular and amorphous material, including fecal pellets and marine snow). The carbonate materials include coccoliths and coccospheres, planktonic foraminifers, pteropod shells, and miscellaneous hard particles such as dinoflagellate-derived thoracospheres. Honjo *et al.* (1982a) reported on the relative contribution of the various biogenic particles from four particle flux experiment –

Table 3. *Fecal pellet data from Pilskaln and Honjo (1987)*

Station	Average mass (μm)	Total CaCO₃ (wt. %)	Average flux (No. m⁻² day)	Average CaCO₃ flux (%)
E	3.0	28	589	1.2
P₁	0.3	38	275	0.3
PB₁	1.5	23	6240	3.1

80 to 90%, had been trapped in 'the amorphous fine fraction (<62 μm), which is believed to come from the disintegration of fragile fecal pellets or other macroscopic aggregates'.

Pilskaln and Honjo (1987) examined the fecal-pellet fraction from three of the stations studied by Honjo *et al.* (1982a), and the same three investigated by Steinmetz (1991), namely E, P₁, and PB₁ (discussed in the following paragraphs). They reported the fecal pellet contribution to the total carbonate flux to be considerably lower than that suggested by Honjo *et al.* (1982a) for the same three stations, ranging between 0.3 to 3.1% (Tables 3 and 4). These

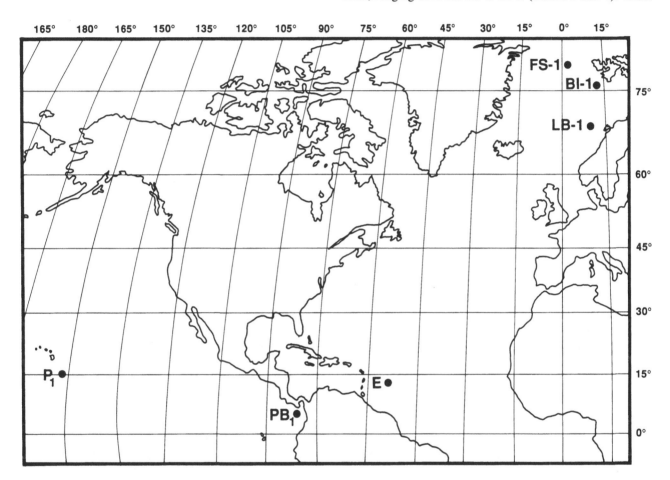

Fig. 4. Locations of sediment-trap stations in Honjo et al. (1982), Pilskaln and Honjo (1987), and Steinmetz (1991) [E, P₁, PB₁]; and Samtleben and Bickert (1990) [LB-l, BI-l, FS-l].

PARFLUX – sediment-trap stations in the Atlantic and Pacific (the same three as examined by Steinmetz (1991) plus the Söhm Abyssal Plain, Station S₂, in the Sargasso Sea, 31.5° N, 56° W) (Fig. 4 and Table 2). They found that 60 to 90% of the total particulate flux was of biogenic origin, and that carbonate particles accounted for 30 to 60% of the flux. The contributions of constituents to the total carbonate flux were as follows: coccoliths: 20–40%, planktonic foraminifers: 15–50%, pteropod shells: 8–12%, and thoracospheres: trace. Honjo *et al.* (1982a, p. 614), realizing that coccoliths found at depth had arrived there in particles considerably larger than individual coccoliths, calculated that 10 to 20% of the coccolith flux was trapped in intact fecal pellets, and the balance,

figures from the later study (Pilskaln and Honjo, 1987) are probably more realistic inasmuch as the earlier figures (Honjo *et al.*, 1982a) appeared to be estimations made from calculations of all biogenic constituents. Moreover, the later study was specifically centered around the fecal-pellet fraction of particle fluxes. The chemical composition of fecal pellets studied by Pilskaln and Honjo (1987) was found to vary by not more than 10% from each trap depth to the next

Table 4. *Free-coccolith fluxes and free-coccolith carbonate fluxes in 63μm size fraction (Steinmetz, 1991), and fecal pellet fluxes and fecal pellet carbonate fluxes (Pilskaln and Honjo, 1987) at three PARFLUX stations.*

Station and depth (m)		Average flux of free-coccoliths (10^6 m^{-2} day^{-1})	Average free-coccolith carbonate flux (mg m^{-2} day^{-1})	Flux of fecal pellets (No. m^{-2} day^{-1})	Flux of fecal pellet carbonate (mg m^{-2} day^{-1})
E:	389	347.46	2.78	604.3	0.20
	988	260.47	2.08	772.3	0.41
	3755	125.39	1.00	498.9	0.37
	5068	531.94	4.26	481.5	0.29
P_1:	378	0.03	Tr	–	–
	978	33.57	0.27	–	–
	2778	529.13	4.23	295.8	0.03
	4280	759.44	6.08	247.5	0.02
	5582	352.52	2.82	281.6	0.03
PB_1:	667	1179.50	9.44	5162.6	1.1
	1268	1165.66	9.32	5174.8	1.4
	2265	1061.03	8.49	–	–
	2869	1113.52	8.91	5857.5	1.4
	3769	515.06	4.12	7704.9	1.7
	3791	423.13	3.38	7302.0	1.5

at any one of the stations, hence, it is reasonable to characterize the fecal-pellet flux characteristics at each station using averages (Tables 3 and 4). Since the carbonate components of the pellets consisted entirely of coccoliths and an occasional coccosphere, and no foraminifers or pteropod shells, or shell fragments, the fecal pellet carbonate flux is a factor in coccolith flux calculations.

In order to translate coccolith fluxes into carbonate fluxes, it is necessary to make an approximation for the average coccolith mass. The values used for the calcite mass of a single coccolith can remarkably affect the final coccolith-carbonate flux values. Steinmetz (1991) used 8×10^{-12} g (or 8×10^{-9} mg), a value suggested by Honjo (1976), for the mass of an average coccolith. Samtleben and Bickert (1990) used 3×10^{-9} mg for the mass of the relatively small species *Emiliania huxleyi* and 1.3×10^{-7} mg for the considerably larger *Coccolithus pelagicus*. Because coccospheres represent such a minor fraction of the coccolith and coccosphere flux (<0.001% of all sinking coccoliths) (Steinmetz, 1991), their contribution can be ignored.

Coccolithophores caught in sediment traps

Surprisingly few sediment-trap studies document the assemblage composition, amount, and preservational condition of coccolithophores. The first was reported by Honjo (1976) on a study of the contents of zooplankton fecal pellets collected in a sediment trap deployed at a depth of 2200 m for two months in the Tongue of the Ocean, Bahamas (Wiebe *et al.*, 1976). He observed that about 80% of the pellets contained the skeletons of phytoplankton (coccoliths and diatoms) and clay-mineral-like particles. Preservation of the coccoliths was found to be excellent.

Pilskaln *et al.* (1989) examined platform-derived and open-ocean particulates caught in a sediment trap deployed at 500 m for two months in the Northwest Providence Channel, northern Bahamas. The total carbonate flux was determined to be 47.6 mg m^{-2} day^{-1}, of which 61% was derived from planktonic carbonate sources (coccolithophores, planktonic foraminifers, and pteropods). The balance of the carbonate flux was from shallow platform-derived sources. Coccoliths (primarily *Emiliania huxleyi*) dominated the <63 μm components, providing a calcite flux of 14 mg m^{-2} day^{-1}. Coccolith calcite within fecal pellets constituted a small contribution (0.34 mg m^{-2} day^{-1} or 0.7%) to the total carbonate flux.

Okada (1989) studied the seasonal variation of coccolithophore flux at a Japanese offshore station. He observed a prominant peak corresponding to a spring bloom. The smallest (0.5×10^9 individuals m^{-2} day^{-1}) and the largest (3.1×10^9 m^{-2} day^{-1}) coccolith fluxes were recorded in the late March and the early June samples, respectively. The calculated average flux was 1.2×10^9 m^{-2} day^{-1}.

Kleijne (1990) examined the species composition, distribution, and condition of preservation of coccolithophores in the Indonesian Seas. Floating sediment traps, deployed at depths between 60 and 100 m at four stations in the eastern Banda Sea, collected 32 species in various stages of preservation. Moored sediment traps samples from 4500 m water depth in the Flores Sea collected 23 species of cocco-lithophores. *Gephyrocapsa oceanica* was found to predominate in both trap types, followed usually by *E. huxleyi*. No flux studies were done on either trap type.

Comprehensive coccolith / sediment-trap studies

To date, only two comprehensive studies of coccoliths caught in sediment traps have been reported. Samtleben and Bickert (1990) examined coccolith and coccosphere species composition, flux, and seasonality from sediment traps deployed at three stations in the Norwegian Sea. Steinmetz (1991) examined coccoliths and coccospheres in three sediment traps deployed at various depths for varying lengths of time on moored vertical arrays in equatorial waters. Since the sediments he studied were collected in traps over a long period of time and do not offer seasonal information, his study will be discussed first as a long-duration trapping study. This will be followed by the work of Samtleben and Bickert (1990) that can be characterized as time-series sediment trapping.

Long-duration sediment trapping

Traps deployed for many months can better characterize the annual flux of a particular region of the ocean. The longest periods sampled in equatorial waters have been between 61 and 112 days in sediment flux studies by Honjo *et al.* (1982a) and Pilskaln and Honjo (1987) in which Steinmetz (1991) examined the coccolithophores.

Steinmetz (1991) examined samples in three PARFLUX Mark II sediment traps deployed in the equatorial Atlantic, central Pacific, and in the Panama Basin (Fig. 4 and Table 2). He studied the <63 μm size fraction in order to characterize the flux of coccoliths and coccospheres, determine the taxa present, and assess their condition of preservation throughout the water column. The results of these determinations varied markedly at each of the three stations, a reflection of the different productivity of the respective water masses. The equatorial Atlantic site (Station E) was located in the western tropical Atlantic at least 750 km from the nearest landmass, the Guyana coast, and within the northwestward-flowing North Equatorial Current. The central Pacific site (Station P_1) was located between the Molokai and Clarion Fracture Zones in the East Hawaii Abyssal Plain and within the main axis of the westward-flowing North Equatorial Current. The Panama Basin site (Station PB_1) was located in the northeast quadrant of the Panama Basin where annually the region is successively influenced by Trade Winds induced upwelling and Calm Belt induced doldrums. Of the three locations, the Panama Basin has the highest level of

primary productivity, exceeding an annual average of 1000 mg C m^{-2} day^{-1} in the euphotic zone. The central Pacific station lies within the area of lowest level of biological productivity of the three regions, exhibiting a productivity of less than 100 mg C m^{-2} day^{-1}. The western tropical Atlantic station displayed a level of primary productivity that was somewhat intermediate between the other two.

Steinmetz (1991) enumerated the coccospheres separately from the coccoliths, since they required different visualizing techniques under the microscope. The Utermöhl or inverted microscope method was used to count coccospheres, and a hemocytometer with improved Neubauer ruling was used to count coccoliths. In general, the results of the coccosphere enumeration show coccosphere flux decreased with depth at all three sites. The flux was lowest at Station P_1, averaging 24 coccospheres m^{-2} day^{-1} for the entire water column. At Station E, the average flux for the entire water column was 4725 coccospheres m^{-2} day^{-1}. The highest fluxes were calculated at Station PB_1. The average coccosphere fluxes for the shallowest trap (667 m) and for the entire water column were, respectively, 23,413 and 8030 coccospheres m^{-2} day^{-1}.

The scanning electron microscope (SEM) was utilized to identify coccolithophore species present, to estimate their relative abundance, and to assess the preservation of the taxa. SEM studies of the samples showed that 56 species were present collectively at all three stations: 50 species at Station E, 35 species at Station P_1, and 26 species at Station PB_1. Preservation of the samples ranged from good to poor in most of the samples. Only at the shallowest trap at each station were all specimens preserved in good condition. In general, Steinmetz found well-, moderately-, and poorly-preserved specimens of the same species in the same trap sample. This substantiated two assumptions that he had made: (1) well-preserved specimens in traps at great depths were rapidly transported there, even through undersaturated waters, and (2) the breakdown of fecal pellets or aggregates proceeds within or just above the traps, exposing coccoliths to partial or complete dissolution.

The coccolith fluxes reported in Steinmetz (1991) are based solely on the free-coccoliths observed in sediment-trap samples, that is, samples consisting of coccoliths most likely freed from marine snow aggregates and from fecal pellets (Table 4). Inasmuch as the fluxes did not include the coccoliths contained in whole fecal pellets, the values represented minimum coccolith-flux values. The present paper incorporates the contribution of coccoliths from whole fecal pellets (Pilskaln and Honjo, 1987), and the numbers likely reflect more accurate coccolith-flux values for the respective sampling stations (Tables 3, 4, and 5). Since these values represent the sum of all coccoliths trapped from the various known sources, the term "total-coccolith" will be used in the flux discussion.

Total-coccolith fluxes differ widely at each of the three

Table 5. *Total-mass flux, biogenic-carbonate flux, and total coccolith-carbonate flux. Total-mass and biogenic-carbonate flux values from Honjo et al. (1982a). Total coccolith carbonate flux is the sum of free-coccolith carbonate flux and fecal pellet carbonate flux (from Table 3).*

Station and depth (m)		Total mass flux (mg m^{-2} day^{-1})	Biogenic carbonate flux (10^6 m^{-2} day^{-1})	Total coccolith carbonate flux (mg m^{-2} day^{-1})	Total coccolith carbonate in biogenic carbonate flux (%)	Total coccolith carbonate in total mass flux (%)
E:	389	69.4	43.5	2.98	6.85	4.29
	988	49.2	27.1	2.49	9.19	5.06
	3755	46.4	26.1	1.37	5.25	2.95
	5068	47.0	23.0	4.5	19.78	9.68
P$_1$:	378	11.4	4.0	Tr	Tr	Tr
	978	7.5	5.4	Tr	Tr	Tr
	2778	17.1	11.7	4.26	36.41	24.91
	4280	16.8	12.0	6.10	50.83	36.31
	5582	11.1	6.8	2.85	41.91	25.68
PB$_1$:	667	114.1	41.2	10.54	25.58	9.24
	1268	104.5	41.0	10.72	26.15	10.26
	2869	158.0	50.9	10.31	20.26	6.52
	3769	179.3	45.2	5.82	12.88	3.24
	3791	179.6	46.9	4.88	10.40	2.72

sites (Table 5, Fig. 5). At Station E, the total-coccolith flux decreases regularly with depth, but it increases sharply at the lowermost trap to its highest value in the profile. The average total-coccolith carbonate flux for the entire column is 2.85 mg m^{-2} day^{-1}. Station P$_1$ has a slightly lower average for the entire column, 2.64 mg m^{-2} day^{-1}, but the profile is substantially different. Instead of decreasing with depth, there is a marked increase in the flux below 2700 m. The average total-coccolith-carbonate flux in the Station PB$_1$ traps, 8.45 mg m^{-2} day^{-1}, is more than three times greater than the average flux in the other two stations. The uppermost three traps show a steady, similar flux averaging 10.52 mg m^{-2} day^{-1}. Below them, there is a drop to about half in the lowermost two traps averaging 5.35 mg m^{-2} day^{-1}.

Utilizing biogenic-carbonate flux data for each sample from Honjo *et al.* (1982a), it is possible to calculate the contribution of total-coccolith carbonate in the <63 μm size-fraction to the biogenic-carbonate flux (Table 5, Fig. 5). This contribution ranges from trace amounts (Station P$_1$: 378 m and 978 m) to 50.83% (Station P$_1$: 4280 m). The relative contribution of total-coccolith carbonate to the biogenic-carbonate flux is lowest at Station E (averaging 10.27% for the entire column), moderate at Station PB$_1$ (19.05%), and highest at Station P$_1$ (25.83%). The contribution of total coccolith carbonate in the total-mass flux for each station (Table 5, Fig. 5) follows a similar ranking: it is lowest at Station E (averaging 5.5%), slightly higher at Station PB$_1$ (6.40%), and about three times higher at Station P$_1$ (17.38%).

The differences in total-mass flux, biogenic-carbonate flux, and total-coccolith flux (Table 5) at the three sediment-trap stations are an indication of differences in surface water productivity. The lowest flux values were recorded at Station P$_1$ in the low-productivity region of the central Pacific. Intermediate flux values occurred at Station E in the equatorial Atlantic. The highest values occurred in the Panama Basin, Station PB$_1$, in a region of coastal upwelling and horizontal advection. There, flux values are approximately double or more than for the other stations for each of the flux parameters measured.

Modification of the coccolith-carbonate flux by calcite dissolution is apparent in the decrease of flux with depth. The only deviation from this trend is seen in the increased coccolith flux with depth for Station ^{1}. This likely represents the record of a coccolithophore bloom which occurred on the surface just before the sediment-trap array was deployed in September 1978. The products of the bloom were trapped "in transit" to the sea-floor. The marked coccolith flux increase in the lowermost trap (5068 m) at Station E is likely due to either resuspension of bottom sediment or the horizontal advection of coccoliths from elsewhere. At Station PB$_1$, the flux of coccoliths correlates well with the findings of Thunell *et al.* (1981) who investigated the sedimentary lysocline at this site. They conducted an *in situ* study of calcite dissolution using foraminifers and noted an increased rate of dissolution in the water column at 2869 m and particularly below that. The coccolith-flux data (Tables 4 and 5)

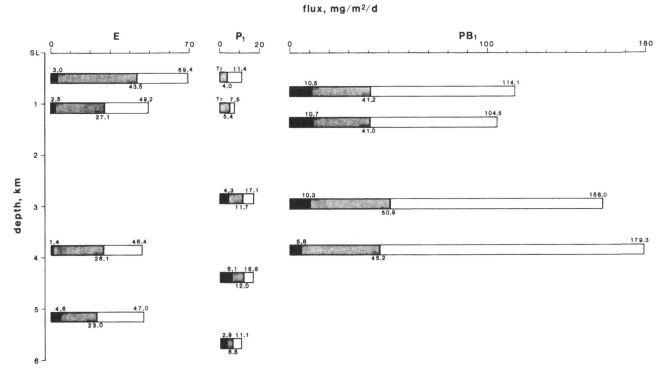

Fig. 5. Fluxes (mg m⁻²day⁻¹) at the three equatorial sediment-trap stations: E, P₁, and PB₁. The total length of each bar represents the total mass flux at the respective depth, the stippled length represents the total biogenic carbonate flux, and the black length represents the total coccolith-carbonate flux. The values are tabulated in Table 5.

show the same trend with depth. The total-coccolith carbonate flux data at Station PB₁ also indicate the proportion of coccoliths that survive the descent to the sea floor, about 51%. (This calculation cannot be made for the other two stations, since fluxes at depth exceed flux near the surface.) Although the coccoliths may survive the trip from the surface, this does not necessarily mean that they will become part of the sediment record. During their exposure on the sea floor, they become susceptible to dissolution. Only a small fraction of the coccoliths will be preserved in the fossil record. Well preserved, or pristine, coccoliths at depth are most likely derived from degraded fecal pellets or marine snow aggregates. Their presence at depth does indicate that fecal pellets and aggregates do provide means of protection and rapid transport to these depths. Okada and Honjo (1973) measured hundreds of coccolithophores per liter suspended at depths of nearly 4000 m in the equatorial Pacific. Fresh, well preserved coccoliths and coccospheres are thus continually replenished at all depths by rapidly descending pellets and aggregates. Free coccoliths caught in sediment traps were released only very recently from these protective, larger particles immediately above the trap mouths, the particles were broken and biodegraded within the traps, or the particles were merely broken during the mechanical handling of the samples. Any exposure of free coccoliths to the undersaturated deep water where the sediment traps were deployed would likely result in their immediate dissolution, so it is reasonable to assume that the pristine free coccoliths observed were from fecal pellets and aggregates broken in field and laboratory handling of the samples (Steinmetz,

1991). Additionally, Honjo (1976) reasoned that free coccoliths would eventually dissolve before they reach the seafloor and would not disturb the biocoenosis-thanatocoenosis correspondence. He calculated that 8% are remineralized (dissolved) within the undersaturated water column; the rest presumably reached the sea floor in fecal pellets.

Steinmetz (1991) found coccoliths in all stages of preservation at all depths, except in the shallowest trap at each of the three stations he studied. He also observed that the ratio of well preserved to poorly preserved (i.e., freshly spilled to nearly dissolved) coccoliths did not change with depth below the shallowest traps. Diversity of the assemblages, however, did, in general, decrease with depth. In Steinmetz' (1991) study, at stations E and PB₁, the number of species in the deepest traps contained half the number in the shallowest. He noted that the evidence of selective dissolution with depth was most apparent among the holococcoliths, those forms composed of small, relatively equidimensional, non-imbricated elements (e.g., *Calyptrosphaera, Corisphaera, Halladosphaera,* and *Homozygosphaera*). He also observed that specimens of these genera were rarely observed below about 3800 m in the sediment-trap samples studied.

Table 6. *Location and logistics of Arctic North Atlantic PARFLUX MARK V sediment traps in Samtleben and Bickert (1990)*

	Lofoten Basin	Bear Island	Fram Strait
Station	LB-1	BI-1	FS-1
Location	69°11' N 11°00' E	75°52' N 11°28' E	78°53' N 1°25' E
Term	8/83–8/84	9/84–8/85	9/84–8/85
Duration	352 days	364 days	359 days
Trap depth	2600 m	1700 m	2000 m
Water depth	3161 m	2119 m	2527 m
Number of samples	12	12	13
Average sampling period	29.3 days	30.3 days	27.6 days

Time-series sediment trapping

Sediment traps deployed for short periods of time, whether for days, weeks, or months, provide valuable insight into oceanographic processes occurring at various depths. It is well established, however, that organisms exhibit a seasonal cyclicity. Production and sedimentation of particles in the ocean are strongly coupled, therefore cycles of organic production in the surface layers are related to cycles of particle fluxes within the water column and on the sea-floor (Honjo and Doherty, 1988). As a result, sediment-trap collections for less than one year, that is, less than one complete seasonal cycle, provide a snapshot of activity in the ocean, and they sample only one element of the cycle. Hence, flux calculations and other determinations do not reveal the true oceanic picture. As demonstrated above, the fortuitous deployment of sediment traps that capture evidence of a surface bloom of coccolithophores, yields a grossly inflated flux and a mistaken characterization at that location if it is falsely assumed to reflect the normal condition.

This is not to say that short-duration deployments do not have merit. On the contrary, most of the qualitative and quantitative information about particle production rates, sinking rates, composition, degradation, and flux was derived from sediment traps placed in various basins for less than one year. Repeated annual samplings of sites are also desirable since the ocean does not operate on simple annual cycles. Such intra-annual phenomena as the El Niño event, monsoons, and Inter-Tropical Convergent Zone anomalies in the tropical and subtropical seas, as well as variability in sea-ice cover in polar regions, are examples of some of the events that influence regional hydrographic conditions, productivity, and

particle fluxes (Deuser *et al.*, 1981; Honjo and Doherty, 1988).

Experience gained from the short-duration sediment-trap experiments has contributed greatly to the design of time-series traps. Honjo and Doherty (1988) discuss and illustrate the considerations necessary in building these very successful sampling mechanisms, the PARFLUX Mark V and Mark VI time-series sediment traps.

Time-series sediment-trap collecting has been performed in the Panama Basin (Honjo, 1982), the Arctic North Atlantic (Wefer and Honjo, 1985; Honjo and Wefer, 1987; Honjo *et al.*, 1988), the Antarctic Bransfield Strait (Wefer *et al.*, 1988), the Arabian Sea (Ittekkot *et al.*, 1987), and the Black Sea (Izdar *et al.*, 1987; Honjo *et al.*, 1987). Although the seasonal variation in the flux of planktonic foraminifers has been investigated from the Panama Basin (Thunell *et al.*, 1983), the northeast Pacific (Sautter and Thunell, 1989), and the Sargasso Sea (Deuser and Ross, 1980), to date, only one such specific study has been published on coccolithophores, the work of Samtleben and Bickert (1990) in the Arctic North Atlantic.

Collection in the West Spitsbergen Current in the Arctic North Atlantic was begun in 1983 in a joint program of the Woods Hole Oceanographic Institution and the Universities of Kiel and Bremen (Honjo *et al.*, 1988). One of the purposes of the study was to characterize annual sedimentation patterns in the Norwegian Current, a relatively warm branch of the North Atlantic Drift that flows northward along the east side of the Norwegian Sea. The three stations, in the Lofoten Basin (LB-1), off Bear Island (BI-1), and in the Fram Strait (FS-1), form a transect through the Norwegian Current from about 69° to 79° N (Fig. 4 and Table 6). Samtleben and Bickert (1990) used continuous-collecting, time-series, PARFLUX Mark V type traps (see Honjo and Doherty, 1988). At each of the three stations, a trap was moored at a single depth for collection. Each was deployed for one year, beginning in the summer, and particulate material was collected at 12 or 13 time intervals, or on the order of once a month. Samtleben and Bickert (1990) examined the fine-fraction (<63 μm) from splits of the monthly samples using the SEM. Results of their enumeration determinations and flux calculations are shown in Fig. 6 and Table 7. They caution that the resulting values represent minimum fluxes because their calculations were based on the small proportion of identifiable coccoliths. Data from their study serve as a very important foundation from which future time-series sediment-trap data and results can be compared.

Among their numerous findings are the following. First, a distinct seasonality could be recognized in the sediments recovered over the course of an annual cycle. This seasonality was particularly evident in the Lofoten Basin, where it was observed in the total flux, the carbonate flux, the coccolith carbonate flux, as well as in the fluxes of the two domi-

Table 7. *Average total mass flux, average total carbonate flux, <1 mm-carbonate flux, <63 µm-carbonate flux,* Coccolithus pelagicus-*carbonate flux, and* Emiliania huxleyi-*carbonate flux at the three Arctic North Atlantic sediment-trap stations, expressed in mg m^{-2} day^{-1} (Samtleben and Bickert, 1990)*

Station	Total mass flux	Total carbonate flux	Carbonate (<1 mm) flux	Carbonate (<63µm)	C. pelagicus flux	E. huxleyi flux
BI-1	77	18.1	14.4	NA	1.92	0.11
FS-1	19*	3	1.75	NA	0.88	0.03
LB-1	64*	31.1	26.4*	18.3	1.51	0.18

* = approximated from graph in Samtleben and Bickert, 1990.
NA = not available.

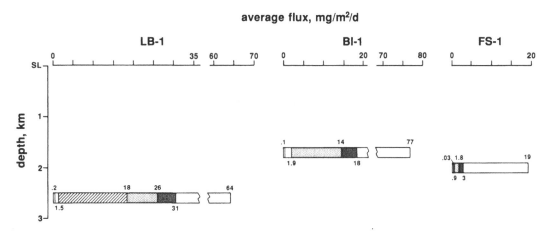

average flux, mg/m²/d

Fig. 6. Average fluxes (mg m^{-2} day^{-1}) at the three Arctic North Atlantic sediment-trap stations of Samtleben and Bickert (1990): BI-l, FS-l, and LB-l. The total length of each bar represents the total mass flux at the respective depth, the finely-stippled length represents the total carbonate flux, the coarsely-stippled length represents the <l mm-carbonate flux, the hashured length represents the <63 µm-carbonate flux, and the vertical bars represent the *Coccolithus pelagicus* and *Emiliania huxleyi* coccolith-carbonate fluxes, respectively right to left. The values are tabulated in Table 7.

nant coccolith components, *Coccolithus pelagicus* and *Emiliania huxleyi*. Second, the seasonality of the annual cycle became less distinct towards the north in sediment traps from Bear Island and the Fram Strait. Similarly, there was a general northward decrease in the coccolith flux. Third, there was a lag of one to two months between the time of phytoplankton productivity in the euphotic zone and the coccolith flux recorded in the sediment traps. This time difference was accounted for by the estimated sinking rates of fecal pellets and macroaggregates (marine snow). Fourth, the species composition of coccolithophores and their relative proportions recovered from the sediment traps differed considerably from those reported from living communities in the eastern Norwegian Sea. Samtleben and Bickert (1990) attributed these differences to the selective destruction of delicate forms, and the resulting enrichment of the more stable forms, during fecal pellet formation. Fifth, resuspension of sediment on the Barents Sea shelf during the winter months greatly obscured seasonal flux trends. And finally, the vertical flux of coccolithophores is controlled by three processes: (l) phytoplankton production, (2) zooplankton grazing and fecal pellet formation, and (3) resuspension of sedimented material. Sedimentation rate provides only an indirect indication of the most important of the processes, phytoplankton productivity.

Coccolith flux variations at depth in the water column

Variations in coccolith flux at depth in the water column result from several possible factors. First, and probably most obvious, is the effect of undersaturation with respect to $CaCO_3$ of the deeper water masses. In such waters, calcite becomes more susceptible to dissolution as a function of its exposure or residence time. Dissolution, of course, would act to decrease the flux. Additionally, Gardner *et al.* (1983) suggested that organic matter decaying in trap samples would cause the pH to be lowered and hasten carbonate dissolution. Sodium azide (NaN_3) is often added to traps to retard microbial activity. Second is the prospect of trapping the products of a surface bloom of coccolithophores as they sink. Steinmetz (1991) documented the predominance of one species, *Umbilicosphaera sibogae*, in the assemblage of coccoliths trapped at depth at Station P_1 in the central Pacific and suggested that the traps intercepted a bloom which had occurred on the surface shortly before the sediment-trap array was deployed. Since the coccoliths of *U. sibogae* had already descended below about 1000 m, the three traps set at depths greater than 2700 m collected material already in transit to the sea-floor. The overall effect on the flux profile was to produce a marked increase at depth. Evidence of blooms of *U. sibogae* have also been reported by Honjo (1982) and Honjo *et al.* (1982a) in the Panama Basin, Station PB_1.

Third, horizontal advective transport may contribute to an increased flux at depth. Bottom nepheloid layers are known to be one form of advective transport. They are believed to be formed by bottom erosion under internal tides and waves. Along continental slopes, these layers may become detached and spread along isopycnal surfaces (Dickson and McCave, 1986). Fine-grained, resuspended sediments, especially clays, silts, and silt-sized coccoliths, are easily incorporated in this form of deep horizontal transport (Honjo *et al.*, 1982b; Izdar *et al.*, 1987). Steinmetz (1991) observed the extinct calcareous nannofossil genus *Discoaster* in the lowermost trap in equatorial Atlantic Station E. Since that trap was 724 m above the sea-floor, he reasoned that horizontal advective transport, likely flowing from the Amazon Cone, was responsible for the fossil material, as well as a markedly increased flux, at depth.

Finally, the hydrodynamic characteristics of the sediment trap itself could affect the flux measured. Honjo (1978) deployed two traps 30 cm apart at 5367 m water depth in the Sargasso Sea. He observed that one trap contained over twice the mass of material trapped as the other in the <10 μm size fraction consisting mostly of clay particles and free coccoliths. He attributed this difference to the increased trapping ability of the rear, or lee trap, in the current vector.

Inasmuch as as these factors have and do affect flux measurements, it is prudent to realize that the coccolith fluxes presented in this paper are baseline approximations for each respective sampling station. They provide values for qualitative, and perhaps semi-quantitative, comparisons, but because of their inherent variability, numerous collections would be required at each station to obtain true average flux values. Time-series sediment trapping (discussed above) provides one of the means to obtain realistic quantitative fluxes at a station.

The fate of coccoliths on the sea-floor

Once a coccolith has survived its descent to the sea-floor, its travels are not likely to have ended. Resuspension of surface sediments is a common phenomenon. It produces a marked increase in particulate flux in the near-bottom zone, and it complicates particulate flux measurements close to the bottom (Dymond, 1984). The common deep-sea 'fluff layer', consisting of a very turbid but fluid layer several centimeters thick suspended over the sediment surface, is easily disturbed by purturbations caused by coring, submersible propulsion, or bottom-equipment placement. The effect on this layer of any kind of sediment collecting tray or chamber is largely unknown. Collecting flux measurements across the sediment-water interface utilizing sediment traps is unlike collecting in the water column. The mere presence of a collecting device is likely to change overlying water concentrations of particulates and dissolved gases and minerals, thereby having an unknown effect on gradients across the sediment-water interface and on flux rates. Finally, logistical and financial considerations also hamper the placement of sophisticated sampling devices that could be deployed long enough to monitor bottom conditions that may affect flux measurements, such as bottom currents, turbulence, and organism movements (Smith, 1984).

Honjo *et al.* (1982a) observed that the composition of the bottom surface sediment rarely resembled that of the particles that settled in the deepest traps. This was true not only for the carbonate in the flux, but also biogenic opal and organic carbon. Although part of this decrease was due to dilution contributed by the large amount of refractory particulate matter in the bottom sediments, solution of the sedimented materials was also responsible. Honjo *et al.* (1982a) suggested that the region approximately 200 m above the sediment surface – what they called the benthic transition layer – was characterized by significant biogenic activity that affected the biogeochemistry of the water, and as a result, it altered the composition of material sinking through it to the bottom. Pilskaln and Honjo (1987) observed that since fecal pellets are found in large numbers in sediment traps at depths greater than 3500 m, this indicates that pellets produced in the surface waters do arrive intact at the benthic boundary layer. However, fecal pellets were completely absent from all site core tops that Pilskaln and Honjo (1987) examined (E, P_1, and PB_1), and they suggested that rapid remineralization of the organic-rich pellets by microbes and detritus feeders was occurring at the sediment–water interface. Cole *et al.* (1987) were able to quantify these perceptions. They placed laboratory-grown, ^{14}C-labelled coccoliths in permeable chambers at the sediment–water interface, 10 cm above the interface, and 1 m above the interface in the Panama Basin for up to one year. By measuring the amount of ^{14}C material remaining, they were able to estimate the potential of near-benthic environments to dissolve fresh surface-derived materials. They found that the mean residence time of this material on the sediments was about 0.7 yr, and that 75–85% of the organic carbon that reaches the deep-sea sediments is dissolved within one year. In conclusion, the flux does not always represent the net accumulation rate of sediment (Honjo, 1986).

Post-depositional processes

Numerous physical, chemical, and biological processes act on particles immediately upon their arrival on the sea floor. The sum total of these processes, that bring about progressive changes in the sediment eventually leading to the formation of sedimentary rock, is called diagenesis. Berner (1980) considered each of these processes in the early stages of diagenesis. Among them are the following: depositional burial, compaction, water flow, benthic boundary diffusion, bioturbation, transfer across the sediment–water interface, equilibrium processes, microbial reactions, homogeneous (single-phase) reactions, and precipitation and dissolution.

Pelagic sediments have been cored at numerous depths worldwide by the Deep Sea Drilling Project and its successor, the Ocean Drilling Program. The results of studies of these materials allow us to follow the progressive diagenetic changes from the sediment–water interface to thousands of meters below the sea floor. Depositional burial and compaction (constituting mechanical compaction) predominate during early stages of burial. These physical processes include simple dewatering and grain reorientation or breakage. Initial values of porosity range from 70 to 95% just below the sediment–water interface (0–1 m). Dewatering starts as soon as mechanical compaction is initiated, and it continues as the constituent grains of sediment are brought closer establishing a grain-supported fabric. Additionally, the dissolution of biogenic carbonate particles is a major process just below the sediment–water interface. The rate and degree of dissolution of these particles is controlled by a number of factors including water depth (i.e., the degree of undersaturation of the water with respect to carbonate), the sedimentation rate (affecting the exposure time of particles to undersaturated waters), bottom currents, skeletal ultrastructure, size, and mineralogy, and any protective films that may be present on the skeletons (Cook and Egbert, 1983). Coccolithophores are significantly more resistant to dissolution than other biogenic carbonate particles (e.g., pteropods and planktonic foraminifers). Several possible reasons have been suggested for this: the purity of the calcium in the skeletal calcite (i.e., the low-Mg nature of the $CaCO_3$), incorporation of cellulose-like material within the skeletal calcite, organic coatings on the surface of the coccoliths, and the relative postion of the optic axis in the elements of the coccolith calcite (Bukry, 1971; Cook and Egbert, 1983).

At porosity levels of 40 to 50% (commonly at about 500–1000 m of burial for pelagic carbonate sediments), mechanical compaction slows and chemical compaction (solution transfer) becomes the major mechanism of poros-

ity reduction (Schlanger and Douglas, 1974; Cook and Egbert, 1983; Scholle et al., 1983). The rate of chemical compaction depends on the chemical composition of the constituents involved, for example, the specific nature of the carbonate sediment (aragonite vs. calcite), variations in sediment composition (clay-rich vs. clay-poor), and the composition of the pore fluids (Mg-rich vs. Mg-poor). Despite the numerous variables, the diagenetic transformation of pelagic carbonate sediment from ooze to chalk to limestone can be predicted with confidence (Scholle, 1977; Scholle et al., 1983).

Information from sediment-trap studies is being used to interpret better the geological record. Hay (1988) investigated the changes in particle flux in the western Black Sea throughout its anoxic past using sediment cores. By measuring and counting varves, and with knowledge gained about coccolithophore blooms from modern sediment-trap studies, he was able to calculate what might be called 'paleofluxes' of the dominant Emiliania huxleyi for the past 5100 years. Thomsen (1989), following similar reasoning, attempted to calculate carbonate production rates by coccolithophores in the Lower Cretaceous Munk Marl Bed in the North Sea. Estimations by both Hay (1988) and Thomsen (1989) fairly approximate modern production rates suggested by Honjo (1976) for the tropical Pacific Ocean.

Future sediment-trap experiments

Interest in sediment-trap investigations continues to be high. Information gained from such oceanographic studies is providing more insight into how oceanic biota regulate carbon fixation and how carbon is cycled through the oceans. Time-series sediment traps play an integral role in calibrating models of physical and biological processes in the ocean. In 1989, multinational, multi-year sediment-trap experiments were begun under the headings of Joint Global Ocean Flux Study (JGOFS) and Biogeochemical Ocean Flux Study (BOFS) (Pain, 1989; SCOR, 1990). Such programs should provide a wealth of information on the ecology, productivity, preservation, and taxonomy of coccolithophores, as well as numerous other microplankton (Young, 1989). The questions of global warming and the greenhouse effect may, in large part, be answered with knowledge gained from sediment-trap studies.

Summary

Our knowledge concerning coccolithophore sedimentation is shown schematically in Fig. 7 and is summarized below.

1. Fecal pellets and other oceanic marine snow (macro-aggregates) are largely responsible for the rapid vertical

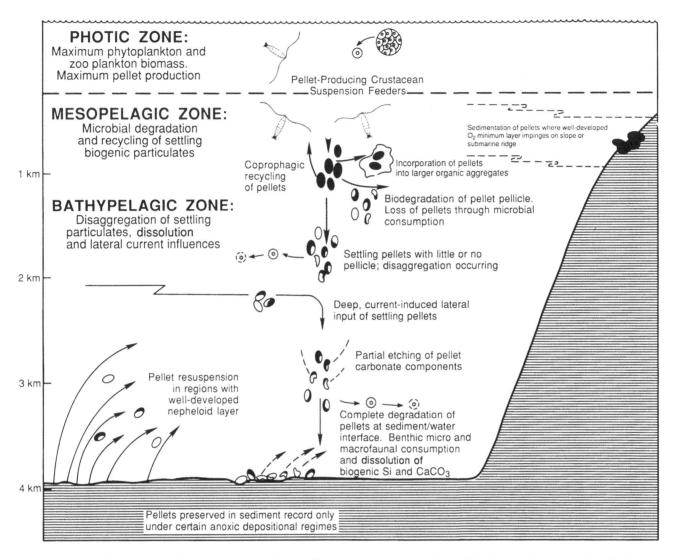

PHOTIC ZONE:
Maximum phytoplankton and zoo plankton biomass. Maximum pellet production

Pellet-Producing Crustacean Suspension Feeders

MESOPELAGIC ZONE:
Microbial degradation and recycling of settling biogenic particulates

Sedimentation of pellets where well-developed O_2 minimum layer impinges on slope or submarine ridge

Coprophagic recycling of pellets

Incorporation of pellets into larger organic aggregates

BATHYPELAGIC ZONE:
Disaggregation of settling particulates, dissolution and lateral current influences

Biodegradation of pellet pellicle. Loss of pellets through microbial consumption

Settling pellets with little or no pellicle; disaggregation occurring

Deep, current-induced lateral input of settling pellets

Partial etching of pellet carbonate components

Pellet resuspension in regions with well-developed nepheloid layer

Complete degradation of pellets at sediment/water interface. Benthic micro and macrofaunal consumption and dissolution of biogenic Si and $CaCO_3$

1 km, 2 km, 3 km, 4 km

Pellets preserved in sediment record only under certain anoxic depositional regimes

transport of the majority of coccospheres and coccoliths through undersaturated waters to the sea-floor.

2. The descent rate of fecal pellets averages 200 m/day, twice that of marine snow aggregates.

3. Coccospheres represent a minor fraction of the coccosphere and coccolith flux (<0.001% of all sinking coccoliths). In equatorial waters fluxes range in the order of tens to thousands of coccospheres $m^{-2} day^{-1}$.

4. Coccolith fluxes in equatorial waters generally range from the tens to the thousands of millions of coccoliths $m^{-2} day^{-1}$. In Arctic Atlantic waters fluxes range in the ones to the tens of millions $m^{-2} day^{-1}$. The differences in total-mass flux, biogenic-carbonate flux, and total-coccolith flux at different locations are an indication of differences in surface-water productivity.

5. The proportion of coccoliths that survive the descent to the sea-floor in the Panama Basin, an area of intense upwelling and high primary productivity, is about 51%. Elsewhere, estimates of survival range as high as 92%.

Fig. 7. A model of coccolithophore sedimentation (modified slightly from Pilskaln and Honjo, 1987).

6. Within the water column, the diversity of coccolith assemblages decreases with depth. Selective dissolution with depth is most apparent among delicate forms such as the holococcoliths.

7. Utilizing time-series sediment traps, seasonality can be detected and measured in total flux, carbonate flux, coccolith flux, and in individual coccolith species.

8. Advection and resuspension can grossly inflate flux values in the water column and near the sea-floor, respectively.

9. Coccolith fluxes represent minimum values because of the inexact nature of trapping and enumerating particles, the unknown loss to dissolution in the trap, and other poorly understood factors. The fluxes do provide baseline approximations for qualitative and perhaps semi-quantitative comparisons between sampling stations.

Reference

Alldredge, A., 1984. Macroscopic organic aggregates (marine snow). Workshop on the Global Ocean Flux Study, Woods Hole, MA, Nat. Acad. Sci., pp. 166–79.

Alldredge, A. L. and Gotschalk, C. C., 1989. Direct observations of the mass flocculation of diatom blooms: characteristics, settling velocities and formation of diatom aggregates. *Deep-Sea Res.*, **36** (2A): 159–71.

Asper, V. L., 1987. Measuring the flux and sinking speed of marine snow aggregates. *Deep-Sea Res.*, **34** (1A): 1–17.

Bernard, F., 1963. Vifesse de chute en mer des amas palmeloides de *Cyclococcolithus*. Ses Consequences pour le cycle vital des mers chaudes. *Pelagos. Bull. Inst. Oceanoqr., Alger.*, **1**: 1–34.

Berner, R. A., 1980. *Early Diagenesis: A Theoretical Approach.* Princeton Univ. Press, Princeton.

Betzer, P. R., Showers, W. J., Laws, E. A., Winn, C. D., DiTullio, G. R., and Kroopnick, P. M., 1984. Primary productivity and particle fluxes on a transect of the equator at 153° W in the Pacific Ocean. *Deep-Sea Res.*, **31** (1A): 1–11.

Bishop, J. K. B., Edmond, J. M., Ketten, D. R., Bacon, M. P., and Silver, W. B., 1977. The chemistry, biology, and vertical flux of particulate matter from the upper 400 m of the equatorial Atlantic Ocean. *Deep-Sea Res.*, **24** (6A): 511–48.

Bukry, D., 1971. Cenozoic calcareous nannofossils from the Pacific Ocean. *San Diego Soc. Nat. Hist., Trans.* **16**(14): 303–28.

Cole, J. J., Honjo, S., and Erez, J., 1987. Benthic decomposition of organic matter at a deep-water site in the Panama Basin. *Nature*, **327** (6124): 703–4.

Cook, H. E. and Egbert, R. M., 1983. Diagenesis of deep-sea carbonates. In *Diagenesis in Sediments and Sedimentary Rocks*, 2, ed. G. Larsen and G.V. Chilingar. pp. 213–288. Elsevier, Amsterdam.

Deuser, W. G. and Ross, E. H., 1980. Seasonal change in the flux of organic carbon to the deep Sargasso Sea. *Nature*, **283** (5745): 364–5.

Deuser, W. G., Ross, E. H., and Anderson, F. R., 1981. Seasonality in supply of sediment in the deep Sargasso Sea and implications for the rapid transfer of matter to the deep ocean. *Deep-Sea Res.*, **28** (5A): 495–505.

Dickson, R. R. and McCave, I. N., 1986. Nepheloid layers on the continental slope west of Porcupine Bank. *Deep-Sea Res.*, **33** (6A): 791–818.

Dunbar, R. B. and Berger, W. H., 1981. Fecal pellet flux to modern bottom sediment of Santa Barbara Basin (California) based on sediment trapping. *Geol. Soc. Amer. Bull.*, Pt. 1, **92** (4): 212–18.

Dymond, J., 1984. Sediment traps, particle fluxes, and benthic boundary layer processes. Workshop on the Global Ocean Flux Study, Woods Hole, MA, Nat. Acad. Sci., pp. 260–84.

Frankenberg, D. and Smith, K. L., 1967. Coprophagy in marine animals. *Limnol. Oceanogr.*, **12**: 443–50.

Gaarder, K. R. and Hasle, G. R., 1971. Coccolithophorids of the Gulf of Mexico. *Bull. Mar. Sci. Gulf Carib.*, **21** (2): 519–44.

Gardner, W. D., 1980a. Field assessment of sediment traps. *J. Mar. Res.*, **38** (1): 41–52.

Gardner, W. D., 1980b. Sediment trap dynamics and calibration: a laboratory evaluation. *J. Mar. Res.*, **38** (1): 17–39.

Gardner, W. D., Hinga, K. R., and Marra, J., 1983. Observations on the degradation of biogenic material in the deep ocean with implications on accuracy of sediment trap fluxes. *J. Mar. Res.*, **41** (2): 195–214.

Hara, S. and Tanoue, E., 1985. Protist [sic] along 150°E in the Southern Ocean: its composition, stock and distribution. *Trans. Tokyo Univ. Fisheries*, no. 6, pp. 99–115.

Hay, B. J., 1988. Sediment accumulation in the central western Black Sea over the past 5100 years. *Paleoceanogr.*, **3** (4): 491–508.

Honjo, S., 1975. Dissolution of suspended coccoliths in the deep-sea water column and sedimentation of coccolith ooze. In *Dissolution of Deep-Sea Carbonates.* ed. W.B. Sliter, A.W.H. Be, and W.H. Berger, Cushman Found. Foraminiferal Res., Spec. Publ., no. 13 : 115–28.

Honjo, S., 1976. Coccoliths: production, transportation and sedimentation. *Mar. Micropaleontol.*, **1** (1): 65–79.

Honjo, S., 1978. Sedimentation of materials in the Sargasso Sea at a 5367 m deep station. *J. Mar. Res.*, **36** (3): 469–92.

Honjo, S., 1980. Material fluxes and modes of sedimentation in the mesopelagic and bathypelagic zones. *J. Mar. Res.*, **38** (1): 53–97.

Honjo, S., 1982. Seasonality and interaction of biogenic and lithogenic particulate flux at the Panama Basin. *Science*, **218** (4575): 883–4.

Honjo, S., 1986. Oceanic particles and pelagic sedimentation in the western North Atlantic Ocean. In *The Geology of North America, vol. M, The Western North Atlantic Region.* ed. P.R. Vogt, and B.E Tucholke, 469–78. Geol. Soc. Amer.,

Honjo, S. and Doherty, K. W., 1988. Large aperture time-series sediment traps; design objectives, construction and application. *Deep-Sea Res.*, **35** (1A): 133–49.

Honjo, S. and Okada, H., 1974. Community structure of coccolithophores in the photic layer of the mid-Pacific. *Micropaleontol.*, **20** (2): 209–30.

Honjo, S. and Roman, M. R., 1978. Marine copepod fecal pellets: production, preservation, and sedimentation. *J. Mar. Res.*, **36**: 45–57.

Honjo, S. and Wefer, G., 1987. Oceanic particle fluxes in the Nordic seas. *Trans. Amer. Geophys. Union*, **68** (50): 1771.

Honjo, S., Connell, J. F., and Sachs, P. L., 1980. Deep-ocean sediment trap: design and function of PARFLUX Mark II. *Deep-Sea Res.*, **27** : 745–53.

Honjo, S., Doherty, K. W., Agrawal, Y. C. and Asper, V. L., 1984. Direct optical assessment of large amorphous aggregates (marine snow) in the deep ocean. *Deep-Sea Res.*, **31** (1A): 67–76.

Honjo, S., Hay, B. J., Manganini, S. J., Asper, V. L., Degens, E. T., Kempe, S., Ittekkot, V., Izdar, E., Konuk, Y. T., and Benli, H. A., 1987. Seasonal cyclicity of lithogenic particle fluxes at a southern Black Sea sediment trap station. In *Particle Flux in the Ocean*, ed. E.T. Degens, E. Izdar, and S. Honjo. Mitt. Geol.-Paläontol. Inst. Univ. Hamburg. Vol. 62, pp. 19–39.

Honjo, S., Manganini, S. J., and Cole, J. J., 1982a. Sedimentation of biogenic matter in the deep ocean. *Deep-Sea Res.*, **29** (5A): 609–25.

Honjo, S., Manganini, S. J., and Wefer, G., 1988. Annual particle flux and a winter outburst of sedimentation in the northern Norwegian Sea. *Deep-Sea Res.*, **35** (8A): 1223–34.

Honjo, S., Spencer, D. W., and Farrington, J. W., 1982b. Deep advective transport of lithogenic particles in Panama Basin. *Science*, **216** (4545): 516–18.

Houghton, J. T., Jenkins, G. J., and Ephraums, J. J. (Eds.), 1990. *Climate Change.* The IPCC Scientific Assessment. Cambridge Univ. Press, Cambridge.

Ittekkot, V., Manganini, S. J., Guptha, M. V. S., Degens, E. T., and Honjo, S., 1987. Particle fluxes in the Arabian Sea. *Trans. Amer. Geophys. Union*, **68** (50): 1772.

Izdar, E., Konuk, T., Ittekkot, V., Kempe, S., and Degens, E. T.,

1987. Particle flux in the Black Sea: nature of the organic matter. In *Particle Flux in the Ocean,* ed. E.T. Degens, E. Izdar, and S. Honjo. Vol. 62 pp. 1–18. Mitt. Geol.-Paläeontol. Inst., Univ. Hamburg.

Johannes, R. E., and Satomi, M., 1966. Composition and nutritive value of fecal pellets of a marine crustacean. *Limnol. Oceanogr.,* 11: 191–197.

Kleijne, A., 1990. Distribution and malformation of extant calcareous nannoplankton in the Indonesian Seas. *Mar. Micropaleontol.,* 16 (3/4): 293–316.

Knauer, G. A., Martin, J. H., and Bruland, K. W., 1979. Fluxes of particulate carbon, nitrogen, and phosphorus in the upper water column of the Northeast Pacific. *Deep-Sea Res.,* 26 (lA): 97–108.

Knauer, G. A., Martin, J. H., and Karl, D. M., 1984. The flux of particulate organic matter out of the euphotic zone. Workshop on the Global Ocean Flux Study, Woods Hole, MA, Nat. Acad. Sci., pp. 136–50.

Lal, D., 1977. The oceanic realm of particles. *Science,* 198 (4321): 997–1099.

Lohmann, H., 1902. Die Coccolithophoridae, eine Monographie der Coccolithen bildenden Flagellaten, zugleich ein Beitrag zur Kenntnis des Mittelmeerauftriebs. *Arch. Protist.,* 1: 89–165.

McCave, I. N., 1975. Vertical flux of particles in the ocean. *Deep-Sea Res.,* 22 (7A): 491–502.

McIntyre, A. and Bé, A. W. H., 1967. Modern Coccolithophoridae of the Atlantic Ocean – 1. Placoliths and cyrtoliths. *Deep-Sea Res.,* 14 (5A): 561–97.

Marshall, S. M. and Orr, A. P., 1956. On the biology of *Calanus finmarchicus* IX. Feeding and digestion in the younger stages. *J. Mar. Biol. Assn. U.K,* 35: 587–603.

Marshall, S. M. and Orr, A. P., 1962. Food and feeding in copepods. *Rapp. Cons. Explor. Mer,* 152: 92–8.

Nishida, S., 1979. *Atlas of Pacific Nannoplanktons.* Micropaleontol. Soc. Osaka, Spec. Pap. no. 3.

Okada, H., 1989. Morphometric and floral variations of calcareous nannoplanktons in relation to their living environment. *INA Newsl.,* 11(2): 87–8.

Okada, H. and Honjo, S., 1973. The distribution of oceanic coccolithophorids in the Pacific. *Deep-Sea Res.,* 20: 355–74.

Paffenhöfer, G.-A. and Knowles, S. C., 1979. Ecological implications of fecal pellet size, production and consumption by copepods. *J. Mar. Res.,* 37 (1): 35–49.

Paffenhöfer, G.-A. and Strickland, J. D. H., 1970. A note on the feeding of *Calanus helgolandicus* on detritus. *Mar. Biol.,* 5: 97–9.

Pain, S., 1989. Ships search for the secret of the carbon cycle. *New Scientist,* 121 (1656): 32.

Pilskaln, C.H., 1982. Fecal pellets in PARFLUX sediment traps. *Trans. Amer. Geophys. Union,* 63 (3): 54

Pilskaln, C. H., 1985. The fecal pellet fraction of oceanic particle flux. unpub. PhD. dissertation, Harvard Univ., Cambridge, Mass.

Pilskaln, C. H., 1988. Zooplankton fecal pellets: a major component of Black Sea surface sediments. *Trans. Amer. Geophys. Union,* 69 (44): 1243.

Pilskaln, C. H. and Honjo, S., 1987. The fecal pellet fraction of biogeochemical particle fluxes to the deep sea. *Global Biogeochem. Cycles,* 1 (1): 31–48.

Pilskaln, C.H., Neumann, A.C., and Bane, J.M., 1989. Periplatform carbonate flux in the northern Bahamas. *Deep-Sea Res.,* 36 (9): 1391–406 .

Roth, P. H., Mullin, M. M., and Berger, W. H., 1975. Coccolith sedimentation by fecal pellets: laboratory experiment and field observations. *Geol. Soc. Amer. Bull.,* 86 (8) : 1079–84 .

Rowe, G. T. and Gardner, W. D., 1979. Sedimentation rates in the slope water of the northeast Atlantic Ocean measured directly with sediment traps. *J. Mar. Res.,* 37 (3) : 581–600.

Samtleben, C. and Bickert, T., 1990. Coccoliths in sediment traps from the Norwegian Sea. *Mar. Micropaleont.,* 16 (1): 3 9 –64 .

Samtleben, C. and Schröder, A., 1990. Coccolithophoriden Gemeinschaften und Coccolithen-Sedimentation im Europäischen Nordmeer. *Zur Abbildung von Planktonzönosen im Sediment. Ber. Sonderforschungsbereich* 313, Univ. Kiel, No. 25.

Sautter, L. R. and Thunell , R. C., 1989 . Seasonal succession of planktonic foraminifera: Results from a four-year time-series sediment trap experiment in the northeast Pacific. *J. Foraminiferal Res.,* 19 (4) : 253–67.

Schlanger, S. O. and Douglas, R. G., 1974. Pelagic ooze-chalk-limestone transition and its implications for marine stratigraphy. In *Pelagic Sediments on Land and under the Sea,* ed. K.J. Hsü, and H.C. Jenkyns, pp . 117–148. Internat . Assn . Sedimentol., Spec . Publ . No . 1.

Scholle, P. A., 1977. Chalk diagenesis and its relation to petroleum exploration: oil from chalks, a modern miracle? *Amer. Assoc. Petrol. Geol. Bull.,* 61 (7) : 982–1009.

Scholle, P. A., Arthur, M. A., and Ekdale, A. A., 1983. Pelagic environment. In *Carbonate Depositional Environments.* Amer. Assoc. Petrol. Geol. Mem. 33. ed. P.A. Scholle, D.G. Bebout, and C.H. Moore, pp. 619–691.

Schrader, H. J., 1971. Fecal pellets: role in sedimentation of pelagic diatoms. *Science,* 174 (4004): 55–57.

SCOR, 1987. *The Joint Global Ocean Flux Study–Background, Goals, Organization, and Next Steps.* Report of the Internat. Sci. Planning and Coordination Meeting of Global Ocean Flux Studies sponsored by the Sci. Comm. on Oceanic Res., Paris.

SCOR, 1990. *The Joint Global Ocean Flux Study–Scientific Plan.* Sci. Comm. on Oceanic Res., Dalhousie Univ., Halifax, 61 pp.

Shanks, A. L. and Trent, J. D., 1980. Marine snow: sinking rates and potential role in vertical flux. *Deep-Sea Res.,* 27 (2A): 137–43.

Silver, M. W., Shanks, A. L., and Trent, J. D., 1978. Marine snow: microplankton habitat and source of small-scale patchiness in pelagic populations. *Science,* 201 (4353): 371–3.

Smith, K. L., Jr., 1984. Measured fluxes across the sediment-water interface. Workshop on the Global Ocean Flux Study, Woods Hole, MA, Nat. Acad. Sci., pp. 325–40.

Soutar, A., Kling, S. A., Crill, A., Duffrin, E., and Bruland, K. W., 1977. Monitoring the marine environment through sedimentation. *Nature,* 266 (5598): 136–9.

Spencer, D. W., Brewer, P. G., Fleer, A. P., Honjo, S., Krishnaswami, S., and Nozaki, Y., 1978. Chemical fluxes from a sediment trap experiment in the deep Sargasso Sea. *J. Mar. Res.,* 36 (3): 493–523.

Spencer, D. W. (Ed.), 1981. *STIE: Sediment Trap Intercomparison Experiment Panama Basin.* WHOI Technical Memorandum No. 1–81.

Steinmetz, J. C., 1991. Calcareous nannoplankton biocoenosis: Sediment trap studies in the equatorial Atlantic, central Pacific, and Panama Basin. In ed. S. Honjo, *Ocean Biocoenosis Series No. 1.* Woods Hole Oceanogr. Inst. Press, 85 pp.

Thomsen, E., 1989. Seasonal variability in the production of Lower Cretaceous calcareous nannoplankton. *Geology,* 17 (8): 715–17.

Thunell, R. C. and Honjo, S., 1981. Planktonic foraminiferal flux to the deep ocean: sediment trap results from the tropical Atlantic and the central Pacific. *Mar. Geol.,* 40 (3/4) 237–53.

Thunell, R. C., Curry, W. B., and Honjo, S., 1983. Seasonal

variability in the flux of planktonic foraminifera: time series
sediment trap results from the Panama Basin. *Earth Planet. Sci.
Lett.*, **64** (1): 44–55.

Thunell, R.C., Keir, R.S., and Honjo S. 1981. Calcite dissolution:
an *in situ* study in the Panama Basin. *Science*, **212** (4495):
659–61.

Wefer, G. and Honjo, S., 1985. Material fluxes under MIZ in the
Fram Strait. *Trans. Amer. Geophys. Union*, **66** (51): 1271.

Wefer, G., Fischer, G., Fütterer, D., and Gersonde, R., 1988.

Seasonal particle flux in the Bransfield Strait, Antarctica. *Deep-
Sea Res.*, **35** (6A): 891–8.

Wiebe, P., Boyd, S. H., and Winget, C., 1976. Particulate matter
sinking to the deep sea floor at 2000 m in the Tongue of the
Ocean, Bahamas, with a description of a new sedimentation
trap. *J. Mar. Res.*, **34** (3): 341–54.

Young, J., 1989. Joint Global Ocean Flux Study – participation of
nannoplankton workers. *INA Newsl.*, **11** (1): 30.

10 Distribution of coccoliths in oceanic sediments

PETER H. ROTH

Introduction

Coccoliths, the minute calcified plates produced by planktonic haptophyte algae, have been a major component of pelagic carbonates for the last 150 million years (Bramlette, 1958; Roth, 1986). Deep-sea carbonate oozes are one of the most widespread sediment types on Earth, covering roughly half of the sea-floor of the world's oceans or one-third of the surface of the Earth (Berger, 1976). Because of widespread occurrence and rapid reproduction and their transfer to the sea floor, coccolithophores act as biological pumps in transferring CO_2 to the deep and bottom waters and thus appear to be significantly affecting the CO_2 concentration of the ocean-atmosphere system (Winter and Briano, 1989).

Coccolithophores are tied to particular water masses (see chapter 8). Coccolith assemblages preserved in sediments serve as excellent proxy indicators of temperature, salinity, and limiting nutrients of surface-waters in which they once lived; they also monitor carbonate dissolution on the sea-floor and in the sediment (McIntyre and McIntyre, 1971; Roth and Berger, 1975). Finally, coccoliths are sensitive indicators of diagenesis in ancient rocks (Roth et al., 1975). The onset of diagenesis may be very early and its effect may be seen in Recent sediments (Roth and Berger, 1975). An understanding of the processes of sedimentation and preservation of coccoliths in Recent sediments is crucial for the understanding of the ancient rock record where coccoliths serve as one of the major environmental guides for the last 200 million years of ocean history.

This chapter focuses on the use of coccolith assemblages as oceanographic indicators, rather than on their properties as components of deep-sea sediments, and on their role in the global carbon cycle.

Previous work

The importance of calcareous nannoplankton as contributors to Recent deep-sea sediments was recognized in the second half of the last century (Huxley, in Dayman, 1858; Wallich, 1861). Subsequent studies showed that the clay-size fraction of deep-sea carbonates consists largely of whole coccoliths and their fragments (Black, 1956; Bramlette, 1958, 1961).

The utility of calcareous nannoplankton as indicators of surface-water conditions (temperature, salinity, concentration of limiting nutrients) and of carbonate dissolution was not fully recognized until the second half of this century (Ushakova 1968, 1970, 1974; Shumenko and Ushakova, 1967; McIntyre and Bé, 1967; McIntyre et al., 1970; Roth and Berger, 1975; Schneidermann, 1977; Roth and Coulbourn, 1982; Dmitrenko, 1985). These studies showed that remains of calcareous nannoplankton preserved in bottom sediments reflect the physical and chemical characteristics of overlying water masses. From the distribution and preservation of coccoliths in bottom sediments it is possible to estimate important chemical and physical parameters of water masses and thus reconstruct circulation patterns of ancient oceans and climates. The first such effort concerned the Pleistocene when the extant species of coccolithophores dominated and only a few exclusively fossil species existed. Coccolith distribution in sediments was used to document changes in surface-water temperature in the North Atlantic since the last glacial (McIntyre, 1967) and helped to estimate surface-water temperature, salinity, and the distribution of water masses during the last glacial maximum 18 thousand years ago (CLIMAP members, 1976, 1981). Much progress was made by applying multivariate statistics to paleoceanographic problems in these studies under the guidance of John Imbrie. The successful application of coccoliths to paleoceanographic reconstructions in the Pleistocene encouraged an extension of such studies to the more distant past (Haq and Lohmann, 1976; Haq, 1980; Roth and Bowdler, 1981; Backman et al., 1986; Roth and Krumbach, 1986).

Biogeographical distribution of coccoliths in surface sediments, and oceanographic interpretations

Only a small percentage of coccoliths produced by the living nannoplankton assemblages are preserved in sediments. The majority of the taxa do not produce skeletal elements that are sufficiently resistant to survive passage through the digestive system of herbivores, descent through the water column, and transition from the bottom waters to the sediment in the benthic boundary layer. Calcareous nannoplankton in sediments occur mostly as individual coccoliths rather than as coccospheres, because the majority of the coccospheres are fragile and disintegrate after organic membranes binding the coccoliths together have been digested by herbivores or metabolized by bacteria during transport to the sea floor, while lying on the sea floor, or after burial in the sediment. Only coccospheres of taxa with interlocking coccoliths, such as *Coccolithus pelagicus* and *Calcidiscus leptoporus*, have a good chance of being preserved in the sediments. Coccospheres of all other Recent coccolith species are exceedingly rare in sediments. The composition of calcareous nannoplankton assemblages preserved in sediments is determined mainly by the flux rate of coccoliths to the sea floor and by the rate of dissolution on the sea floor and within the sediment. Flux rate of coccoliths is a function of the relative abundance of a particular species in the living assemblage, the number of coccoliths per coccosphere, the rate of reproduction of the taxon under consideration, and other factors (see chapter 9).

Numerous secondary processes, none of them well understood, affect the composition and preservation of nannoplankton assemblages that are ultimately buried as a major component of pelagic sediments. These processes include differential predation, transfer to the sea floor as fecal pellets or as large aggregates (often referred to as "marine snow"), dissolution, resuspension, winnowing, and redeposition in the benthic boundary layer before final burial. Secondary alteration of coccolith assemblages continues even after burial. The composition and preservation of coccoliths may be affected by passage through the digestive system of deposit-feeding organisms, by chemical fronts in sediments, and finally by heat and pressure due to an ever increasing overburden. The final burial assemblage incorporated into the permanent rock record is often a mere shadow of the living assemblage, and it mirrors only the major patterns of surface-water conditions and carbonate dissolution on the sea floor. Sediment assemblages are spatially and temporally averaged over thousands of years except in varved sediments that accumulated under anoxic conditions in an environment that is generally not conducive to the preservation of coccoliths. Patchiness and small temporal fluctuations are obscured by processes of sediment mixing and redeposition.

It is therefore a great challenge to reconstruct the initial composition of the preservable component of living coccolith assemblages. First, it is important that assemblages that have been greatly affected by preservational changes are recognized and excluded from reconstructions of surface-water conditions. Second, an attempt needs to be made to estimate the amount of distortion of the environmental signal by secondary processes of preferential predation, transport, dissolution, and diagenesis and to make an attempt to remove it. This is very difficult and has not been seriously attempted. Preservational studies of coccolith assemblages indicate a critical water depth below which coccolith assemblages are so severely altered by dissolution that preservational changes overshadow paleoenvironmental differences. Only samples from locations shallower than this critical depth are used to determine distribution patterns and relate them to physical and chemical parameters in the surface waters. A level of 3000 m was chosen for the Pacific by Roth and Coulbourn (1982). Paleoenvironmental reconstructions based on calcareous nannoplankton remains may be meaningless if preservational changes are not considered.

Differential dissolution is more important than any other process of secondary alteration of coccolith assemblages in Recent sediments (McIntyre and McIntyre, 1971; Schneidermann, 1977; Roth and Berger, 1975; Roth and Coulbourn, 1982). Therefore, dissolution is discussed in greater detail later in this paper. Examples of the state of preservation of coccolith sediment assemblages and important indicator species are shown in Figs 1 and 2.

Biogeographic distribution patterns of coccoliths in sediments reflect the distribution of the living coccolithophore species in the water masses (McIntyre and Bé, 1967), proving rapid transport of coccoliths from the surface waters to the sea floor (Roth et al., 1975; Honjo, 1976). Thus, sediment assemblages provide good proxy records of the environmental conditions that control the distribution and production of calcareous nannoplankton in surface-waters and dissolution of their calcitic remains on the sea floor. The first step in a study of sediment assemblages of calcareous nannoplankton generally involves simple mapping of relative abundances of the various calcareous nannoplankton taxa. In order to minimize the effects of species-differentiated preservation, only samples from cores recovered at relatively shallow water depth, this is, well above the calcite compensation depth (CCD), and if possible above the lysocline, should be used (Roth and Coulbourn, 1982). The observed biogeographic distribution patterns reflect major environmental parameters, especially temperature, concentration of limiting nutrients, and to a lesser extent salinity of surface waters. Reduction of census data using various sophisticated mathematical techniques is discussed below.

Open-ocean sediments

Much of the data on the biogeography of Recent sediment assemblages was published in the late 1960s to early 1980s. Relatively little new information has appeared in the last few years, except for some small-scale, regional studies.

The density and nature of distribution data on calcareous nannoplankton in sediments are variable. Presence–absence data of the most important species have been determined for all the major ocean basins but overall coverage is not very dense (e.g., Ushakova, 1974). Quantitative data have been published for some parts of the oceans, and for some of the

A

B

Fig. 1.A. Well preserved tropical sediment assemblage of coccoliths with a proximal view of *Umbellosphaera tenuis* (lower left), *Gephyrocapsa oceanica* (center), with an *Umbellospheara* sp. in distal view, a small specimen of *Umbilicosphaera sibogae* just above it, a stem of *Rhabdosphaera clavigera* just next to it, and three specimens of *Emiliania huxleyi* along the right edge of the picture. Isolated plates of the deep-water species *Flabellosphaera profunda* occur to the left of the specimen of *G. oceanica* and above the specimen of U. *tenuis*. Several fragments of *Umbellosphaera, Emiliania huxleyi,* and unidentifiable coccoliths occur mainly in the upper left-hand and lower right-hand quadrants. Minor amounts of clay are present as small platy crystals. Sample from the top of core NEL-H-S, in the North Pacific near Hawaii, at 21° 33' N, 157° 23' W 2873 m depth.

B. Moderately preserved high-latitude assemblage, dominated by *Gephyrocapsa caribbeanica* (three specimen in the center), with *Emiliania huxleyi* and a stem of *Rhabdosphaera clavigera* (to the right of the center of the picture). A fragment of *Umbellosphaera tenuis* (below center) and numerous elements of *Florisphaera profunda* (especially to the left of the three *Gephyrocapsa* specimens. This sample from the top of core DWD 70, taken by Scripps Institution of Oceanography on the East Pacific Rise in the South Pacific at 48° 29' S, 113° 17' W, is from a shallower water depth than the better preserved tropical assemblages.

Scale bar on both micrographs represents 1 μm

marginal seas. The relative abundances of coccoliths in sediment assemblages from some three dozen trigger-weight cores from the Atlantic were compared to the distribution patterns of living coccolithophores in the water column (McIntyre and Bé, 1967). Major patterns in ocean water were displayed in the sediments and the potential of coccoliths for paleoclimatologic research was clearly indicated. A more detailed study of the biogeography of coccoliths in sediments of the North Atlantic was performed by Geitzenauer *et al.* (1977). An extensive data set on calcare-

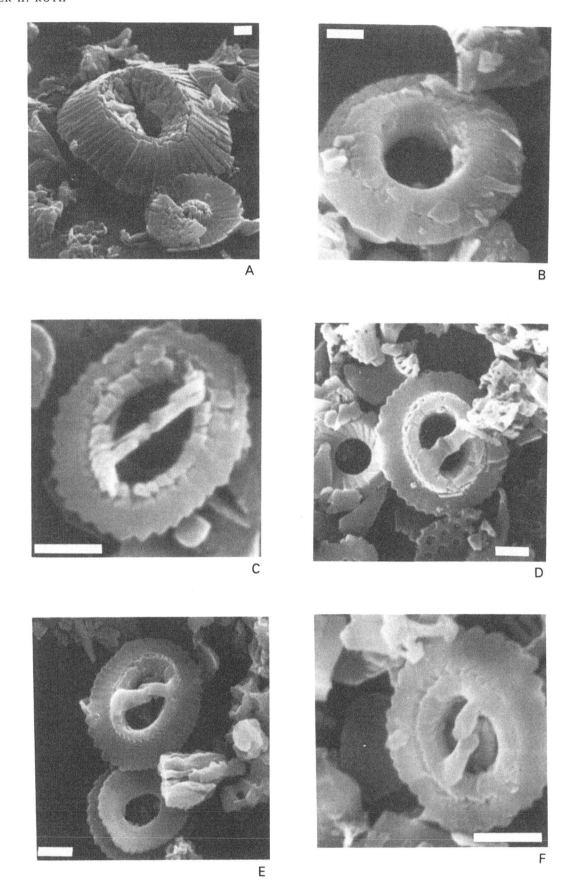

ous nannoplankton in the South Atlantic and other parts of the world's oceans gathered for the CLIMAP project by numerous scientists under the leadership of Andrew McIntyre at Lamont-Doherty still awaits publication.

Indian Ocean sediments have received less study than the other two major oceans. The relative abundances of the most abundant species of coccolithophores, *Gephyrocapsa oceanica* and *Calcidiscus leptoporus*, and of the less common species *Helicosphaera carteri*, and presence–absence data for the less abundant species *Ceratolithus cristatus, Rhabdosphaera clavigera, Syracosphaera pulchra, Oolithotus antillarum , Neosphaera coccolithomorpha*, and *Umbellosphaera tenuis* in sediments from the western Indian Ocean are given by Dmitrenko (1985). The preservation and distribution of the most common coccolith taxa (*Emiliania huxleyi, G. oceanica, Calcidiscus leptoporus*, and *Umbilicosphaera sibogae*) and the relationship of coccolith assemblages and oceanographic parameters in a more restricted area in the southwestern Indian Ocean off southern Africa are discussed in detail by Fincham and Winter (1989).

The composition and preservation of coccoliths in surface sediments of the Pacific has received the most study of all oceans and seas. Presence–absence data by McIntyre *et al.*

Fig. 2. A. *Coccolithus pelagicus* (center and small specimens of *Calcidiscus leptoporus* below (proximal view) and on the upper right (distal view). Notice that the proximal shield with its c-axis in the plane of the shield has been more damaged by dissolution than the distal shield with the c-axes of the elements perpendicular to the shield because, as Baine (1940) has shown, dissolution is more rapid in the direction of the c-axis than perpendicular to it. From the top of core DWD35-HG Downwind Expedition of Scripps Institution of Oceanography, 44° 21' S, 127° 14' W, depth of 4675 m.
B. *Umbilicosphaera sibogae*, a taxon that is most common in tropical waters of high salinity and moderately high nutrient concentration. From the top of core TP 97 (Trans Pac Expedition of Scripps Institution of Oceanography, 1953) , 32° 02' N, 139°25' E, depth of 1340 m.
C. *Gephyrocapsa ericsonii* a species of *Gephyrocapsa* with a highly arched bridge which occurs along the periphery of the Pacific and in the equatorial Pacific but is too rare to be reliably related to water-mass characteristics, according to Roth and Coulbourn, (1982). From the top of core SCAN 83 PG of Scripps Institution of Oceanography, taken at 2° 41' N, 130° 29' W, depth of 4400 m.
D. *Gephyrocapsa oceanica* in the center, with a specimen of *Umbellosphaera sibogae* to the left. From the top of core SCAN 96PG of Scripps Institution of Oceanography, 1° 29' N, 113° 52' W, depth 3856 m.
E. *Gephyrocapsa caribbeanica*, upper specimen in distal view, lower one in proximal view, from the same sample as Fig. 4.
F. *Gephyrocapsa caribbeanica* with an unusual strongly offset central bridge. From the top of core SCAN 91G, 0° 6' N, 86° 52' W, depth of 2707 m.

Scale bar on all micrographs represents 1 μm.

(1970, Figs. 1–4; 6–8) delineate the major biogeographic patterns of important species. Comprehensive studies of the distribution and dissolution of coccoliths in Recent sediments of the Pacific by Roth and Berger (1975, Figs. 4 and 5) and Roth and Coulbourn (1982, Figs. 3 through 18) contain biogeographic maps of most important species. Relationships are shown between relative abundance of all major taxa and the two most important environmental parameters: temperature and dissolved phosphate of surface-waters. Multivariate statistical techniques were also used to group coccolith taxa and relate their distribution patterns to oceanographic parameters. Relative abundances of major species and the distribution of coccolith assemblages in a different set of samples from the Pacific Ocean were a part of a study by Geitzenauer *et al.* (1976; 1977).

Although the geographic coverage of published data on the distribution of coccoliths in bottom sediments is uneven, it is possible to discuss the major biogeographic patterns for the most important taxa and to relate the distribution patterns to paleoenvironmental parameters. The most abundant species in Recent sediments include *E. huxleyi,* a cosmopolitan, eurytopic taxon with a wide distribution, *G. oceanica,* a warm-water species, *G. caribbeanica,* a cosmopolitan form with a preference for colder waters, and *Coccolithus pelagicus,* definitely a cold-water species. All three taxa are relatively resistant to dissolution, especially *Gephyrocapsa* and *C. pelagicus*. Published relative abundance patterns appear reliable even if information on the depth of particular samples is not taken into consideration. Differences in relative abundances between the Atlantic and the Pacific oceans are definitely due to differences in acidity of their bottom waters.

Uniform species concepts are crucial if one wants to compare data from different sources as is done on Figs 3 to 5. There is excellent agreement among most authors on the taxonomy of *E. huxleyi* and *C. pelagicus*. This is not the case for the genus *Gephyrocapsa* whose taxonomy is in a constant stage of flux (see chapter 6). Therefore, it is virtually impossible to compare data from different investigators, which obscures the paleoenvironmental information that can be extracted from distribution patterns of various taxa. Numerous workers (McIntyre *et al.*, 1970; Roth and Berger, 1975; Roth and Coulbourn,1982) have demonstrated that the bridge-angle is the best criterion to distinguish two large and abundant morphotypes, the cosmopolitan species *G. caribbeanica* and the warm-water species *G. oceanica*. Using this species concept, it is possible to extract paleoenvironmental information from spatial and temporal distribution patterns of these two taxa (see Rahman and Roth, 1990, for further discussion and references). Other taxonomic schemes used to subdivide the genus *Gephyrocapsa* into species and subspecies (e.g., Pujos-Lamy, 1977; Bréhéret, 1978; Samtleben, 1980; Ghidalia, 1988; Biekart, 1989) have neither resulted in a more stable taxonomy, nor clarified phylogenetic relationships of this complex and highly

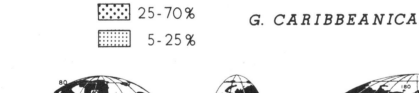

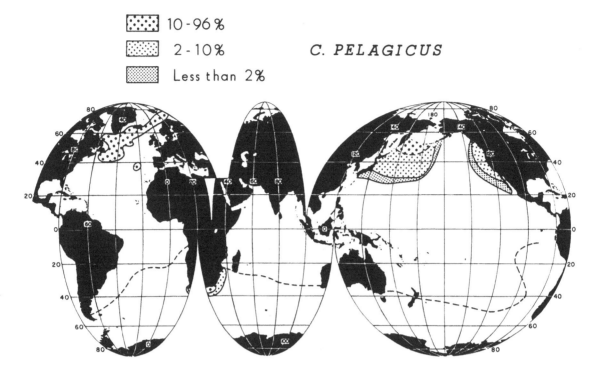

Fig. 3. Relative abundance of *Coccolithus pelagicus* in Recent surface sediments of the oceans. Compiled by Atiur Rahman from the literature listed in the Reference list. Dotted line marks the northern limit of occurrence of this species in the sediments according to McIntyre *et al.*, (1970).

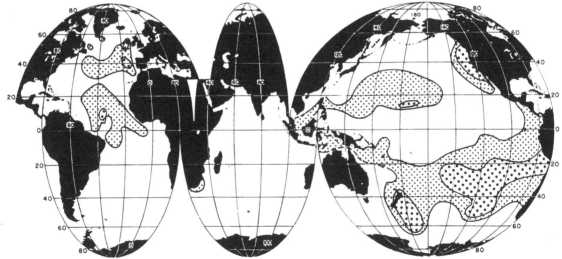

Fig. 4. Relative abundance of *Gephyrocapsa caribbeanica* in Recent surface sediments of the oceans. Compiled by Atiur Rahman from the literature listed in the Reference list.

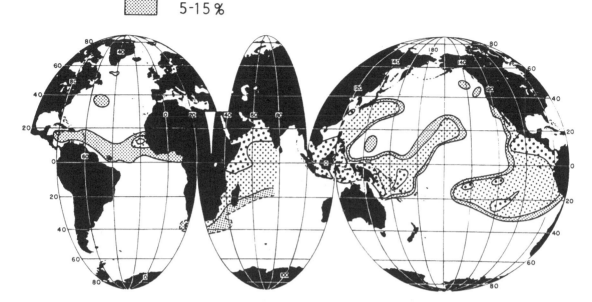

Fig. 5. Relative abundance of *Gephyrocapsa oceanica* in Recent surface sediments of the oceans. Compiled by Atiur Rahman from the literature listed in the Reference list.

variable group, nor improved the usefulness of the resulting taxa as paleoceanographic indicators or biostratigraphic markers.

Winter *et al.* (chapter 8) discuss the environmental preferences and biogeography of living coccolithophores in the water column. The following is a brief summary of the environmental affinities of several important species as inferred from studies of their distribution patterns in surface sediments. Species are divided into major taxa that usually comprise more than 10% of an assemblage and are thus less affected by preservational changes than species that are a very minor component of the total assemblage. The major sources of information for these discussions are the most recent compilations of distribution patterns on an ocean-basin-wide scale by Roth and Berger (1975), Geitzenauer *et al.* (1977), Roth and Coulbourn (1982) and Dmitrenko (1985).

Emiliania huxleyi is a eurytopic species that can tolerate very low salinities (10 and 20‰ in the Sea of Azov and the Black Sea respectively, according to Bukry, 1974), high but not the highest salinities (40 to 45‰ in the Gulf of 'Aqaba, according to Winter, 1982) and very low temperatures (2°C, according to McIntyre and Bé, 1967). It is abundant and relatively uniformly distributed in all major ocean basins. It may dominate assemblages of nannoplankton in sediments, comprising up to 82% of the total assemblage in the North Atlantic and 67% in the Pacific (Geitzenauer *et al.*, 1977), but due to its eurytopic nature its distribution patterns do not convey much paleoenvironmental information.

Coccolithus pelagicus is a cold-water species that prefers high nutrient concentrations (Fig. 1). It is most abundant in

the southern Labrador Sea, the southern part of the Norwegian Sea, the northwestern North Pacific, and off the coast of South Africa. This species has a long geologic record from the lower Tertiary to the Recent and thus may sometimes be reworked from older sediments (Roth and Coulbourn, 1982).

The genus *Gephyrocapsa* tolerates inner-shelf conditions and often outnumbers *E. huxleyi* (Chen and Sieh, 1982; Okada, 1983, 1989; Zhang and Siesser, 1986). However, in the warm, hypersaline environment of the Red Sea and the Gulf of 'Aqaba, *E. huxleyi* strongly dominates over *G. oceanica*. Winter (1982) explained this pattern by assuming that the latter species is a pelagic rather than neritic form that requires high nutrient concentrations. The relationship of high nutrient concentration with high abundance of *G. oceanica* is not very strong in the open ocean as shown by the results of a canonical correlation analysis of data from the North Pacific (Roth and Coulbourn, 1982, pp. 20–2).

Gephyrocapsa caribbeanica has a preference for cold, nutrient-rich waters (Fig. 2). It is most abundant in the Pacific in the Transitional Region underlying the California Current, the southern South Pacific between 20° and 65° S latitude underlying the Circum-Antarctic Current and the Peru Current. In the Atlantic it is most abundant in the northeastern North Atlantic. Because it is a dissolution-resistant species, its overall relatively enhanced abundance in the

Pacific compared to the Atlantic may be due to increased dissolution.

Gephyrocapsa oceanica prefers warm waters of high to moderate fertility (Fig. 3). It can tolerate the highest salinities (45–51‰) of any of the major coccolith species (Winter, 1982). It is most abundant in the tropical west Pacific, the Panama Basin, the South China and Banda Seas, the northwestern Indian Ocean, and the equatorial Atlantic. Its higher abundances in marginal seas compared to the open ocean indicates that it is rather eurytopic, tolerating relatively large fluctuations in physical and chemical conditions, as long as the water temperatures are relatively high. Its relatively lower abundance in the Atlantic compared to the Pacific is due to less intensive dissolution and thus reduced secondary enrichment of this solution-resistant form.

Distribution patterns of other taxa are less striking, because they are either less common or more eurytopic than the major taxa discussed above, but they still reflect environmental patterns to a certain extent. Remarks on their distribution patterns, are summarized in the following paragraphs. Detailed illustrations of the patterns are shown in the most recent biogeographic compilations on coccoliths in Recent sediments (Geitzenauer *et al.*, 1977; Roth and Berger, 1975; Roth and Coulbourn, 1982; Dmitrenko, 1985). *Umbilicosphaera sibogae* appears to favor warm, salty and relatively nutrient-rich waters. *Helicosphaera carteri* is a warm-water species that rarely exceeds 10% of the total assemblage. *Rhabdosphaera clavigera* is a rather solution-resistant form; *Umbellosphaera tenuis* and *U. irregularis,* two relatively delicate species, and the very dissolution-prone *Discosphaera tubifera* are all characteristic of central water masses, that is, relatively warm, salty but nutrient-depleted waters. Other taxa constitute such a small percentage of the total assemblage that even minor changes in preservation appear to alter their relative abundance greatly.

The deep-living species *Florisphaera profunda* is found in sediments as individual plates which are very simple in shape but recognizable in the light and electron microscope (Okada, 1983). Because of its simple shape it has been largely overlooked and it is probably still underreported. It serves as an excellent indicator of depth of deposition. Its relative abundance is just a few percent in shelf sediments but may reach 70% on continental slopes and in deep basins (Okada, 1983, 1989). Its greatest concentration appears to mark the thermocline and thus could serve as a proxy indicator of this level in the oceanic sedimentary record (Molfino and McIntyre, 1987).

Marginal-sea sediments

Coccolithophores can tolerate more extreme and less stable conditions than calcareous and siliceous microzooplankton (radiolarians and planktonic foraminifers), but not as well as siliceous phytoplankton (diatoms). Coccoliths thus occur in Recent sediments of marginal seas, even fairly close to the shore, and in fairly restricted basins with extreme and unstable environmental conditions. Dilution by terrigenous material and dissolution caused by oxidation of organic matter in sediments are the major secondary factors that affect the concentration and composition of coccolith assemblages in sediments from shallow marginal seas.

Although more accessible for sampling than deep ocean basins and thus quite extensively studied, coccolith assemblages in sediments from shelf and marginal seas are more difficult to interpret because of a lack of thorough documentation of oceanographic parameters of the overlying waters. Few of the studies provide coccolith census data that are needed to test thoroughly the effects of environmental conditions on burial assemblages of coccolithophores. Most investigations have simply reported relative abundances or average composition from a relatively small number of samples. The oceanography of the waters overlying the sampling stations is generally not discussed in sufficient detail. Statistical analyses are entirely missing and quantitative assessment of the influence of various environmental parameters and of preservational effects on coccolith assemblages are not possible. Thus, only some of the major patterns of coccolith distribution going from the least restricted to the most restricted marginal-marine environments are discussed below.

Sediment assemblages of coccoliths from large marginal seas with good connections to the open oceans, such as the Gulf of Mexico, are virtually identical to open-ocean assemblages (Pierce and Hart, 1979). This also seems to be the case for the Mediterranean (Bartolini, 1970; Violanti *et al.*, 1987), although modern studies of nannofossils in Recent sediments do not yet adequately define species distribution patterns and relate them to oceanic parameters. This situation ought to be remedied because of the unique oceanography of the Mediterranean which may provide a model for the mode of bottom-water formation and halothermal circulation which prevailed when the Earth was in a green-house stage during the Mesozoic; see Roth, 1989.

Coccoliths in sediments from marginal seas and continental margins of the western Pacific have received considerable study (Burns, 1975; Conley, 1979; Chen and Sieh, 1982; Okada, 1983, 1989; Wang and Samtleben, 1983; Varol, 1985; Wang and Chen, 1985; Zhang and Siesser, 1986). Compared to open-ocean assemblages, diversity is reduced and the assemblages are clearly dominated by one or two taxa, namely *Gephyrocapsa oceanica* and *Emiliania huxleyi*. As the shore is approached, one or both of these two taxa completely dominate in sediment assemblages, and the diversity of coccoliths in sediments drops off rapidly. Which of the two major taxa dominates seems to depend on various factors, including surface-water fertility, temperature, and salinity. Winter (1982) feels that *G. oceanica* prefers high nutrient concentrations, but it was shown by Roth and Coulbourn (1982) that the three major oceanographic parameters (temperature, salinity, and phosphate concentration) are quite highly correlated and that the interrelationship of these parameters and relative-abundance patterns of coccoliths in sediment assemblages are complex and not yet well understood.

Malformed *E. huxleyi* are common in the water column of marginal seas (Okada and Honjo, 1975) but are rarely observed in the sediment (Wang and Samtleben, 1983). Perhaps these malformed coccoliths occur only during particular seasons or are a relatively recent phenomenon and thus would become increasingly rare in older sediments because of downward mixing. Alternatively, perhaps malformed coccoliths are more easily dissolved than normal ones.

Restricted marginal seas display more extreme environmental conditions and thus more unusual nannoplankton assemblages. The back-reef lagoons off Belize are dominated by *E. huxleyi*, with smaller numbers of *G. oceanica*, and very rare *Umbellosphaera tenuis, U. irregularis, Umbilicosphaera sibogae, Rhabdosphaera clavigera,* and *Braarudosphaera bigelowii*. Conspicuously absent are *Calcidiscus leptoporus* and *Scyphosphaera* spp. (Scholle and Kling, 1972; Kling, 1975). In the Great Barrier Reef region, another coral-reef environment, *E. huxleyi* dominates in the southernmost part and *G. oceanica* in the central and northern part and in the Torres Straits. All other species combined comprise less than 20% of the total assemblage. This is in great contrast to the Coral Sea, where taxa other than *G. oceanica* and *E. huxleyi* make up a total of more than 20% and reach a high of over 45% of the total assemblage (Conley, 1979). Diversity of coccolith assemblages preserved in sediments is generally only half as high in sediments of the Great Barrier Reef and Torres Straits as in the Coral Sea.

In diatomaceous sediments from the Gulf of California, coccolith assemblages are dominated by *G. oceanica* (50–80% of the total assemblage) and *E. huxleyi* (12–30%), and all other species comprise percentages in single digits (A. Winter, pers. comm.).

Two marginal seas are particularly interesting because they represent two end-members. The Black Sea is a good example of a marginal basin with estuarine circulation, having surface-water outflow and deep-water inflow and stagnant bottom waters, thus acting as a 'nutrient trap'. The Red Sea and Gulf of Aqaba have lagoonal circulation with surface-water inflow and deep-water outflow and thus are 'nutrient deserts'.

Calcareous nannoplankton in sediments of the Black Sea have received very little study. Because of its low surface-water salinity (around 17–20‰) it supports a low diversity nannoflora completely dominated by *E. huxleyi*. This cosmopolitan and highly eurytopic species is the major component of white laminae in the sediments that appear to represent summer blooms, but also dominates in the intervening gray laminae that appear to have been deposited in the winter when input of detrital sedimentary particles was greater. Other species encountered in Recent sediments of the Black Sea include *B. bigelowii*, known for its great tolerance for extreme environmental conditions (Roth, 1974), *G. caribbeanica, Anacanthoica acanthos, Syracosphaera mediterranea,* and *S. pirus*, together with reworked Cretaceous and Eocene age nannofossils (Bukry, 1974; Toker, 1984). High abundance of *E. huxleyi* in the high-salinity waters of the Gulf of Aqaba (Winter *et al.*, 1979) and of *B. bigelowii* in the Persian Gulf (Roth, 1974) suggest that these species may not be useful paleosalinity indicators.

Coccolith assemblages in Recent sediments of the Gulf of Aqaba are dominated by *E. huxleyi*, which gradually increases from a few percent in the Gulf of Aden to 25–29% in the southern Red Sea, 52 to 71% in the northern Red Sea, and 85% in the Gulf of 'Aqaba. *Gephyrocapsa oceanica* shows the opposite trend, with 40–60% in the Gulf of Aden, 49% in the southern Red Sea, 10–15% in the central to northern Red Sea, and 3% in the Gulf of 'Aqaba (Winter *et al.*, 1979). *Gephyrocapsa ericsonii* is very common in the Gulf of 'Aqaba, and *C. leptoporus* is also more common in the surface sediments of the Red Sea than in the Gulf of 'Aqaba (Winter *et al.*, 1979). The deep-water species *Florisphaera profunda* occurs in the water column of the southern Red Sea and should also occur in the underlying sediments. This deep-water coccolithophore is absent from the Gulf of 'Aqaba (Winter *et al.*, 1979) although its seafloor depth is 1800 m and thus deep enough for a species that is most common in the water column of the open-ocean between 200 and 1000 m. Unusually high salinities and low nutrient concentrations may be the cause for its absence.

Data reduction using multivariate statistics and oceanographic reconstructions

Much can be learned about the environmental conditions responsible for the distribution of coccoliths from a comparison of maps displaying the relative abundance of particular taxa in sediments and maps of surface-water temperature, salinity, and inorganic phosphate concentration. Simple plots of the relative abundance of particular species versus temperature or phosphate concentration further aid in the interpretation of distribution patterns as related to oceanographic conditions (see e.g., Roth and Coulbourn, 1982, Fig. 19). Inspection of biogeographic maps for a larger number of coccolith species (e.g., Geitzenauer *et al.*, 1977, Figs. 1–18; Roth and Coulbourn, 1982, Figs. 3–18) reveals that numerous species have similar distribution patterns and thus respond to the same environmental parameters. This was already recognized by McIntyre and Bé (1967), who proposed four coccolithophore floral zones for the Atlantic. Clearly, there is much redundancy in the use of numerous coccolith species in paleoenvironmental analysis. Thus, for large data sets with good geographic coverage, more elaborate methods of information handling are appropriate.

A whole range of mathematical techniques exist, including simple exploratory techniques that display relationships among samples or species, such as cluster analysis, to more powerful multivariate statistical techniques, such as principal component, factor, and canonical correlation analysis, all of which tend to remove redundancy in the data set and may establish quantitative relationships among the composition of coccolith assemblages and environmental parameters (Roth and Berger, 1975; Roth and Coulbourn, 1982). Most of the techniques used in these studies were developed initially for the social sciences and later adapted for geology. Multivariate statistical techniques are well described in various textbooks (e.g., Cooly and Lohnes, 1971; Tatsuoka, 1971; Jöreskog et al., 1976; Johnston, 1980; Davis, 1986) and need not be discussed here in detail. Statistical packages containing all the necessary programs for main-frame computers (e.g., Norusis, 1982, 1985; Dixon, 1985), and more interactive statistical packages which can be run on microcomputers are widely available.

Statistical techniques have been used successfully in studies of coccoliths in sediments in an effort to: (1) reduce the redundancy of information in raw counts, and (2) quantify relationships between nannofloral composition and environmental conditions. Cluster analysis was applied to all coccolith counts from Recent sediments extending over a wide range of water depths and latitude in the South Pacific. Nine clusters were identified and related to both water mass and water depth of deposition of the sediment (Roth and Berger, 1975). Factor analysis was used by Geitzenauer et al. (1976) for coccolith census data from surface sediments of the Pacific Ocean in order to derive a transfer equation to estimate surface-water temperatures and salinities for the Pleistocene. Unfortunately, two important and often dominant species, *Gephyrocapsa caribbeanica* and *Emiliania huxleyi*, were excluded from the analysis because the latter evolved during the late Pleistocene and the former showed a marked decline during the late Pleistocene.

Many different statistical techniques (factor analysis, canonical correlation, stepwise discriminant analysis) were applied to coccolith census data from Recent sediments of the North and Equatorial Pacific by Roth and Coulbourn (1982). Factor analysis proved useful in reducing the number of variables (taxa) to a more manageable number of new and mutually independent variables (factors) that could be related more easily to oceanographic conditions. The geographic distribution of the factor assemblages of Roth and

Coulbourn (1982) agrees well with that of Geitzenauer et al., (1976) for those factors that are dominated by the same species. However, some differences in the two studies occurred because of the above mentioned exclusion of important species by the latter authors and because of differences in sample coverage.

Roth and Coulbourn (1982) were the first and so far the only authors to use canonical correlation analysis to relate factor assemblages to physical and chemical parameters of surface water (Fig. 6 and Table 1). Canonical correlation analysis is a powerful technique that explores the relationship between two sets of independent variables (Buzas, 1979; Roth and Coulbourn, 1982). Discriminant analysis shows that only about a third of the taxa are needed to characterize water masses. So far only the coccolith sediment assemblages from the Pacific Ocean have been thoroughly analyzed with a number of different multivariate statistical techniques. The North Atlantic coccolith census data have been analyzed using factor analysis.

A purely mechanical interpretation of the results of multivariate statistical analyses by statisticians and non-specialists may result in misleading results. A careful inspection of the raw data, a basic understanding of the biology and taxonomy of coccolithophores and of the effects of preservational changes on coccolith assemblages and some insight into the oceanography of the area of interest are important for the interpretation of the results of a statistical analysis. Not even the most elaborate statistical techniques will extract much useful information from poor data sets.

Dissolution of coccoliths

About four-fifths of all the carbonate fixed by organisms in surface-waters is dissolved (Broecker, 1974). Dissolution of nannoplankton remains becomes more and more pronounced as water depth increases (see Plates 1 and 2 in Roth and Berger, 1975). The degree of dissolution can be determined visually and can be expressed semiquantitatively (Roth and Thierstein, 1972) or quantitatively by the use of dissolution rankings and indices (Roth and Berger, 1975), or by carefully inspecting similarity matrices (Roth and Coulbourn, 1982). Calcareous nannoplankton remains are very sensitive indicators of carbonate dissolution, especially at shallow water depth where planktonic foraminifers are less sensitive (Roth and Berger, 1975). Thus dissolution of calcareous nannoplankton is not necessarily a nuisance; it provides important information on the carbonate chemistry of bottom waters.

The question on how carbonate dissolution affects coccolith distribution in sediments in the ocean and how the preservation of coccoliths may be useful in estimating the amount of carbonate dissolution and the position of various important dissolution horizons in the oceans is addressed

Table 1

Correlation coefficients among Varimax Factor loadings of Appendix B and among selected environmental variables (values less than |0.5| not shown)

Variable number	Variable name	Factor 1	Factor 2	Factor 3	Factor 4	Temperature	Salinity	Nutrients	Oxygen
1	Factor 1 (*G. oceanica*)	1.000							
2	Factor 2 (*G. caribbeanica*)	—0.618	1.000						
3	Factor 3 (*U. sibogae*)			1.000					
4	Factor 4 (*E. huxleyi*)				1.000				
5	Temperature	0.638	—0.905			1.000			
6	Salinity		—0.562	0.668			1.000		
7	Nutrients					—0.671		1.000	
8	Oxygen	—0.674	0.605			—0.569			1.000

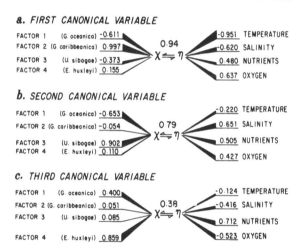

Fig. 6. Three independent, ranked, canonical correlations relating four coccolith factor assemblages to four environmental variables. a. The first canonical variables accounts for 89% of the variance within the data and the relationship is significant at the 0.001-level. Loadings show that as the contribution of Factor 1 to the sediment decreases and that of Factor 2 increases, sea-surface winter temperature and salinity decreases while nutrients increase. b. The second canonical variable accounts for 63% of the variance within the data and describes a relationship significant at the 0.01-level. Loadings suggest that that as the contributions of Factor 3 to the sediment increases salinity also increases. c. The third canonical variable accounts for nearly 15% of the variance within the data and describes a relationship that has a good chance of being fortuitous. (After Roth and Coulbourn, 1982.)

here. The relationship between properties of the water column and sedimentation of carbonates has been reviewed in numerous publications (see Berger, 1976; Broecker and Peng, 1982, for review and references).

The important effect of dissolution on coccolith assemblages was recognized by McIntyre and McIntyre (1971), who determined the relative strength of modern heterococcoliths using ultrasonic vibration. They concluded that *Emiliania huxleyi, Calcidiscus leptoporus, Gephyrocapsa oceanica,* and *Umbilicosphaera sibogae* are the most resistant forms. Berger (1973), using the rather limited data of McIntyre and Bé (1967) and Schneidermann (1977), who did not present a thorough quantitative analysis of his data, postulated that partial dissolution of coccolith assemblages introduces a 'cold' aspect due to removal of the more delicate tropical species and the relative enrichment in the more

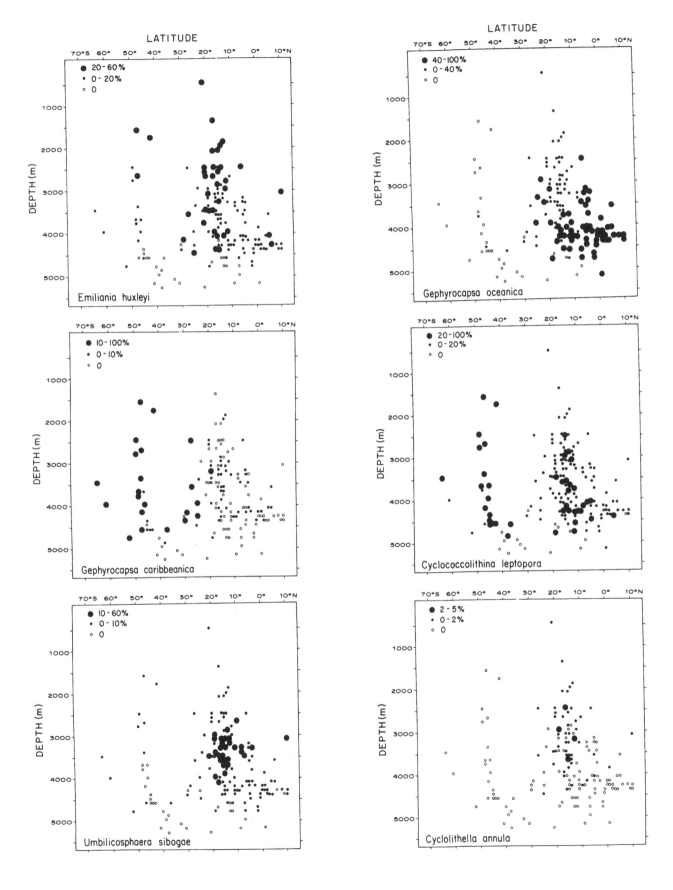

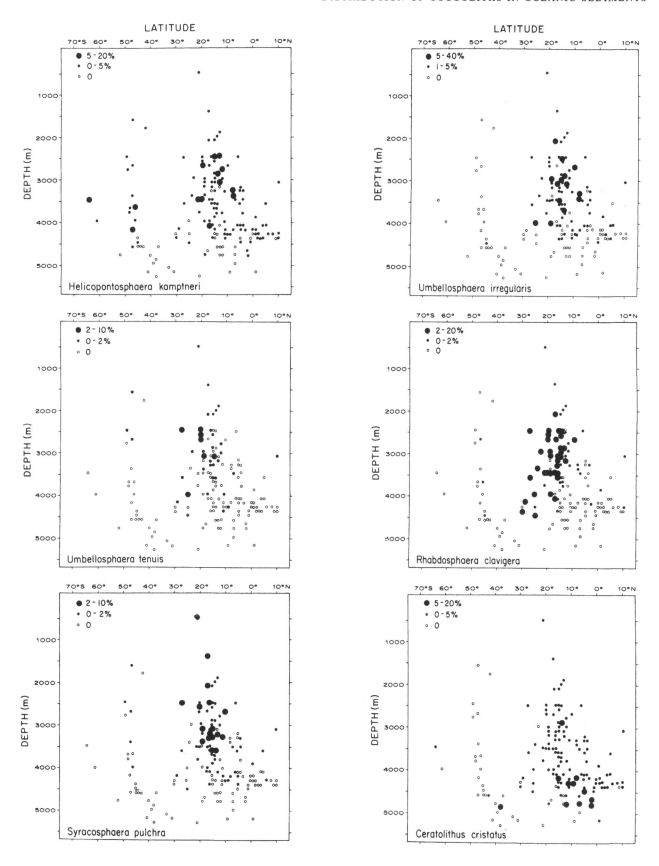

Fig. 7. Depth distribution of important coccolith species in surface sediments of the South Pacific. (From Roth and Berger, 1975.)

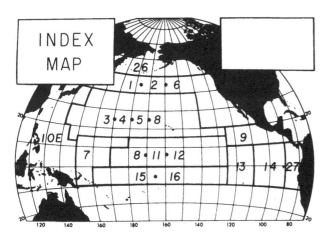

Fig. 8. Index map of areas used in Figs. 7 and 8. (From Roth and Coulbourn, 1982.)

igera) are more susceptible to dissolution than most placoliths (*Gephyrocapsa, Emiliania, Coccolithus, Calcidiscus*). Very rare species may appear in different positions in the dissolution rankings because of chance recovery (e.g., *Ceratolithus cristatus* appears among the most robust taxa in the South Pacific but among the more delicate forms in the equatorial Atlantic).

Roth and Coulbourn (1982) described the patterns of dissolution of coccoliths for different geographic areas. Samples were arranged according to increasing depth for each geographic area. The amount of carbonate dissolution can be expressed by dissolution indices, which are the sum of the product of rank position and relative abundance for all species in an assemblage (Roth and Berger, 1975). Percent similarity between samples provides another good measure of dissolution. It was used together with similarity indices to evaluate coccolith dissolution with depth in the North and

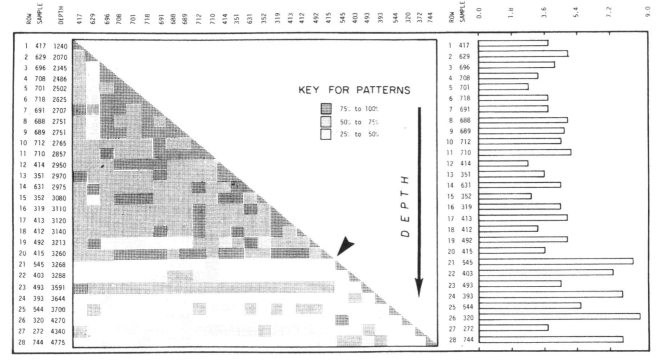

Fig. 9. Patterned similarity matrix and solution indices of coccolith assemblages from the Transition Region of the North Pacific (Areas 1.2.6). The computation of similarity and dissolution indices are discussed in the original paper. Arrows mark major breaks in the pattern. (From Roth and Coulbourn, 1982.)

robust cold-water species. Roth and Berger (1975) showed that is not the case and that sediment assemblages preserve a good record of temperature even in highly dissolved samples, except at high latitudes. Mechanical stability rankings established experimentally by McIntyre and McIntyre (1971) and depth distribution plots by Roth and Berger (1975) for the South and Central Pacific and by Schneidermann (1977) for the central Atlantic are comparable (Fig. 7). Caneoliths (e.g., *Syracosphaera pulchra*) and cyrtoliths (e.g. *Discosphaera tubifera, Rhabdosphaera clav-*

Central Pacific (Roth and Coulbourn, 1982). Different regions were distinguished that did not cross major boundaries of water masses (Fig. 8). Dissolution rankings differ somewhat from one area to the next (see Roth and

212

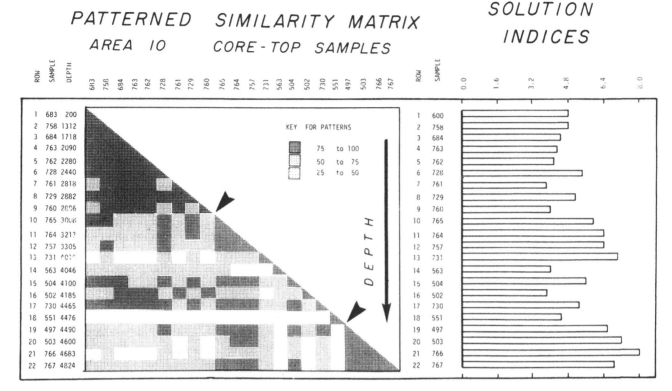

Fig. 10. Patterned similarity matrix and solution indices of coccolith assemblages from the Western Pacific. (From Roth and Coulbourn, 1982.)

Coulbourn, 1982, Table 6). There are some differences in similarity patterns, but some general trends can be observed. Going from shallow to deep samples in a particular region, great similarity is noticed among shallow samples. Partially dissolved samples are quite similar among themselves, but show low similarity when compared to shallow samples. Assemblages from close to the calcareous nannoplankton compensation depth (CNCD) contain only a few species and chance recovery of the same few in several samples is unlikely. Thus, such samples from very great depths display very low similarities among themselves and between samples from shallower water-depths. Major breaks in similarity patterns and changes in solution indices occur at slightly different levels in different regions but dissolution seems to be first registered between 3000 and 3500 m (Figs. 9 and 10). The CNCD in the North Pacific is bowl-shaped, reaches its greatest depth of 5500 m in the central equatorial Pacific and shallows towards the margins of the basin to about 3300 m. Even moderately dissolved samples retain their characteristic coccolith assemblages and can be assigned to the correct water mass. Thus, the basic signature of surface-water masses is recognizable in spite of dissolution if a sufficiently large number of samples are available.

The two major dissolution horizons in the ocean are the lysocline and the calcite compensation depth. Planktonic foraminifers show major changes at the lysocline and thus it is a distinctive boundary. This is not the case for the coccolith lysocline which is difficult to recognize and thus not a very useful level. Berger (1973) analyzed data of McIntyre

and Bé (1967) and postulated a coccolith lysocline for the Atlantic. Subsequent studies of Roth and Berger (1975) and Roth and Coulbourn (1982) showed that it is difficult to define a coccolith lysocline. The reason for this is that even well preserved coccolith assemblages are dominated by solution-resistant species rather than delicate ones, as is the case for planktonic foraminifers (Roth and Berger, 1975). Coccolith assemblages do not record slight degrees of dissolution very well but provide a very robust signal of surface-water conditions of the modern and Cretaceous oceans (Roth and Coulbourn, 1982; Roth, 1986, 1989).

Conclusions and perspective

Coccoliths are a major component of pelagic sediments, but it is surprising how little quantitative information is available in the published literature about the relative proportion of coccoliths in sediments from various settings. Speculations on the effects of redepositon and winnowing have not been substantiated by data. More work is needed along those lines. With scanning electron microscopes and coulter counters, this task is now much easier than it was in the 1950s when the last estimates were published (Black, 1956; Bramlette, 1958).

Recent studies indicate that the high rate of carbon fixation of coccolithophores and rapid transport of coccolithophore remains to the sea floor may have important effects on the carbon dioxide concentration of the surface oceans and thus the atmosphere. Therefore, in order to understand better possible implications for the greenhouse effect, more attention ought to be given to production, transport, preservation, and burial of coccolithophores and their skeletons.

The general relationship between the composition and preservation of coccolith sediment assemblages and the oceanography of overlying waters is reasonably understood for the Pacific Ocean and the North Atlantic. Regional coverage of published data for the other ocean basins and for most of the marginal seas is still inadequate. Recent studies have clearly shown that coccoliths in sediments delineate overlying water masses very well and thus serve as proxy indicators of the distribution of major water masses. The complex relationship and the relatively high degree of correlation between basic physical and chemical parameters of surface waters (e.g., Roth and Coulbourn, 1982, Table I) makes it difficult to interpret coccolith data in terms of a single environmental parameter. Dissolution and diagenesis complicate oceanographic interpretations of coccolith assemblages in sediments to a certain degree. However, major oceanographic signals are preserved in coccolith sediment assemblages, and accurately reflect spatial and temporal changes in ocean circulation.

Considerable work is needed to gain a better understanding of the coccolith assemblages in marginal marine settings. Detailed sampling combined with careful oceanographic data gathering is needed. Quantitative studies combined with thorough statistical analyses should result in a better understanding of factors responsible for the distribution and preservation of coccoliths in marginal seas and on continental margins. Such insight would be very helpful in understanding nannoplankton distribution in epicontinental seas which were much more widespread during high sea-level stands in the past. Examples of such marginal environments in which coccoliths thrived are the chalk seas of the Cretaceous and the poorly oxygenated basins in the boreal realm during the late Jurassic where cyclic black shale sequences rich in coccoliths accumulated and gave rise to important petroleum source rocks.

Recent analogues are few, and they remain largely unstudied. The study of coccolith sedimentation in marginal-marine settings is a major challenge for nannoplankton specialists. It will, however, require close cooperation with physical, chemical and biological oceanographers if real progress is to be made in obtaining a thorough understanding of the processes and products in the pelagic sedimentary environment.

Models of production, transport, and preservation of coccoliths in sediments need to be refined. Geographic coverage of samples studied quantitatively in all ocean basins and marginal seas should be improved. Complete oceanographic data from the waters overlying the sample stations are needed to provide a better understanding on the factors affecting production, transport and preservation of coccoliths in sediments and thus make coccoliths in sediments even better oceanographic proxy indicators.

Appendix – *Methods of study*

Calcareous nannoplankton in sediments received relatively little attention prior to the introduction of the electron microscope. The majority of taxa in sediment assemblages can be identified in a high-power optical microscope with oil-immersion objectives with a magnification of × 40 to 100 and both polarized light, phase-contrast or other contrast-enhancing illumination systems. Electron microscopy is necessary to identify smaller forms and to determine the detailed ultrastructure of the majority of taxa. Electron microscopy has greatly contributed to a better understanding of the architecture and taxonomy of coccoliths and related forms, often referred to as nannoliths. Electron microscopy, however, is very time-consuming, especially if used for quantitative study where one needs to count at least 300 specimens per sample to determine the relative abundance of all taxa that comprise more than 1-2% of the total assemblage with 90 to 95% confidence. Combining light and electron microscopy saves much time and still produces reproducible results (Roth and Berger, 1975; Roth and Coulbourn, 1982).

Techniques of sample preparation for the study of coccoliths in sediments are relatively simple and considerably faster than preparation techniques for other microfossils. Great care must be taken to avoid contamination of samples. Disposable materials such as toothpicks and siphon straws should be used wherever possible. All non-disposable glassware and utensils should be washed in hydrochloric acid, followed by a distilled water rinse, and finally an ammonia rinse to remove all traces of acid that would otherwise dissolve coccoliths.

Preparation techniques are described briefly here, and the reader is referred to the published literature for more detailed descriptions (e.g., Taylor and Hamilton, 1982; Perch-Nielsen, 1985). Light microscope slides should be prepared of all samples to evaluate the overall abundance and preservation of coccoliths and to decide if the sample is suitable for detailed study and to decide which of the more elaborate preparation methods may be most suitable. Light microscope preparations are best made in the following way. A tiny amount of sediment (about the size of a grain of rice) is mixed with a drop of water on a microscope cover glass. The resulting suspension is spread evenly across the cover glass with a flat wooden tooth pick and dried on a hot plate.

When the suspension is completely dry the cover glass is attached to a glass microscope slide with a mounting medium that is acid free and has an index of refraction around 1.5. Traditional mounting media, such as natural Canada Balsam (slightly acidic) or some sort of synthetic resin, require heat curing. More recently developed optical adhesives are much more convenient to use, because they cure under ultraviolet light, but they are considerably more expensive and some have a shelf-life of only a few months.

Light microscope slides may be used to perform a census of all but a few species of calcareous nannoplankton. In most cases, it is desirable to supplement light microscopy with electron microscopy for detailed study of the ultrastructure and preservation changes, and especially to identify with certainty and perform reliable counts of some of the small forms, such as *Emiliania huxleyi*, or forms with specific features which are beyond the resolution of the light microscope, such as the distinctive surface texture on the distal side of *Umbellosphaera tenuis* and the central structure of *Syracosphaera*. Small coccoliths may be counted in the light microscope as groups (e.g., *Gephyrocapsa* spp.–*Emiliania huxleyi* group and *Umbellosphaera* spp.), which may be split up later into species under the scanning electron microscope (see Roth and Berger, 1975; Roth and Coulbourn, 1982).

Both types of electron microscope (transmission electron microscope and the scanning electron microscope) are used for the study of coccoliths, each requiring different preparation techniques. The transmission electron microscope has a higher resolution than the scanning electron microscope (about 2 Å for the former, about 100 Å or somewhat better for the latter), but transmission electron microscopy requires elaborate sample preparation that involves the preparation of carbon replicas. This method is not described here, because transmission electron microscopy now is used only rarely for coccolith studies and because numerous good descriptions of these techniques have been published (Noel, 1965; McIntyre *et al.*, 1967).

Scanning electron microscopy requires only minor preparation, especially if the samples are rich in coccoliths and do not contain other components, as is generally the case for deep-sea carbonates. A dilute suspension of sediment in water is prepared by adding a small amount of raw sediment (a few mm^3 in 1 ml of water) in a small glass vial, shaking the mixture until all the sediment is dispersed, removing a few drops using a siphon straw and placing them in the center of a cover glass that has been attached to a specimen stub with silver paint, and spreading the suspension with a toothpick. The suspension on the stub is left to dry and then coated with gold in a sputter coater or a vacuum coater. If quantitative studies are envisioned it is best to avoid using an ultrasonic cleaner, because differential destruction of particular taxa adds bias. Short centrifugation or gravity settling to remove clay from the sediment samples also may distort the composition of nannofossil assemblages. If such techniques are used where supernatants are discarded, but which may be necessary for samples particularly rich in clay minerals that tend to mask coccoliths in the electron microscope, great care is required to avoid the loss of small coccoliths. Sample suspensions stored for more than just a few hours should be buffered to avoid dissolution by the carbonic acid produced by absorption of carbon dioxide from the air and oxidation of organic matter. A pH of about 9 is sufficient to protect against dissolution and provides maximum dispersion of carbonate particles. However, routine use of sodium carbonate-bicarbonate buffers is discouraged, because small encrustations may form and obscure ultrastructural details or disturb the optical images.

Methods of information handling, display, and data reduction vary in their degree of sophistication and effort. Many authors, especially in the early days of coccolith studies but even to this day, simply list the presence of particular taxa. In some cases relative abundances are expressed either verbally or by using some measure of relative abundance such as four or five categories of relative abundance or the number of specimens per field of view, in some cases using standardized slide preparation. Determination of numbers of coccoliths per gram of sediment is very difficult because of the small sample size and inhomogeneities of samples and prepared slides. More sophisticated data reduction use multivariate statistical technique groupings of taxa that respond in similar ways to particular environmental conditions and thus facilitate environmental reconstruction.

Acknowledgments

I am grateful to the funding agencies that supported my research and to the institutions that have employed me and thus made it possible for me to work on these wonderful little plants for over twenty years. Research summarized in this paper was supported by grants of the National Science Foundation to Scripps Institution of Oceanography (GA-36697; OCE 76-84029 A02; OCE 78-25587) and to the University of Utah (OCE 78-25844; OCE 82-14704). All samples used for illustrations were supplied by Scripps Institution of Oceanography and were obtained with the help of Tom Walsh, then Curator of the Core Laboratory at SIO. Over the years I have profited from discussions with Wolfgang Berger, David Bukry, Stefan Gartner, William Hay, Andrew McIntyre, Barbara Molfino, Bill Siesser, Hans Thierstein, and Amos Winter, and with my former students Sharon Alley and Angela Gorman. My doctoral student Atiur Rahman, not only prepared Figs. 1 through 3 and reviewed earlier versions of this paper, but he also constantly challenged me in my own thinking about the sedimentation of cocolith remains and thus contributed much to this review. I thank all these colleagues for their help.

This paper is dedicated to the memory of two of my former coauthors and friends, Bill Coulbourn and Sy Schlanger, who taught me much about statistics and carbonates and who both died prematurely in 1990 when I was in the midst of writing this review.

References

Backman, J., P. Pestiaux, H. Zimmerman, and D. Hermelin, 1986. Palaeoclimatic and palaeoceanographic development in the Pliocene North Atlantic: Discoaster accumulation and coarse fraction data,. In *North Atlantic Palaeoceanography*, ed. C.P. Summerhayes and N.J. Shackleton, pp. 231–42. Geol. Soc. Spec. Pub. No. 21. Blackwell Scientific Publications, London.

Bartolini, C., 1970. Coccoliths from sediments of the western Mediterranean. *Micropaleontol.*, **16**: 129–54.

Berger, W.H., 1973. Deep-sea carbonates: evidence for a coccolith lysocline. *Deep-sea Res.*, **20**: 917–21.

Berger, W.H., 1976. Biogenous deep-sea sediments: production, preservation and interpretation. In *Chemical Oceanography*, ed. J.P. Riley and R. Chester, vol. 5, pp. 265–383.

Biekart, J.W, 1989. The distribution of calcareous nannoplankton in Late Quaternary sediments collected by the Snellius II Expedition in some southeast Indonesian basins. *Proc. K. Neder. Akad. Wetensch.*, Ser. B., **92**: 77–141.

Black, M., 1956. The finer constituents of globigerina ooze. *Int. Geol. Cong., 20th, Mexico, 1956.* Resumen de los trabajos presentados, p. 173.

Bramlette, M.N., 1958. Significance of coccolithophores in calcium carbonate deposition. *Geol. Soc. Amer. Bull.*, **69**: 121–6.

Bramlette, M.N., 1961. Pelagic sediments. In *Oceanography* ed. M. Sears, Amer. Assn. Advancement Sci. (Washington, D.C.), Pub. No. 67: 345–66.

Bréhéret, J., 1978. Formes nouvelles quaternaires et actuelles de la famille des Gephyrocapsaceae (Coccolithophorides). *C. R. Acad. Sci. Paris*, Serie D., **287**: 599–601.

Broecker, W.S., 1974. *Chemical Oceanography*. Harcourt, Brace, Jovanovich, New York.

Broecker, W.S. and Peng, T-H., 1982. *Tracers in the Sea.* Lamont-Doherty Geol. Observatory, Palisades, New York.

Bukry, D., 1974. Coccoliths as paleosalinity indicators -evidence from the Black Sea. In *The Black Sea, Its Geology, Chemistry and Biology.* ed. E.T. Degens and D.A. Ross, Amer. Assn. Pet. Geol. Mem., **20**: 353–63.

Burns, D.A., 1975. The abundance and species composition of nannofossil assemblages in sediments from continental shelf to offshore basin, western Tasman Sea. *Deep-Sea Res.*, **22**: 425–31.

Buzas, M.A., 1979. Quantitative biofacies analysis. In *Foraminiferal Ecology and Paleoecology*, ed. J.H. Lipps, W.H. Berger, M.A. Buzas, R.G. Douglas, and C.H. Ross. SEPM Short Course No. 6, Houston, Texas, Soc. Econ. Paleontol. & Mineral., Tulsa, Oklahoma, pp. 11–20.

Chen, M.P. and Sieh, K.S., 1982. Recent nannofossil assemblages from Sunda shelf to abyssal plain, South China Sea. *Nat. Sci. Counsel*, ROC(A), **6**(4): 250–85.

CLIMAP project members, 1976. The surface of the ice age earth. *Science*, **191**: 1131–7.

CLIMAP project members, 1981. *Seasonal Reconstructions of the Earth's Surface at the Last Glacial Maximum.* Geol. Soc. Amer. Map and Chart Series, MC-36.

Conley, S.M., 1979. Recent coccolithophores from the Great Barrier Reef-Coral Sea region. *Micropaleontol.*, **25**:20–43.

Cooley, W.W., and Lohnes, P.R., 1971. *Multivariate Data Analysis*. Wiley, New York.

Davis, J. C., 1986. *Statistics and Data Analysis in Geology*, 2nd edition. Wiley, New York.

Dayman, J., 1858. *Deep sea soundings in the north Atlantic Ocean between Ireland and Newfoundland, made in HMS Cyclops.* Eyre & Spottiswoode, London.

Dixon, W.J. (Ed.), 1985. *BMDP Statistical Software Manual*: to accompany the 1988 software release. Univ. California Press, Berkeley.

Dmitrenko, O. B., 1985. Zakonomernosti raspredileniya izvestkovogo nannoplanktona v donnyx osadkakh severo-zapanoy chasti indiyskogo okeana (The regularities of distribution of calcareous nannoplankton in bottom sediments of north-west part of Indian Ocean). *Okeanologia*, **25**: 483–8.

Fincham, M.J. and Winter, A., 1989. Paleoceanographic interpretations of coccoliths and oxygen-isotopes from the sediment surface of the southwest Indian Ocean. *Mar. Micropaleontol.*, **13**: 325–51.

Geitzenauer, K.R., Roche, M.B. and McIntyre, A., 1976. Modern Pacific coccolith assemblages: derivation and application to late Pleistocene paleotemperature analysis. In *Investigation of Late Quaternary Paleoceanography and Paleoclimatology.* R.M. Cline and J.D. Hays. Geol. Soc. Amer. Mem., **145**: 423–48.

Geitzenauer, K.R., Roche, M.B., and McIntyre, A., 1977. Coccolith biogeography from North Atlantic and Pacific surface sediments. In *Oceanic Micropaleontology*, ed. A.T.S. Ramsey, pp. 973-1008. Academic Press, New York..

Ghidalia, M.-J., 1988. Les Gephyrocapsaceae (Coccolithophorées) des coupes de Stuni et de Vrica, Italie, stratotypes de la limite Plio-Pleistocene. Étude systématique et intérêt stratigraphique. *Cahier de Micropaléontologie*, N.S., **3**(1): 5–49.

Haq, B.U., 1980. Biogeographic history of Miocene calcareous nannoplankton and paleoceanography of the Atlantic Ocean. *Micropaleontol.*, **26**: 414–43.

Haq, B.U. and Lohmann, G.P., 1976. Early Cenozoic nannoplankton biogeography of the Atlantic Ocean. *Mar. Micropaleontol.* **1**: 119–94.

Honjo, S., 1976. Coccoliths: production, transportation and sedimentation. *Mar. Micropaleontol.*, **1**: 65–79.

Johnston, R.J., 1980. *Multivariate Statistical Analysis in Geography*. Longman, Essex.

Jöreskog, K.G., Klovan, J.E., and Reyment, R.A., 1976. *Geological Factor Analysis*. Elsevier, Amsterdam.

Kling, S.M., 1975. A lagoonal coccolithophore flora from Belize (British Honduras). *Micropaleontol.*, **21**: 1–13.

McIntyre, A. 1967. Coccoliths as paleoclimatic indicators of Pleistocene glaciation. *Science*, **158**: 1314–17.

McIntyre, A. and Bé, A.W.H., 1967. Modern Coccolithophoridae of the Atlantic Ocean I. Placoliths and Cyrtoliths. *Deep-Sea Res.*, **14**: 561–97.

McIntyre, A. and McIntyre, R., 1971. Coccolith concentrations and differential solution in oceanic sediments. In *The Micropalaeontology of Oceans.* ed. B.M. Funnell B.M. and W.R. Riedel, pp. 253–61. Cambridge Univ. Press, London.

McIntyre, A., Bé, A.W.H. and Preikstas, R., 1967. Coccoliths and the Pliocene-Peistocene boundary. *Progress in Oceanography*, **4**: 3–25.

McIntyre, A., Bé, A.W.H. and Roche, M.B., 1970. Modern Pacific Coccolithophorida: a paleontological thermometer. *Trans. N.Y. Acad. Sci.*, **32**: 720–31.

Molfino, B. and McIntyre, A., 1987. Fluctuations of the equatorial Atlantic thermocline. *EOS*, **68**: 1330.

Noel, D., 1965. *Sur les coccoliths du jurassique européen et d'afrique du nord*. Editions du *CNRS (Paris)*.

Norusis, M. J., 1982. *SPSS Introductory Guide: Basic Statistics and Operations*. McGraw-Hill, New York.

Norusis, M. J., 1985. *SPSS-X advanced Statistics Guide*. McGraw-Hill, New York.

Okada, H.,1983. Modern nannofossil assemblages in sediments from coastal and marginal seas along the Western Pacific Ocean. *Utrecht Micropaleontol. Bull.*, **30**: 171–87.

Okada, H., 1989. Morphometric and floral variations of calcareous nannoplankton in relation to their living environment. *INA Newsl.*, **11**(2): 87–8.

Okada, H. and Honjo, S., 1975. Distribution of coccolithophores in marginal seas along the western Pacific Ocean. *Mar. Biol.*, **31**: 271–85.

Perch-Nielsen, K., 1985. Mesozoic calcareous nannofossils. In *Plankton Stratigraphy*, ed. H.M. Bolli, J.B. Saunders and K. Perch-Nielsen, pp. 329–426. Cambridge University Press, Cambridge.

Pierce, R.W. and Hart, G.F., 1979. Phytoplankton of the Gulf of Mexico. Taxonomy of calcareous nannoplankton. *Geoscience and Man*, vol. 20. School of Geoscience, LSU, Baton Rouge.

Pujos-Lamy, A.1977. *Emiliania et Gephyrocapsa* (nannoplancton calcaire): biométrie and intérêt biostratigraphique dans le Pleistocène supérieur marin des Açores. *Rev. Esp Micropaleontol.*, **9**: 69–84.

Rahman, A. and Roth, P. H., 1990. Late Neogene paleoceanography and paleoclimatology of the Gulf of Aden region based on calcareous nannofossils. *Paleoceanogr.*, **5**(1): 97–107.

Roth, P.H., 1974. *Calcareous Nannoplankton from the Northwestern Indian Ocean, Leg 24, Deep Sea Drilling Project*, pp 969–94. Init. Repts. DSDP, v.24, U.S. Govt. Printing Office, Washington.

Roth, P.H., 1986. Mesozoic paleoceanography of the North Atlantic and Tethys Oceans. In *North Atlantic Palaeoceanography*, ed. C.P. Summerhayes and N.J. Shackleton, pp. 299–320. Geol. Soc. Spec. Publ. no. 21. Blackwell Scientific Publications, London.

Roth, P.H., 1989. Ocean circulation and calcareous nannoplankton evolution during the Jurassic and Cretaceous. *Palaeogeogr., Palaeoclimatol., Palaeoecol.*, **74**: 111–26.

Roth, P.H. and Berger, W.H., 1975. Distribution and dissolution of coccoliths in the south and central Pacific. In *Dissolution of Deep-Sea Carbonates*, ed. W.V. Sliter, A.W.H. Bé, and W.H. Berger, pp. 87-113. Cushman Found. Foraminiferal Res., Spec. Publ. 13.

Roth, P.H. and Bowdler, J.L., 1981. Middle Cretaceous calcareous nannoplankton biogeography and oceanography of the Atlantic Ocean. *Soc. Econ. Paleontol. Mineral. Spec. Pub.*, no. 32, pp. 517–46.

Roth, P.H. and Coulbourn, W.T., 1982. Floral and solution patterns of coccoliths in surface sediments of the North Pacific: *Mar. Micropaleontol.*, **7**: 1–52.

Roth, P.H. and Krumbach, K.R.,1986. Middle Cretaceous calcareous nannofossil biogeography and preservation in the Atlantic and Indian oceans: *Mar. Micropaleontol.* **10**: 235–66.

Roth, P.H. and Thierstein, H.R., 1972. *Calcareous Nannoplankton – Leg 14, Deep Sea Drilling Project*, pp. 421–86. Init. Repts., DSDP, vol. 14, U.S. Govt. Printing Office, Washington.

Roth, P.H., Mullin, M.M., and Berger, W.H., 1975. Coccolith sedimentation by fecal pellets: laboratory experiments and field observations. *Geol Soc. Amer. Bull.*, **86**: 1079–84.

Roth, P.H., Wise, S.W. and Thierstein, H., 1975. Early chalk diagenesis and lithification: Sedimentological applications of paleontological approaches. *9th International Sedimentological Congress, Nice, France, 1975*, **7**: 187–99.

Samtleben, C., 1980. Die Evolution der Coccolithophoreen-Gattung *Gephyrocapsa* nach Befunden im Atlantik. *Paläont. Z.*, **54**: 91–127.

Schneidermann, N., 1977. Selective dissolution of Recent coccoliths in the Atlantic Ocean. In *Oceanic Micropaleontology*, ed. A.T.S. Ramsey, pp. 1009–53. Academic Press, New York.

Scholle, P. A. and Kling, S.A., 1972. Southern British Honduras: Lagoonal coccolith ooze. *J. Sediment. Petrol.*, **42**: 195–204.

Shumenko, S.I. and Ushakova, M.G., 19 67. Kokkolity v donnykh osadkakh Tikhogo okeana [Coccoliths in bottom sediments of the Pacific Ocean]. *Dokl. Akad. Nauk SSSR*, **176**: 932–4.

Tatsuoka, M. M., 1971. *Multivariate Analysis: Techniques for Educational and Psychological Research*. Wiley, New York.

Taylor, R. and Hamilton, G., 1982 Techniques. In *A Stratigraphical Index to Calcareous Nannofossils*. ed. A.R. Lord, pp. 11–15. Ellis Horwood, Chichester.

Toker, V., 1984. Nannoplankton and uranium concentration relations in the Black Sea deposits. *Bull. Mineral Res. Explor. Instit. Turkey*, **99/100**: 100–6.

Ushakova, M.G., 1968. Kokkolithy vo vzvesi i v poverkhnostom sloe osadkov Tikhogo i Indiyskogo okeanov [Coccoliths in suspension and in surface layers of the sediment in the Pacific and Indian Ocean]. In *Osnovnye problemy micropaleontologii i organogenogo osakonakopleniya v okeanakh i moryakh*, pp. 96–104. Nauka, Moscow. 7th Congr. INQUA, Paris, 1969.

Ushakova, M.G., 1970. Coccoliths in suspension in the surface layer of sediment in the Pacific Ocean. In *The Micropalaeontology of Oceans*, ed. B. Funnel and W. Riedel, pp. 245–51. Cambridge University Press, London.

Ushakova, M.G., 1974. Biogeograficheskaya zonalnost karbonatnogo nannoplanktona v poverkhnostom sloe donnikh osadkov Tikhogo Atlanticheskogo i Indiyskogo okeanov [Biogeographic zonation of calcareous nannoplankton in the uppermost layers of the bottom sediments of the Pacific, Atlantic, and Indian oceans]. In *Mikropaleontologiya okeanov i morey*, pp. 106–17. Nauka, Moscow.

Varol, O., 1985. Distribution of calcareous nannoplankton in surface sediments from intertidal and shallow marine regimes of a marginal sea: Jason Bay, South China Sea. *Mar. Micropaleontol.*, **9**: 369–74.

Violanti, D., Parisi, E. and Erba, E. 1987. Fluttuazioni climatichi durante il Quaternario nel Mar Tirreno, Mediterraneo occidentale (carota PC-l9 Ban 80). *Riv. Ital Paleontol.Strat.*, **92**: 515–70.

Wallich, G.C., 1861. Remarks on some novel phases of organic life, and on boring powers of minute annelids, at great depths in the sea. *Ann. & Mag. Nat. Hist.*, Ser. 3, **8**: 52–8.

Wang, P. and Chen, X., 1985. Distribution of calcareous nannoplankton in the East China Sea. In *Marine Micropaleontology of China*, Wang Pixian, *et al.*, pp. 218-228. Ocean Science Press, Bejing; Springer, Berlin.

Wang, P. and Samtleben, C., 1983. Calcareous nannoplankton in surface sediments of the East China Sea. *Mar. Micropaleontol.*, **9**: 249–59.

Winter, A., 1982. Paleonvironmental interpretation of Quaternary coccolith assemblages from the Gulf of Aqaba (Elat), Red Sea. *Rev. Esp. Micropaleontol.*, **14**: 291–314.

Winter, A. and Briano, J., 1989. Coccolithophores and global CO_2. *INA Newsl.*, **11**(2): 109–110.

Winter, A., Reiss, Z., and Luz, B., 1979. Distribution of living coccolithophore assemblages in the Gulf of Elat 'Aqaba. *Mar. Micropaleontol*, **4**: 197–223.

Zhang, J. and Siesser, W.G., 1986. Calcareous nannoplankton in continental-shelf sediments, East China Sea. *Micropaleontol.*, **32**: 271–81.

11 Stable isotopes in modern coccolithophores

JOHN C. STEINMETZ

Introduction

The stable-isotope composition of calcareous microfossils preserved in deep-sea sediments has served as a major tool in both paleoceanographic and paleoclimatic reconstructions. Oxygen isotopic variations have been widely used as indicators of oceanic paleotemperature and paleosalinity and of past fluctuations in sea level and glacial ice volumes. Carbon isotopic ratios, although less well studied, have provided insight into changes in oceanic fertility, circulation patterns, and the global carbon budget. The majority of this research has utilized both benthic and planktonic foraminifers. Recently, however, calcareous nannoplankton have been examined as an alternative microfossil group which preserves similar information in the deep-sea record. This paper reviews the development of isotopic studies of Recent coccolithophores, for both oxygen and carbon, and summarizes the state of knowledge to date.

The isotopic composition of calcareous marine organisms is a function of the $\delta^{18}O$, $\delta^{13}C$, and temperature of the ambient waters in which the organisms lived (Anderson and Arthur, 1983). Numerous factors complicate this seemingly simple relationship. Among them are the following:

1. There exists an oxygen isotopic fractionation between $CaCO_3$ and water as temperature of the environment changes. No such fractionation is evident in carbon isotopes.
2. Oxygen or carbon isotopic equilibrium with ambient water is rarely recorded in biogenic carbonate precipitation. Some physiological control is usually imposed by the organism. The magnitude of this 'vital effect' is different for each different group of organisms and often for each different species within a group.
3. Isotopic variations within the oceans are due to the interaction with the hydrologic cycle (evaporation, precipitation, runoff), freezing effects, and the mixing of water masses. Because contributions from these are much smaller than the overall mass of sea-water, isotopic variations in the ocean are small over short time intervals (less than 1000 years). For longer time periods of several thousands or tens of thousands of years, glacial–interglacial cycles become an important factor in the hydrologic cycle. The 'glacial effect', resulting from the preferential removal of ^{16}O from the ocean during glacial episodes and its deposition as isotopically light ice on land, has a marked effect on the oxygen isotopic composition of the world ocean (Anderson and Arthur, 1983).

The interested reader is referred to Anderson and Arthur (1983) for a detailed discussion of these factors.

Finally, diagenesis, or post-depositional changes, of marine sediments often results in the modification of the isotopic composition of carbonate skeletons. Interstitial solutions play an important role in the dissolution and recrystallization of marine carbonates (Milliman, 1966, Anderson and Schneidermann, 1973; Killingley, 1983).

Oxygen isotope studies

The impetus to examine the stable isotopic composition of calcareous marine microfossils derives from a classic paper by Emiliani (1955) in which he attempted for the first time to determine oceanic temperature fluctuations of the Quaternary utilizing the isotopic record of planktonic foraminifers. Although the actual temperature values are debatable, this work initiated the application of oxygen-isotope analysis as a powerful tool in paleoceanography and paleoclimatology. The 'paleotemperature method' is based on the fact that sea-water and $CaCO_3$ precipitated in equilibrium with sea-water differ in their $^{18}O/^{16}O$ ratios. (This is illustrated graphically by the heavy line in Fig. 1.) The carbonate precipitated is enriched in ^{18}O relative to sea-water, but the difference between the carbonate and the seawater ratios decreases with increasing temperature. This difference and the relationship between the temperature of precipitation and the isotopic composition of the carbonate is given by the following equation:

$$\text{Temperature, } T = 16.5 - 4.3\,(\delta - A) + 0.14\,(\delta - A)^2$$

$$\delta = \frac{(^{18}O/^{16}O)_{sample} - (^{18}O/^{16}O)_{standard\ carbonate}}{(^{18}O/^{16}O)_{standard\ carbonate}} \times 1000$$

$$A = \frac{(^{18}O/^{16}O)_{water} - (^{18}O/^{16}O)_{standard\ water}}{(^{18}O/^{16}O)_{standard\ carbonate}} \times 1000$$

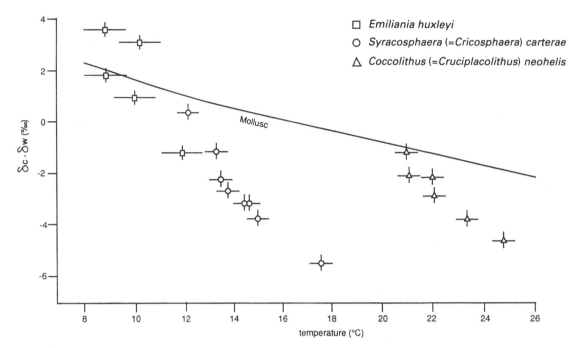

Fig. 1. Results of the oxygen isotopic analyses on cultured coccolithophores of Dudley and Goodney (1979). δC - δW values are δ^{18}O VS. PDB of coccolith calcite minus δ^{18}O vs. SMOW of water of growth media. The heavy line fits the equilibrium paleotemperature equation that was derived experimentally from CaCO$_3$ precipitated by molluscs grown in controlled environments. Molluscs precipitate CaCO$_3$ at oxygen isotopic equilibrium with ambient waters (Epstein *et al.*, 1953).

The constants were determined experimentally, and A must be measured in the modern ocean (Craig and Gordon, 1965) and estimated for the past. See Appendix to this chapter for further discussion.

Planktonic foraminifers have been used in the majority of the oxygen isotopic analyses of the past 35 years to derive information on Recent and past ocean surface and near surface conditions. Nevertheless, numerous factors about these calcareous microfossils keep them from being the "ideal" paleotemperature microfossil: the magnesian calcite in their tests renders them more susceptible to dissolution and recrystallization at depth than nearly pure calcite, the depth habitat of many species is below the surface waters and hence exposed to cooler temperatures, calcification occurs at different depths and different temperatures in the water column as the foraminifers migrate in their life cycle, and vital effects differ among species to a largely unknown degree.

Calcareous nannoplankton, although perhaps not being the ideal paleotemperature microfossil either, should prove to be a useful counterpart to the foraminifers. Coccolithophores may more accurately reflect sea-water surface temperatures (and paleotemperatures) for several reasons. First, they are predominantly found in the upper photic layer (with maximum population densities occurring between 50 and 100 m depth in most oceanic areas). Secondly, they do not migrate vertically during their life cycles, nor do they undergo stages of calcification at various depths during their lives. Thirdly, since their skeletons are composed of nearly pure calcite, they exhibit a higher resistance to dissolution and are less susceptible to recrystallization (either of which can affect the isotopic composition) than planktonic foraminifers. Fourthly, the vital effects of

many species can be determined through laboratory culture experiments. To their detriment, in their regular application to paleotemperature analyses, is the fact that monospecific oceanic samples of coccolithophores have not yet been possible to obtain (Okada and Honjo, 1973; Berger, 1973; Sliter *et al.*, 1975; Honjo, 1977; Wise, 1977).

The following section reviews the progressive development of knowledge in the study of the oxygen and carbon isotopes of calcareous nannoplankton. The fossil record is mentioned only as it is necessary to understand isotopes as applied to Recent coccolithophores.

Interest in the study of stable isotopes of coccoliths was first shown in a paper by Anderson and Cole (1975). They examined the oxygen and carbon isotopes of the coccolith size fraction (<44 µm) of Pleistocene sediment from a Caribbean and an east Pacific core, and they compared the isotopic composition of that fraction with those of planktonic foraminifers. They discovered that the oxygen isotopic compositions of the coccolith fraction varied systematically with the oxygen isotopic compositions of the planktonic foraminifers from the same depths in the cores. Scanning electron microscopy had shown that the coccolith fraction and foraminiferal samples were well preserved and that the

<44 μm material was contaminated by only trace amounts of non-coccolith carbonate (i.e., broken foraminiferal tests). Anderson and Cole reasoned that the systematic covariation, or correlation, between the isotopic compositions of the coccoliths and planktonic foraminifers reflected the temperature and the isotope composition of the water in which they lived. Moreover, Anderson and Cole concluded that the consistent oxygen isotopic enrichment of the coccoliths (by +2 to +3 permil in the Caribbean), with respect to the planktonic foraminifers, indicated a 'biological isotope effect' and that this controlled the isotopic composition of coccoliths. Hence, they had demonstrated that the oxygen-isotope composition of coccoliths could be used in paleotemperature or paleoclimatic investigations.

At nearly the same time that Anderson and Cole were examining the stable isotope signals in Pleistocene coccoliths, Margolis et al. (1975) were analyzing the stable isotope record of calcareous nannofossils through the entire Cenozoic. They determined the $\delta^{18}O$ and $\delta^{13}C$ values from well-preserved nannofossils also contained in the <44 μm fraction of sediment from Deep Sea Drilling Project (DSDP) Leg 29 cores collected in the Southern Ocean. Published a month after Anderson and Cole's article appeared in the *Journal of Foraminiferal Research*, Margolis et al. (1975) reported very similar results (in the 15 August 1975 issue of *Science*): (1) the $\delta^{18}O$ profile through the Cenozoic for calcareous nannofossils closely paralleled both the planktonic and benthic foraminiferal profiles for most of its length, and (2) nannofossil $\delta^{18}O$ values from DSDP Site 277 are equal to or slightly more negative than those of associated planktonic foraminifers. They concluded that calcareous nannofossils deposit $CaCO_3$ at or near equilibrium with oceanic surface waters and that well-preserved, polyspecific samples of calcareous nannofossils could be used to estimate surface water paleotemperatures.

The oxygen-isotope content of cultured coccoliths was first reported by Dudley and Goodney in 1979. This study was in direct response to the works just discussed. Dudley and Goodney reasoned that since calcareous nannofossils were too small to permit selection of enough individual specimens for a monospecific analysis, all isotopic work in sediment samples must be done on polyspecific assemblages. Their study was undertaken to determine if coccolithophores secrete $CaCO_3$ coccoliths in isotopic equilibrium with the sea-water in which they grow or if a species-specific vital effect exists. Three species, *Cricosphaera carterae*, *Coccolithus* (= *Cruciplacolithus*) *neohelis*, and *Emiliania huxleyi* were cultured at constant temperatures over a range of 8 to 28 °C. They clearly showed (Fig. 1) that there was a definite dependence between oxygen-isotope composition and temperature of nutrient-limited growth. Nitrogen is the critical limiting factor to algal growth in the marine environment (Ryther and Dunstan, 1971); hence, growth controlled in the laboratory by nitrate limitation appears to closely

approach natural conditions (Dudley and Goodney, 1979). Dudley and Goodney concluded that because nutrient-limited growth was probably the most common condition in the marine environment, it followed that naturally occurring coccoliths could be expected to show a similar temperature dependence of isotopic composition.

Dudley and Goodney's (1979) work was quickly followed by additional laboratory culture-isotope studies of the previously analyzed species *E. huxleyi*, as well as species newly isolated and cultured for the first time: *Calcidiscus* (= *Cyclococcolithus*) *leptoporus*, *Gephyrocapsa oceanica*, and *Thoracosphaera heimii* (Dudley et al., 1980).[1] This time all four species were not only prevalent in Recent oceans, but they also were common in Pleistocene sediments.

Continuous batch cultures at various temperatures were analyzed for their oxygen isotopic composition. The $\delta^{18}O$ values of the four species showed a strong temperature dependence which in no case corresponded to equilibrium precipitation of $CaCO_3$. The $\delta^{18}O$ values of the coccoliths of *E. huxleyi* and *G. oceanica* were approximately 1‰ positive relative to calcium carbonate precipitated at equilibrium, whereas the isotopic compositions of coccoliths from *C. leptoporus* and *T. heimii* were approximately 2.5‰ depleted in ^{18}O relative to equilibrium. They concluded that the fractionation of oxygen isotopes during the secretion of the calcite in all species studied was due to vital effect influences and that the calcification processes may vary among different taxa. The close grouping of the isotopic data for *E. huxleyi* and *G. oceanica* is reasonable in light of the two species' suspected phylomorphogenetic lineage (McIntyre, 1970; Dudley et al., 1980).

Dudley et al. (1980) reasoned that the presence of a vital effect in the fractionation of oxygen isotopes in coccolithophores did not preclude their use for paleotemperature analysis, as long as the magnitude of the effect was known over the temperature range in question. Despite the fact that coccolith calcite was not precipitated in equilibrium, the oxygen isotopic values of an assemblage reflected the temperature and isotopic ratio of the water in which the coccolithophores had lived. They also understood that variables such as light intensity, nutrients, and salinity may play a role in regulating fractionation. Goodney et al. (1980) investigated the oxygen isotopes of Recent nannoplankton from core tops and also concluded that their $\delta^{18}O$ composition did not in all cases indicate deposition in isotopic equilibrium with oceanic surface waters, but that a definite temperature-dependence trend existed.

Interest in the applications of oxygen isotopes of calcareous nannofossils for paleotemperature analysis continued

---

[1] The taxonomic position of thoracospheres was largely undecided when Dudley et al. submitted their paper to *Nature* in January 1980. Thoracospheres had been classified either as coccolithophores or dinoflagellates, until Brand, Blackwelder, and Guillard, upon examining cell ultrastructure of *Thoracosphaera*, observed the presence of a typical dinoflagellate nucleus (Note added in proof of Dudley et al., 1980).

221

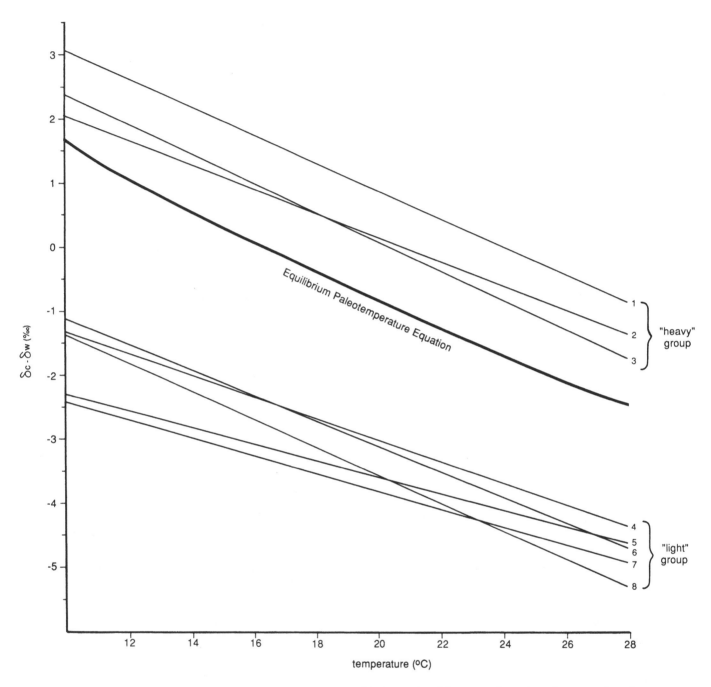

Fig. 2. Results of the oxygen isotopic analyses on cultured coccolithophores of Dudley et al. (1986). The eight thinner lines are plots of the least squares fits of the culture data in Table 1.

with Anderson and Steinmetz' (1981, 1983) and Steinmetz and Anderson's (1984) examination of Quaternary core P6304-4 from the Caribbean. Emiliani (1955, 1966, 1972) had earlier reported on the isotopic record for the planktonic foraminifers *Globigerinoides sacculifer* (Brady) in this and adjacent cores. Comparing the isotopic records of the nanno-fossils and of the foraminifer (Fig. 3), and examining the species composition of well preserved nannofossil assem-blages downcore to isotope stage 14, Anderson and Steinmetz (1981, 1983) and Steinmetz and Anderson (1984)

made several conclusions:

1. The excellent correlation between global Quaternary biostratigraphic data and the isotopic record are firm evi-dence for the reliability and precision of the stratigraphi-cal information in the $\delta^{18}O$ record of nannofossils.

2. Nannofossils are enriched in ^{18}O relative to *G. sacculifer*

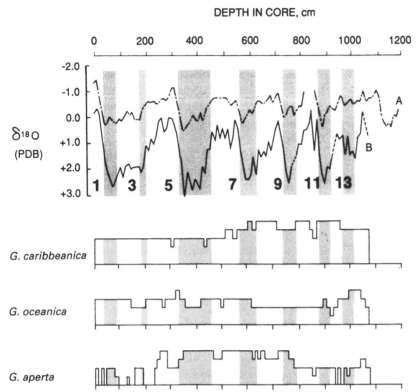

DEPTH IN CORE, cm

Fig. 3. Results of the oxygen isotopic analyses on the Pleistocene Caribbean core P6304-4. A = Emiliani's (1972) oxygen isotope data of *Globigerinoides sacculifer*. B = Steinmetz and Anderson's (1984) oxygen isotope data on the coccolith-size fraction (3-25 μm). The odd numbers beneath curve B identify the interglacial isotope stages (Emiliani, 1966). The shaded areas are glacial stage intervals. The lower half of the figure shows the relative frequency of the three most common species of *Gephyrocapsa*.

by 1 to over 2‰, consistent with data on the oxygen isotopic fractionation of *G. oceanica* grown in culture experiments (Dudley *et al.*, 1980) (*Gephyrocapsa* is the dominant taxon in the 3–25 μm fraction examined throughout the core).

3. The glacial–interglacial δ¹⁸O fluctuations of the nanno-fossil record are not attributable to variations in the abundance of the dominant (in numbers and volume) taxon, *Gephyrocapsa*, or the most common species of this genus, *G. aperta*, *G. caribbeanica*, and *G. oceanica*.

4. The nannofossil isotopic record is a more reliable indicator of temperature and δ¹⁸O changes in surface sea-water, because selective dissolution or deep calcification during interglacials reduced the amplitude of the signal for *G. sacculifer*. This conclusion was later supported by evidence from the Venezuela Basin presented by Showers and Margolis (1985).

5. The isotopic temperature preserved in nannofossils is a weighted average of the respective vital effects of the various species constituting the assemblage. Despite species-dependent non-equilibrium isotopic fractionation in coccoliths, the variations in species abundance did not obscure the paleoenvironmental significance of the isotope signal from coccolith assemblages (Anderson and Steinmetz, 1981).

These studies demonstrated that a clear understanding of the 'vital effect' in coccolith formation was necessary to per-mit more reliable use of coccoliths as paleotemperature or paleoclimatic tools. In 1986, Dudley *et al.* reported on further laboratory studies in order to investigate biological fractionation during coccolith formation. In addition to examining four previously cultured species (*Calcidiscus leptoporus, Cricosphaera carterae, E. huxleyi, and G. oceanica*), they investigated four species heretofore not analyzed (*Reticulofenestra* (= *Crenalithus*) *sessilis, Syracosphaera pulchra, Umbilicosphaera hulburtiana, and U. sibogae*). As before (Dudley and Goodney, 1979; Dudley *et al.*, 1980), monospecific algal cultures were grown and acclimated for two to three weeks at selected temperatures (12, 16, 20, 24, 28 ˚C). Coccoliths, as well as water of the growth medium, from more than 150 samples were analyzed for oxygen isotopes. The results clearly showed that biological fractionation occurs during the formation of $CaCO_3$ in all species studied. Interestingly the species fell into two distinct groups: those which were isotopically light or heavy with respect to $CaCO_3$ precipitated at equilibrium (see Fig. 2).

The 'heavy' group consisted of *R. sessilis, E. huxleyi*, and *G. oceanica*. They proved to be about 1‰ heavy (enriched in ¹⁸O relative to equilibrium) and indicated a significant temperature dependence in oxygen isotopic composition. Additionally, the relationship between temperature and isotopic composition for the three species was similar and with no significant differences among them at warmer temperatures.

The 'light' group consisted of *C. leptoporus, C. carterae, S. pulchra, U. hulburtiana*, and *U. sibogae*. Their oxygen

Table 1. *Equations for the least squares fit to coccolith oxygen isotope data ($\delta^{18}O$ coccolith vs. PDB minus $\delta^{18}O$ growth media vs. SMOW) (from Dudley* et al., *1986).*

'heavy group'		
1	*Emiliania huxleyi*	$Y = 5.27 - 0.22\,X$
2	*Gephyrocapsa oceanica*	$Y = 3.95 - 0.19\,X$
3	*Reticulofenestra sessilis*	$Y = 4.69 - 0.23\,X$
'light group'		
4	*Cricosphaera carterae*	$Y = 0.38 - 0.17\,X$
5	*Syracosphaera pulchra*	$Y = -1.00 - 0.13\,X$
6	*Umbilicosphaera sibogae*	$Y = 0.89 - 0.20\,X$
7	*Calcidiscus leptoporus*	$Y = -1.02 - 0.14\,X$
8	*Umbilicosphaera hulburtiana*	$Y = 0.83 - 0.22\,X$

isotope values were about 2.5‰ light (depleted in ^{18}O relative to equilibrium). A marked temperature dependence was also shown, although significant differences in the isotopic composition of the species was exhibited at the temperature extremes.

Perhaps the most important contribution of the work by Dudley *et al.* (1986) to paleotemperature analysis was the determination of equations for least squares fit to their data (Table 1). Using these equations together with the relative contribution of each species, they suggested it should be possible to decipher the relative importance of each species' contribution to the bulk coccolith isotopic signal.

This idea was developed in a paper by Paull and Thierstein (1987) in which they attempted to identify the 'major signal carriers' in the fine fraction (<38 μm) and understand the effect of compositional variations on isotopic ratios. They employed an automated decanting device to separate fine fraction particles into as many as 13 subfractions from Recent and glacial samples from three sites. Although about half the subfractions consisted of mixtures of coccoliths of various species because of their size range overlap, taxonomically distinct subsamples were also produced in each initially polyspecific <38 μm sample. These subsamples were those dominated by small foraminifers and foraminiferal fragments, thoracosphaerids, *G. oceanica*, *Emiliania huxleyi* and *G. oceanica*, and *Florisphaera profunda*.

Paull and Thierstein (1987) examined the ratios of $\delta^{18}O$ and $\delta^{13}C$ in the subfractions and compared them to expected equilibrium ratios calculated from temperature and salinity values at each of two depths from the sites in the western Pacific, South Atlantic, and Caribbean. Comparison of the oxygen isotopic values predicted by culture experiment data (Dudley and Goodney, 1979; Dudley *et al.*, 1980, 1986) with those measured in the Recent subfractions showed little correspondence: predicted values for the large-size subfrac-

tions were outside the range of the measured values, and although the discrepancies were less in the small subfractions for two of the sites, there was no overlap in the values for the equatorial Pacific samples[2].

They concluded that compositional changes may strongly influence the isotopic content of polyspecific samples, and that any mixture of the size components will have intermediate isotopic values according to the relative proportions of the mixture.

Dudley and Nelson (1989) conducted a similar investigation and supported this conclusion. They examined the oxygen isotope values of calcareous nannofossils from mid-to-late Quaternary cores in the southern Tasman Sea and explored the possibility of adjusting the nannofossil $\delta^{18}O$ values for vital effects. They assigned 'species-specific correction factors' to each of the major nannofossil contributors in the cores as follows:

(a) gephyrocapsids, −1.2‰
(b) *E. huxleyi*, −1.6‰
(c) *C. leptoporus*, +2.5‰
(d) *Umbilicosphaera* sp., +2.5‰
(e) *Syracosphaera* sp., +2.2‰
(f) *Pseudoemiliania lacunosa* (= *Emiliania ovata*), −1.6‰

(This final value was tentatively assigned based on the taxonomic affinity of this species to *E. huxleyi*). A weighted correction factor (Σ f) for each sample was then calculated based on the following formula:

$$\Sigma\,f = (-1.2a - 1.6b + 2.5c + 2.5d + 2.2e - 1.6f)/\,100$$

where a, b, c, d, e, and f represented estimates of the weight

---

[2] Paull and Thierstein (1987) noted that the 'heavy' group of Dudley *et al.* (1986) (*E. huxleyi* and *G. oceanica*) consisted of coccolithophores with small placoliths, and the 'light' group (*T. heimii*, *C. leptopora*, *H. carteri*) consisted of nannofossils with large calcite plates. They (p. 427) apparently and mistakenly assumed *Cricosphaera carterae*, cultured by Dudley *et al.* (1986), to be *Helicosphaera carteri*, probably based on the similar species epithets. The two species, in fact, bear little resemblance to each other. *Cricosphaera carterae*, originally described as *Syracosphaera carterae* Braarud and Fagerland (1946), was described as a coccolith oval about 2 μm long. Braarud *et al.* (1953) published electron micrographs of *S. carterae* showing the oval consisted of about thirteen weakly-joined, anvil-shaped segments with a 'column-like base and a flattened upper part' resembling 'an elliptical balustrade' (p. 132). Each segment is about 0.25 μm at its largest dimension. In contrast, *H. carteri* consists of a large elliptical central shield surrounded by a spiralling flange constructed of radial elements (Tappan, 1980).

The mistake in nomenclature means that characterizing the 'light' group as 'nannofossils with large calcite scales' (Paull and Thierstein, 1987, p. 427) is not meaningful, and that their figure 4, showing the ranges of both predicted and measured $\delta^{18}O$ values for the three sample sites, must be viewed with this constraint. Characterising the 'heavy' group with 'small placoliths' is still possible, although with respect to *C. carterae*, all 'small' forms are not necessarily 'heavy'. I suggest that since each group size consists of a heterogeneous group of species with different sized structural elements, that the large and small designation of Paull and Thierstein (1987) be avoided and that the light and heavy characterizations be used instead. In spite of the nomenclatural problem, Paull and Thierstein's observations are not seriously affected because *C. carterae* was not present in their samples.

percent of calcite contributed by the six species listed above.

Dudley and Nelson then calculated a 'vital effect adjusted' (vea) $\delta^{18}O$ nannofossil value from the following expression:

$$\delta^{18}O_{vea} = \delta^{18}O_{raw} + \Sigma f$$

where $\delta^{18}O_{raw}$ was the analytical result for $\delta^{18}O$ for bulk calcareous nannofossils in 4–25 μm sample fractions. With the exception of *Coccolithus pelagicus*, which was found in some of the samples and for which culture and correction factor information does not currently exist, all of the species listed above represented >5% of the flora in any one sample.

In comparing the raw and the vea nannofossil $\delta^{18}O$ data, the correspondence in isotope stages and shifts were excellent from about 0 back to 1660 Ka. However, with few exceptions, the effect of the 'correction' was to displace the record to lighter $\delta^{18}O$ values, typically 0.5–1.0‰.

Carbon isotopes studies

The isotopic carbon composition of marine surface waters is controlled by the relative importance of biologically and thermodynamically controlled processes. Specifically it is a function of rates of primary productivity, temperature, the rate of upwelling of deeper water, and accompanying rates of CO_2 exchange with the atmosphere. Since these factors vary over space and time as a function of local conditions and watermass mixing, and they are everywhere different within certain tolerances, it is not possible to theoretically calculate equilibrium water-isotope values (Margolis *et al.*, 1982; Anderson and Arthur, 1983). Some generalizations, however, can be made. At low latitudes, biological processes are more important than thermodynamic ones (i.e., those affecting residence time and atmospheric gas exchange as a function of wind speed) (Kroopnick, 1985; Oppo and Fairbanks, 1989). Tropical and mid-latitude surface waters, especially the more stably stratified parts of the oceans such as the central gyres, are enriched in ^{13}C. This is because phytoplankton preferentially fractionate ^{12}C and incorporate it in their cells. Their subsequent sinking removes ^{12}C from the total dissolved carbon reservoir at the surface (Anderson and Arthur, 1983). Generally these waters have $\delta^{13}C$ values around 2‰ (PDB), but values are highly variable due to biological and air–sea mixing processes (Kroopnick, 1985). Upwelling of ^{12}C-enriched intermediate waters at the circumpolar and equatorial divergences partly explains the lighter $\delta^{13}C$ values at high southern latitudes and at the equator (Anderson and Arthur, 1983).

As organic matter sinks in the water column and is oxidized, it adds isotopically light carbon to that which is already present. The result is a $\delta^{13}C$ value near 0.0‰ (Shackleton and Vincent, 1978). At great depths in the water column, the dissolution of carbonate adds carbon to that derived from the oxidation of organic carbon (Kroopnick *et al.*, 1977).

Carbon isotopes are fractionated during the production of organic matter and to a lesser extent during the formation of biogenic $CaCO_3$. The organic carbon produced as a result of photosynthesis is isotopically lighter than the dissolved organic carbon; typically the $\delta^{13}C$ of marine organic carbon is about –26 to –30‰ (Sackett *et al.*, 1965). The $\delta^{13}C$ of calcite is relatively insensitive to changes in temperature, about 0.035‰ per degree C) (Anderson and Arthur, 1983). The productivity of ocean surface-waters is thus recorded in the $\delta^{13}C$ of organic carbon and biogenic carbonate. With a thorough understanding of the distribution of carbon and its isotopes in the modern oceans, paleoproductivity can be estimated from a record of past oceanic $\delta^{13}C$ values (Kroopnick, 1985).

Only one study has been made on the carbon isotopic composition of Recent calcareous nannoplankton (Goodney *et al.*, 1980). As seen earlier with oxygen isotopes, the impetus for this investigation had largely come from a desire to understand the variations in the isotopic composition of nannofossils. Earlier analyses of fossil marine sediments had suggested that the carbon isotopic composition of biogenic carbonate may provide paleoceanographic information (Broecker and Broecker, 1974; Margolis *et al.*, 1975; Kroopnick *et al.*, 1977; Williams *et al.*, 1977). Margolis *et al.* (1975) noted that the $\delta^{13}C$ profile for calcareous nannofossils paralleled that of the planktonic and benthic foraminiferal curves. The progressive increase (or enrichment) of $\delta^{13}C$ values from the benthic and planktonic foraminifers to the nannofossils suggested that carbon isotopes reflected the depth of growth of the organisms and that the $\delta^{13}C$ of the surface waters was preserved in the nannofossils. Similarly, after examining isotopic data from Cretaceous and Cenozoic Deep Sea Drilling Project cores, Kroopnick *et al.* (1977) provisionally concluded that nannofossil $\delta^{13}C$ values are a better indicator of surface sea-water than planktonic foraminiferal values.

Research first started with evidence to suggest that $\delta^{13}C$ values of well preserved nannofossils appeared to reflect their depth of growth when compared with $\delta^{13}C$ profiles from modern oceans (Margolis *et al.*, 1975). Anderson and Cole (1975) investigated the stable isotopes in the <44μm fraction from a Caribbean core and from an east Pacific core, both of Pleistocene age. Isotopic carbon ratios in the Caribbean core ranged from +0.71 to +1.97‰, and in the Pacific core from +0.20 to +0.67‰. Since the assemblages in both cores contained essentially the same species, Anderson and Cole could not suggest any systematic reason for the differences of carbon isotopes between the two localities. Inasmuch as carbon-isotope composition data of deep-sea biogenous carbonate were quite limited, they hesitated to speculate beyond suggesting that the differences might be due to biological 'vital effects', variations in the isotopic composition of particulate carbon in the ocean, or some other cause. Berger *et al.* (1978) examined Holocene and late Pleistocene sediments recovered from a box core taken in the western equatorial Pacific. For the coccolith size fraction 0.45–20 µm, values ranged from –1.10 to +0.45‰. After comparing $\delta^{13}C$ values with those calculated from equilibrium precipitation, they concluded that the nannofossils recovered were probably deposited between about –3 and –4‰ from equilibrium. Goodney *et al.* (1980) examined the carbon isotopes of Recent nannoplankton in Indian Ocean core tops. They found the values to range from –1.33 to –0.33‰ for the >15 to <44-µm size fraction, and from –0.77 to +0.94‰ for the <15 µm size fraction. Additionally, they found that $\delta^{13}C$ data plotted as a function of latitude did show regular variations. Although they could not explain the range in carbon isotopic values or the latitudinal variations, they did suggest that the values may be influenced by complex interrelationships between differences in surface water $\delta^{13}C$ of CO_2 or species diversity changes as a function of temperature, latitude, and productivity. They also suggested that metabolic CO_2 ('vital effects') might explain the departure of the $\delta^{13}C$ from equilibrium, or to changes in the local rate of productivity which can control the $\delta^{13}C$ of the dissolved bicarbonate reservoir.

Despite evidence indicating a vital effect in the fractionation of the carbon isotopes during the deposition of coccoliths, the isotope-culture studies by Dudley and Goodney (1979) or Dudley *et al.* (1980, 1986) presented data only on oxygen isotope fractionation. To date, no data on carbon isotope-culture studies have been published. As a result, we must rely on inferences made from the analyses of plankton tows and core tops. In these cases, analytical results are based on multi-species, and even multi-organism, assemblages.

Vital effects

Organisms which precipitate $CaCO_3$ with either oxygen or carbon isotope ratios similar to those of inorganically precipitated carbonate are considered to be in isotopic equilibrium with their surrounding medium. Those organisms depositing carbonate out of equilibrium with their environments are said to exhibit 'vital effects' on the fractionation of the isotopes. That is, the physiological control of biochemical processes taking place within the organism influences the fractionation of isotopes.

The mechanism by which organisms fractionate the isotopes is complex and not wholly understood. Culture and ultrastructure studies of coccolithophores by de Jong *et al.* (1983) and van der Wal (1984) have indicated that even within this group, there exists a variety of mineralizing systems. Not only may coccoliths be formed in different areas of the cell, but the composition of their organic constituents

may also substantially differ. Furthermore, their research suggests that the calcifying system itself may vary within one species, differing in morphology and production of coccoliths depending on the life cycle of the alga.

In spite of the enormous morphological variety of coccoliths and the apparent variability of the calcifying systems, there are some common features that can be discerned (de Jong *et al.*, 1983; van der Wal, 1984): (1) Coccoliths of both *Emiliania huxleyi* and *Cricosphaera* (= *Hymenomonas*) *carterae*, for example, are formed in intracellular vesicles in close proximity to the nucleus. The vesicle membrane plays an important role in regulating mineralization by controlling the intracellular pH and ion concentrations. (2) Coccoliths are associated with soluble acid polysaccharides, molecules of remarkable complexity thought to fulfill an important role in the regulation of $CaCO_3$ crystallization. (3) Coccolith formation is strongly dependent upon the availability of light.

Because coccolithophores are highly autotrophic organisms, photosynthesis is probably the most important mechanism responsible for energy production (van der Wal, 1984). Bicarbonate is known to be the main source of carbon in coccolithophores and in coccolith formation (Paasche, 1964). Calcification and photosynthesis are thought to be mutually dependent according to the following reaction (Sikes *et. al.*, 1980; Sikes and Wilbur, 1982):

$$2HCO_3^- \rightleftharpoons CO_3^{2-} + CO_2 + H_2O$$

calcification photosynthesis

Mechanisms of isotope fractionation

Swart (1983) wrote a thorough review of carbon and oxygen fractionation processes from which much of the following discussion was developed. Despite the wide use of oxygen isotopes in paleotemperature studies, considerably less attention has been paid to the fate of oxygen metabolic and photosynthetic processes than that of carbon. Oxygen may be derived from three sources: H_2O and CO_2 used during photosynthesis, or O_2 consumed during photorespiration (Swart, 1983). During photosynthesis, the H_2O largely determines the isotopic composition of the oxygen even though much of the oxygen itself involved is derived from CO_2. The CO_2 rapidly comes into equilibrium with the much larger amount of H_2O. Photosynthesis itself results in only a slight fractionation of the oxygen.

The role of photorespiration in the fractionation of oxygen isotopes remains largely understudied and unresolved, but it is generally assumed to produce a depleted ratio. It is important to realize, however, that the overwhelming influence on oxygen-isotope composition of the skeleton is exerted by seawater (Swart, 1983). The laboratory culture experiments of

Dudley *et al.* (1980, 1986) have demonstrated species-dependent fractionation factors ranging from + 1.6 to –2.5‰.

In the conversion of inorganic carbon to organic carbon in photosynthesizing organisms, there is a conspicuous fractionation of the two stable isotopes, ^{12}C and ^{13}C. All common photosynthetic pathways preferentially fix ^{12}C, primarily because of a kinetic effect in the first carbon-fixing carboxylation reaction. The heavy carbon remaining is retained in the surface reservoir mostly as dissolved bicarbonate and is fixed as skeletal carbonate (Schidlowski, 1988). Since there is little isotope fractionation in carbon assimilation, respiration, or excretion, the carbonate retains nearly the same $\delta^{13}C$ ratio as the bicarbonate in the reservoir (Goreau, 1977).

Conclusions

1. Coccolithophores secrete coccoliths out of isotopic equilibrium with sea-water.
2. The oxygen isotope vital effect of a coccolithophore can be determined through isotope-culture studies at various temperatures.
3. The $\delta^{18}O$ value of a coccolith varies as a function of temperature and the vital effect of the species.
4. The oxygen isotopic signal of a polyspecific assemblage of coccoliths (e.g., a bulk sample from a plankton tow or a deep-sea ooze) is a reflection of the relative proportions of the component $\delta^{18}O$ values of the species making up the assemblage.
5. The $\delta^{13}C$ value of a coccolith is a reflection of the carbon isotopic composition of the surface waters where it was secreted; hence, it records surface water productivity.

Future studies

Much research obviously remains to be done on the stable isotopes of coccolithophores, so that a better understanding of Recent and fossil samples can be derived. Among the questions that need to be answered are the following:

- What are the precise metabolic pathways of carbon and oxygen in coccolithophore photosynthesis?
- How do such factors as light intensity, nutrients, and salinity play a role in regulating isotope fractionation?
- How do calcification processes in coccolithophores vary among the different taxa? Can knowledge of calcification processes in other closely or distantly related organisms help in understanding those in coccolithophores?
- Are there ecophenotypic similarities between coccolithophores with similar isotopic signatures? Can the signatures be used to determine phylogenetic lineages?
- Is it safe to extrapolate information learned about modern representatives to ancestors in the fossil record? If so, how far back in geological time can the extrapolations be made?
- Is any or all of the $\delta^{13}C$ signal incorporated in coccoliths the result of vital effects?
- How is the global carbon cycle affected by coccolithophore productivity and preservation?

Appendix – Conventions

The isotopic fractionation between two substances (sample and standard) is described in terms of the following simple formula:

$$\delta = \left[\frac{R_{sample}}{R_{standard}} - 1 \right] \times 1000$$

the δ notation represents per ml (‰) deviations from the isotopic standard, PDB, prepared from the rostrum of the belemnite *Belemnitella americana* from the Cretaceous Peedee Formation of South Carolina. The PDB standard is defined as 0 per mil for carbon and oxygen. R_{sample} is either $^{13}C/^{12}C$ or $^{18}O/^{16}O$ in the sample and $R_{standard}$ is the corresponding ratio in the standard. The isotopic composition of water is reported as per ml deviations of the sample from Standard Mean Ocean Water (SMOW), a hypothetical water close to average ocean water (Craig, 1957). To relate $\delta^{18}O$ values of calcite on the PDB and SMOW scales, the following expression (Craig, 1961) is used:

$$\delta^{18}O_{SMOW} = 1.03086 \, \delta^{18}O_{PDB} + 30.86$$

In other words and since δ vs. PDB = 0, PDB Calcite is +30.86 on the SMOW scale (Anderson and Arthur, 1983).

Since the sample supply of PDB has been exhausted, calibrations are normally done through analysis of Natural Bureau of Standards (NBS) samples. For example, NBS-20 is a homogenized sample of Jurassic Solnhofen limestone from southern Germany;

$$\delta^{18}O_{(NBS-20 \, vs. \, PDB)} = -4.14‰.$$

Mass spectrometric instrument precision of at least +/0.01‰ is currently normal. A positive δ value indicates enrichment in the heavy isotope, relative to the standard. It is colloquially referred to as 'heavy'. In graphs, by convention, positive values are plotted to the left, negative values to the right.

The results of culturing experiments are reported as δC–δW vs. temperature. δC–δW values are $\delta^{18}O$ vs. PDB of coccolith $CaCO$ minus $\delta^{18}O$ vs. SMOW of water of growth media. Hence, any effects caused by the variations in isotopic composition of the medium are accounted for by subtracting the $\delta^{18}O$ value of the medium from the $\delta^{18}O$ value of the coccolith carbonate (Dudley and Goodney, 1979).

References

Anderson, T. F. and Arthur, M. A., 1983. Stable isotopes of oxygen and carbon and their application to sedimentologic and paleoenvironmental problems. In *Stable Isotopes in Sedimentary Geology*. ed. M.A. Arthur, T.F. Anderson, I.R. Kaplan, J. Veizer, and L.S. Land, pp. 1-151. SEPM Short Course No. 10.

Anderson, T. F. and Cole, S. A., 1975. The stable isotope geochemistry of marine coccoliths: a preliminary comparison with planktonic foraminifera. *J. Foram. Res.*, **5** (3): 188–92.

Anderson, T. F. and Schneidermann, N., 1973. Stable isotope relationships in pelagic limestones from the central Caribbean: Leg 15, Deep Sea Drilling Project. In Edgar, N.T., Saunders, J.B. *et al.*, Init. Repts. DSDP, vol. 15, pp. 795-803. U.S. Govt. Printing Office, Washington.

Anderson, T. F. and Steinmetz, J. C., 1981. Isotopic and biostratigraphical records of calcareous nannofossils in a Pleistocene core. *Nature*, **294** (5843): 741–4.

Anderson, T. F. and Steinmetz, J. C., 1983. Stable isotopes in calcareous nannofossils: potential application to deep-sea paleoenvironmental reconstructions during the Quaternary. In ed. J.E. Meulenkamp, *Reconstruction of Marine Paleoenvironments*. Utrecht Micropaleontol. Bull., **30**: 189–204.

Berger, W. H., 1973. Deep-sea carbonates: evidence for a coccolith lysocline. *Deep-Sea Res.*, **20** (10): 917–21.

Berger, W. H., Killingley, J. S. and Vincent, E., 1978. Stable isotopes in deep-sea carbonates: Box Core ERDC-92, west Equatorial Pacific. *Oceanol. Acta*, **1** (2): 203–16.

Braarud, T. and Fagerland, E., 1946. A coccolithophoride in laboratory culture, *Syracosphaera carterae*, n. sp. *Avh. Norske Vidensk-Akad. Oslo*, no. 2: 3–9.

Braarud, T., Gaarder, K. R., Markali, J. and Nordli, E., 1953. Coccolithophorids studied in the electron microscope. Observations on *Coccolithus Huxleyi* and *Syracosphaera Carterae*. *Nytt Mag. Botanikk.*, **1**: 129–34.

Broecker, W. S. and Broecker, S., 1974. Carbonate dissolution on the western flank of the East Pacific Rise. In *Studies in Paleo-Oceanography*, ed. Hay, W.W. pp. 44–57. SEPM Spec. Publ. 20.

Craig, H., 1957. Isotopic standards for carbon and oxygen and correction factors for mass-spectro-metric analysis of carbon dioxide. *Geochim. Cosmochim. Acta*, **12**: 133–49.

Craig, H., 1961. Isotopic variations in meteoric waters. *Science*, **133** (3465): 1702–3.

Craig, H. and Gordon, L. I., 1965. Deuterium and oxygen-18 variations in the ocean and the marine atmosphere. In *Stable Isotopes in Oceanographic Studies and Paleotemperatures*, ed. E. Tongiorgi, pp. 9–130. Consiglio Nazionale delle Richerche, Laboratorio di Geologia Nucleare, Pisa.

de Jong, E. W., Wal, P. van der, Borman, A. H., de Vrind, J. P. M., van Emburg, P., Westbroek, P. and Bosch, L., 1983. Calcification in coccolithophorids. In *Biomineralization and Biological Metal Accumulation*, ed. P. Westbroek, and E.W. de Jong, pp. 291-301. D. Reidel, Dordrecht.

Dudley, W. C. and Goodney, D. E., 1979. Oxygen isotope content of coccoliths grown in culture. *Deep-Sea Res.*, **26** (5A): 495–503.

Dudley, W. C. and Nelson, C. S., 1989. Quaternary surface-water stable isotope signal from calcareous nannofossils at DSDP Site 593, southern Tasman Sea. *Mar. Micropaleontol.*, **13** (4): 353–73.

Dudley, W. C., Blackwelder, P. L., Brand, L. E. and Duplessy, J. C., 1986. Stable isotopic composition of coccoliths. *Mar. Micropaleontol.*, **10** (1): 1–8.

Dudley, W. C., Duplessy, J. C., Blackwelder, P. L., Brand, L. E., and Guillard, R. R. L., 1980. Coccoliths in Pleistocene-Holocene nannofossil assemblages. *Nature*, **285** (5762): 222–3.

Emiliani, C., 1955. Pleistocene temperatures. *J. Geol.*, **63** (6): 538-578.

Emiliani, C., 1966. Paleotemperature analysis of Caribbean cores P6304-8 and P6304-9 and a generalized temperature curve for the last 425,000 years. *J. Geol.*, **74** (2): 109–26.

Emiliani, C., 1972. Quaternary paleotemperatures and the duration of the high-temperature intervals. *Science*, **178** (4059): 398–401.

Epstein, S., Buchsbaum, R., Lowenstam, H. A. and Urey, H. C., 1953. Revised carbonate water isotopic temperature scale. *Geol. Soc. Amer. Bull.*, **64**: 1315–26.

Goodney, D. E., Margolis, S. V., Dudley, W. C., Kroopnick, P. and Williams, D. F., 1980. Oxygen and carbon isotopes of Recent calcareous nannofossils as paleoceanographic indicators. *Mar. Micropaleontol.*, **5** (1): 31–42.

Goreau, T. J., 1977. Carbon metabolism in calcifying and photosynthetic organisms: theoretical models based on stable isotope data. *Proc. 3rd. Int. Coral Reef Symp., Miami*, **2**: 395–401.

Honjo, S., 1977. Biogenic carbonate particles in the ocean: do they dissolve in the water column? In *The Fate of Fossil Fuel CO_2 in the Ocean*, ed. N.R. Anderson, and A. Malahoff, pp. 269–94. Plenum, New York.

Killingley, J. S., 1983. Effects of diagenetic recrystallization on $^{18}O/^{16}O$ values of deep-sea sediments. *Nature*, **301** (5901): 594-597.

Kroopnick, P. M., 1985. The distribution of C-13 of ΣCO_2 in the world oceans. *Deep-Sea Res.*, **32** (1): 57–84.

Kroopnick, P.M., Margolis, S.V. and Wong, C.S., 1977. 13c variations in marine carbonate sediments as indicators of the CO_2 balance between the atmosphere and oceans. In *The Fate of Fossil Fuel CO_2 in the Ocean*. ed. N.R. Andersen, and A. Malahoff, pp. 295-321. Plenum Press, New York.

McIntyre, A., 1970. *Gephrocapsa protohuxleyi* sp. n. A possible phyletic link and index fossil for the Pleistocene. *Deep-Sea Res.*, **17**(1): 187–90.

Margolis, S. V., Kroopnick, P. M., Goodney, D. E., Dudley, W. C. and Mahoney, M. E., 1975. Oxygen and carbon isotopes from calcareous nannofossils as paleoceanographic indicators. *Science*, **189** (4202): 555–7.

Margolis, S. V., Kroopnick, P. M. and Showers, W. J., 1982. Paleoceanography: the history of the ocean's changing environments. In *The Environment of the Deep Sea*, ed. W.G. Ernst and J.G. Morin Rubey Vol. 2, pp. 18-54. Prentice Hall, Englewood Cliffs.

Milliman, J. D., 1966. Submarine lithification of carbonate sediments. *Science*, **153** (3739): 994–7.

Okada, H. and Honjo, S., 1973. The distribution of oceanic coccolithophorids in the Pacific. *Deep-Sea Res.*, **20** (4A): 355–74.

Oppo, D. W. and Fairbanks, R. G., 1989. Carbon isotope composition of tropical surface water during the past 22,000 years. *Paleoceanogr.*, **4** (4): 333–51.

Paasche, E., 1964. A tracer study of the inorganic carbon uptake during coccolith formation and photosynthesis in the coccolitho-phorid *Coccolithus huxleyi. Physiol. Plant.*, Suppl. 3: 1–82.

Paull, C. K. and Thierstein, H. R., 1987. Stable isotope fractionation among particles in Quaternary coccolith-size deep-sea sediments. *Paleoceanogr.*, **2** (4): 423–9.

Ryther, J. H. and Dunstan, W. M., 1971. Nitrogen, phosphorus, and euthrophication in the coastal marine environment. *Science*, **171** (3975): 1008–13.

Sackett, W. M., Eckelmann, W. R., Bender, M. L. and Bé, A. W. H., 1965. Temperature dependence of carbon isotopic composition in marine plankton and sediments. *Science*, **148** (3667): 235–7.

Schidlowski, M., 1988. A 3000-million-year isotopic record of life from carbon in sedimentary rocks. *Nature*, **333** (6171): 313–18.

Shackleton, N.J. and Vincent, E., 1978. Oxygen and carbon isotope studies in recent Foraminifera from the southwest Indian Ocean. *Mar. Micropaleontol.*, **3** (1) : 1–14 .

Showers, W. J. and Margolis, S. V., 1985. Evidence for a tropical freshwater spike during the last glacial–interglacial transition in the Venezuela Basin: $\delta^{18}O$ and $\delta^{13}C$ of calcareous plankton. *Mar. Geology*, **68** (1/4): 145–65.

Sikes, C. S. and Wilbur, K. M., 1982. Functions of coccolith formation. *Limnol. Oceanogr.*, **27** (1) : 18–26.

Sikes, C. S., Roer, R. D. and Wilbur, K. M., 1980. Photosynthesis and coccolith formation: inorganic carbon sources and net inorganic reaction of deposition. *Limnol. Oceanogr.*, **25** : 248–61.

Sliter, W. V., Bé, A. W. H. and Berger, W. H., (Eds.) 1975. Dissolution of Deep-Sea Carbonates. Cushman Found. Foram. Res., Spec. Publ. 13, 1–159.

Steinmetz , J. C. and Anderson, T. F., 1984 . The significance of isotopic and paleontologic results on Quaternary calcareous nannofossil assemblages from Caribbean core P6304–4 . *Mar. Micropaleontol.*, **8** (5) : 403–24.

Swart, P. K., 1983. Carbon and oxygen isotope fractionation in scleractinian corals: a review. *Earth-Sci. Reviews*, **19**: 51-80.

Tappan, H., 1980. *The Paleobiology of Plant Protists*. W. H. Freeman, San Francisco.

Wal, P. van der, 1984. Calcification in two species of coccolithophorid algae. *GUA Papers in Geology, Amsterdam*, Ser. 1, No. 20, 112 pp.

Williams, D. F., Sommer, M.A. and Bender, M.L., 1977. Carbon isotopic compositions of Recent planktonic foraminifera of the Indian Ocean. *Earth Planet. Sci. Lett.*, **36**: 391–403.

Wise, S. W., Jr., 1977. Chalk formation: early diagenesis. In ed. N.R. Anderson, and A. Malahoff, *The Fate of Fossil Fuel CO₂ in the Ocean*, pp. 717–39. Plenum, New York.

Index

Author

Species

SEM micrograph for the species is indicated in bold.

Subject